EXPERIMENTAL PHYSICAL CHEMISTRY

FREDERICK A. BETTELHEIM

PROFESSOR OF CHEMISTRY, ADELPHI UNIVERSITY,
GARDEN CITY, LONG ISLAND, NEW YORK

with a section on Electronics contributed by

DONALD T. OPALECKY, Adelphi University

1971 □ W. B. SAUNDERS COMPANY · PHILADELPHIA · LONDON · TORONTO

SAUNDERS GOLDEN SERIES

W. B. Saunders Company: West Washington Square
Philadelphia, Pa. 19105

12 Dyott Street
London, WC1A 1DB

1835 Yonge Street
Toronto 7, Ontario

Experimental Physical Chemistry

SBN 0-7216-1690-9

Print No.:   1   2   3   4   5   6   7   8   9

To my son

ADRIEL

PREFACE

This book is intended as a laboratory manual for undergraduate physical chemistry courses. It can be used concurrently with most physical chemistry textbooks, or it can be used as an independent textbook for a separate physical chemistry laboratory course on the junior or senior level.

The basic idea that guided the writing of this textbook was that the laboratory part of a physical chemistry course must be more than mere illustration of the principles learned in lecture. The experiments are intended to be a bridge to research, by use of physical measurements to yield quantitative data. With this purpose in mind, most of the experiments contain the seed of novelty, the unknown, and thereby stimulate interest in interpretation.

To achieve this, the book contains a large number of experiments that investigate only *one system* by use of a variety of techniques. These are designated by asterisks. In the past 12 years' experience, this approach has proved to be useful. Applying experiments to one system (for example, dimethylsulfoxide-water) provides a deeper understanding of the principles. Studying the thermodynamic and structural properties of the same system and extending it to kinetics, electrochemistry, and other fields, the student obtains an overall view of the relationships of different physical measurements. Moreover, with each successive measurement of the same system, he is acquiring a research orientation in the sense that he can appreciate the way in which certain properties can be determined by use of different techniques and the manner in which special techniques may fill a gap in our knowledge of the system under investigation. From the pedagogic point of view, this is also cumulative. With each successive experiment the student becomes more and more involved, his anticipation is enhanced, and his enthusiasm constantly grows.

Such an approach obviously is not limited to only the dimethylsulfoxide-water system. Other binary systems of water, acetone, dioxane, methyl alcohol, dimethylformamide, dimethyl sulfoxide, benzene, toluene, and xylene have been used in my laboratory and could be investigated by different groups in one class or by different classes

v

from year to year. However, for the sake of interest to be generated and of identification of the student with the experiments, it is recommended that one group of students investigate one system most of the year.

A large number of binary systems can be used, enabling each student or group of students to work on one system. This distribution of binary systems encourages student involvement and discourages the exchange of data among students or the passing on of information from former students.

Not all physical chemical principles can be applied to one system only, without belaboring the underlying philosophy and without large expense for laboratory equipment. Therefore, another group of experiments is also provided in which physical chemical principles are applied to a material with which the specific principle can be investigated conveniently and economically. The instructor's judgment should provide the proper balance between the experiments selected to investigate one system in depth and the experiments more suited to other systems. Obviously, the limitations of laboratory equipment will largely determine the experiments selected.

Most of the experiments recommended here can be performed with a simple assemblage of available standard parts or commercially available instruments. In some cases the design of the equipment is kept relatively simple with the obvious drawback that some sacrifices must be made regarding accuracy of results. However, once the students become familiar with the techniques and limitations, sufficient latitude may be allowed for the students to improve the design of the equipment.

Most of the experiments can be finished in a four- to five-hour laboratory period, assuming that the students have read the experiments in advance. However, this is only the first task. In experiments that are designed to generate interest in research, evaluation of the experimental part is even more important than performance of the experiment. Hence, emphasis on well written laboratory reports, especially on interpretation of results, is in order.

All the experiments in this book are designed to raise questions; in order to answer them, the student has to go beyond the straightforward application of textbook information to the problem. Logical extensions of the basic principles and analogies drawn from the literature are sometimes necessary for full appreciation of the results.

To facilitate the student's involvement in the outcome of the experiments, at the beginning of each chapter a general description of the applicable theory is given. In the experimental description, detailed instructions are given to enable the student to perform the laboratory part within four to five hours. However, I tried to avoid instructions that would pinpoint the outcome of the experiment. (For example, in the boiling point–composition diagram experiment the students are not told the temperatures at which the samples should be taken, although from previous measurements on the system they may anticipate a maximum boiling point between 60 and 90 weight per cent DMSO, and they may take more samples in this range.) One purpose of this experiment is for the student to find whether an azeotrope exists, and if it does, at what composition this may be. Although one may sacrifice accuracy of results by this approach, it is more than compensated by the implantation of a sense of discovery.

Since the number of binary systems that can be assigned to students is large, it is beyond the scope of this book to give references to all physical chemical measurements on binary systems, especially because many of these systems have not been fully investigated. References are provided that illustrate the specific phenomenon. The students should be encouraged to do their own literature research pertaining to the binary system they investigate.

It is recommended that the final week of the course be left open to allow the students to summarize their experiences working on one binary system with many different physical chemical techniques. This should take the form of an essay or discussion report based on the laboratory reports.

Most of the experiments are complete and self-contained in the sense that each provides all the information necessary to understand the phenomena and perform the experiment. I tried to avoid references to previous experiments because my experience has taught me that the simple act of leafing back and forth in a book distracts many students from their task. Therefore, in this book one finds identical descriptions of phenomena or technique in two different experiments in instances in which both of them are not likely to be performed or they are not dependent on each other, e.g., Experiments 8 and 9, and 28 and 47. On the other hand, in experiments that depend upon each other, e.g., Experiments 48 and 49, such repetition was eliminated.

This striving for self-contained experiments is also the reason no separate chapters are provided on general techniques, such as vacuum systems and optics. All this information is provided in the specific experiments where it is needed. One exception is a general chapter on electronics written by Dr. D. Opalecky. It is felt that, although the descriptions of the instruments in each experiment are sufficiently detailed, a basic knowledge of the electronic components must be provided in a separate chapter; otherwise each experiment would be too voluminous.

One may decry the absence of instructions in glass blowing, especially with reference to vacuum techniques. This information was purposely omitted from the book because physical chemistry laboratory periods are too short for constructing apparatus, and the vacuum systems are usually built and maintained by the instructional staff. Another reason is that glass blowing is an art. Many students, even with two or three laboratory periods of glass blowing instructions, do not advance beyond the stage of satisfactorily fusing two Pyrex tubes.

Two other aspects of this book may be emphasized. Of the 49 experiments given here, the professor in charge of the course may select 20 or 30 experiments of special significance for chemical engineers taking physical chemistry. Experiments on atomic and molecular structure, solids, phase equilibria, kinetics, electrochemistry, and especially macromolecules may provide a core particularly useful to train engineers for experiences they will encounter in industry. Because a large part of the chemical industry now deals with polymers, the chapter on macro-molecules covers essentially all the major theoretical and experimental aspects of polymer chemistry one encounters in industrial laboratories, which are neglected in classic physical chemistry courses.

The second aspect to be emphasized relates to the fact that increasing use is made of physical chemical techniques in biochemistry and medical sciences. Some universities provide a one-semester physical chemistry course for those who are not majoring in chemistry, especially pre-medical students. More and more medical schools are presenting physical biochemistry or biophysics as a laboratory course and some experiments in this book may be used for such a course. Experiments that were performed in a one-semester course for pre-medical students in my lab are as follows: heat of mixing (Exp. 6), heat of combustion (Exp. 7), temperature dependence of solubility of solids (Exp. 10), infrared spectrum (Exp. 12), electronic spectra of charge transfer complexes (Exp. 13), dipole moment and dielectric constant (Exp. 15), proton magnetic resonance (Exp. 16), density and partial molal volumes of solutions (Exp. 19), viscosity of liquids (Exp. 20), inversion of sucrose – kinetics of a pseudo–first order reaction by polarimetry (Exp. 28), kinetics of enzyme reactions – activity of tyrosinase (Exp. 32), radio-isotope dilution technique (Exp. 34), conductance of electrolytes (Exp. 35), dissociation constant determination by potentiometric titrations (Exp. 38), zone (gel) electrophoresis of proteins (Exp. 42), osmotic pressure (Exp. 43), molecular weight distribution by gel filtration (Exp. 44), macromolecular properties as determined by light scattering (Exp. 45), and helix-coil transition as determined by polarimetry and viscosity measurements (Exp. 47).

This book is the result of crystallization of ideas during a 12-year period. Many people have contributed by their helpful suggestions and criticism. Among those, I am especially thankful to my former graduate assistants and to the students who were the first to try new experiments under classroom conditions. I would like to acknowledge the help of my colleagues Drs. D. Opalecky, D. Purins, R. Rudman, and S. Windwer. Furthermore, I would like to thank Dr. A. Gordus, University of Michigan, Dr. W. Brennen, University of Pennsylvania, Dr. M. Nichol, UCLA, Dr. H. Pardue, Purdue University, Dr. R. N. Porter, University of Arkansas, Dr. R. B. Solo, State University of New York, Stony Brook, and Dr. C. Trapp, University of Louisville, who reviewed the manuscript or parts of it and helped to put it in the final form with their constructive criticism and suggestions. In addition I would like to thank Miss Grace Dey for the typing of this manuscript. Last but not least, I am very grateful to the publisher and staff and especially to Mrs. Deanna St. Aubin, who as editor of this book was of immense help.

F. A. BETTELHEIM

Garden City, N.Y.

CONTENTS

IX KINETICS

X RADIOACTIVITY

XI ELECTROCHEMISTRY

XII SURFACE PHENOMENA

XIII MACROMOLECULES

XIV ELECTRONICS

by Donald T. Opalecky

XV APPENDIX

I

GENERAL PROCEDURES

LABORATORY WORK
AND REPORTS

The purpose of a laboratory course in physical chemistry, whether part of a general physical chemistry course or separate, is to familiarize the student with experimental approaches to the abstract concepts. Furthermore, since the theoretical part of an introductory physical chemistry course by its very nature emphasizes the limiting laws (ideal cases), the task falls upon the laboratory course to acquaint the student with the behavior of real systems. This dual aim can be achieved only: (1) if the student is well prepared before he starts the experiment; i.e., he is acquainted with the theoretical aspect of the phenomenon under investigation; (2) if he spends the time allotted for the laboratory in full concentration; i.e., he will be aware and notice if gross errors are committed and observe details that may lead to the explanation of unexpected results; (3) if he spends sufficient time to digest and sum up his experiences in the form of a laboratory report.

Let us consider these three points as prerequisites for the successful completion of this course. In regular physical chemistry courses the laboratory course is given simultaneously with the lectures, but in some colleges there may be a lag of half a year between the lectures and the laboratory course. In both cases, the student may have to perform experiments before being exposed to the theory of the phenomena under observation. This usually occurs because many physical chemistry experiments require elaborate setups.

For some experiments, such as determining the viscosity of gases or measuring density and partial molal volume, perhaps 10 to 15 identical setups can be provided so that a class of 20 to 30 students working in pairs can conduct the same experiment at the same time. But, usually a rotating schedule is in order. Even if a department is well equipped and has ten infrared spectrophotometers for students' use it is usually uneconomical to purchase a large number of identical instruments. For this reason, a rotating schedule is quite common in which one group does an experiment that will be done in succeeding weeks by other groups. This rotating schedule inevitably results in certain experiments being performed before the subject has been covered in lecture.

Since the laboratory period is usually limited and since students quite often work in groups of two or three, it is essential both for execution of the experiment and for meaningful cooperation between students that everyone knows in advance what the experiment is about and how it is carried out. For that purpose, a sufficiently detailed theoretical discussion is given in each experiment in this book, and it should be read and understood by the student before the experiment starts. He also should read the experimental procedures in advance so that the stepwise approach in the execution is ingrained.

The importance of advanced preparation cannot be emphasized sufficiently. A student who fails in this duty is more prone to laboratory accidents; by his fumblings and gropings he not only misses important observations but also delays his partners and other students in their performance.

Many experiments given in this book can be performed in a two to three hour laboratory period. Some may require four to five hours, depending upon the skill and advanced preparation of the student.

Regardless of whether the experiment is performed by one person or a group of students, *meticulous notes should be kept of the primary data obtained.* For example, in the conductance measurements of an electrolyte, accurate record of the concentrations of the solutions is of utmost importance. The primary data that should be recorded include the tare weight, the quantity of electrolyte, the dilution procedures, and the temperature of the thermostat. Triplicate conductance readings for each solution must be recorded, making sure that they are given for the correct concentrations. Keeping a meticulous record of primary data is even more important when the students work in groups because the responsibilities under such conditions are divided and the entire group is affected by the sloppiness of one member.

The division of labor in groups is also important. Some have more ability to make up solutions accurately than do others. Similarly, in taking some visual measurements, such as cathetometer readings or matching field intensities in the polarimeter, some students may perform better than others. Such division of labor develops naturally within a group. However, the instructor should make certain that each member of the group learns the basic techniques by taking at least some of the readings in the polarimeter, and so forth.

Most instructors request that students cease collaborating beyond taking the primary data, and each one should be responsible for interpreting the data obtained and writing the laboratory report. (A comparative reading of the laboratory reports of the members of a group usually discloses whether such collaboration has occurred.)

In the performance of the experiment, the usual safety precautions have to be observed. Most of the experiments in this book are quite safe, and the greatest hazards are damaging the instruments. However, some, especially those that involve high vacuum or high pressure systems (second virial coefficient, critical constants and vapor sorption on solids), require special precautions. Wearing safety glasses is necessary, as well as avoiding oral contact with organic chemicals and breathing in vapors that may cause ill effects, such as spilled mercury. In this respect, there should be special emphasis on working with dimethylsulfoxide. Although it is not poisonous, it has an unpleasant odor and *prolonged exposure to its vapor may cause headaches.* Also, because it penetrates the skin, if accidental spills occur it should be washed off quickly with water.

These hazards may also be minimized by the economy of the experiments. The student group that performs a number of experiments in the dimethylsulfoxide-water system should prepare sufficient amounts of the solutions in the first experiments on this binary system (usually the density and partial molal volume experiment) for use throughout the year This way the test solutions do not have to be prepared for each experiment, a saving in preparation time as well as in exposure time to the organic reagent. Other organic compounds of the binary systems (dimethylformamide, acetone, dioxane, and others) require the same safety precautions, and the solutions can also be prepared for the whole term or year in the first experiment.

The special safety measures that must be observed with radioactive isotopes (Exp. 33 and 34) and x-rays (Exp. 22 and 48) are described in these experiments and should be observed meticulously.

The part of the experiment involving calculation, organization, and interpretation of results is the most demanding of intellect and time. This part must be the work of the *individual* student and must show *his* understanding and capacity for thinking. (The format of this part is the laboratory report.) Many students spend as much or more time on the laboratory report than on the actual performance of the experiment.

Throughout the years, a relatively uniform format, consisting of four parts, has evolved regarding laboratory reports in most colleges.

1. The first part is a brief introduction, describing the purpose of the experiment and the general approach. No detailed theory or experimental technique is given because these are provided in this book. Only one or two paragraphs of summation of the purpose are required.

2. In the second part the primary data is reported. Organization is necessary in this part, because the data in the original notebook is recorded as it is gathered. On the other hand, in the laboratory report, reorganization is necessary to present the data in logical sequence or even in tabulated form.

3. The third part includes the calculations, the analysis of errors, and the presentation of the derived data in graphic or analytical form. In calculating the derived data, slide rule accuracy is usually not sufficient. Desk calculators for students' use are available in most departments. Moreover, some colleges provide computer facilities, and usually preliminary training in computer programming is also given.

Although it is beyond the scope of this book to provide such an introduction to programming, in the appendix a simple program is presented as developed for one with no previous computer training. Most calculations in the physical chemistry laboratory are not unduly long, and the use of computers may not be economical; there are a few experiments, however, in which the use of computers saves time (Exp. 22 and 45, for example). Programs for these two experiments are provided in the Appendix.

Beyond the calculations of the secondary data, analysis of errors is also in order. Part II of this book deals in detail with treatment of experimental data, and the student should read these pages carefully. In calculating standard deviations or fitting with least squares, desk calculators or computers may be used.

4. The last part is the interpretation of the data calculated. This is an intellectual exercise in which the student has to demonstrate his ability to assimilate the pertinent facts from literature and compare his data numerically with those reported or to observe certain trends that may show that certain laws are obeyed or deviations from such laws have occurred. The analysis of both systematic errors and random errors also has to be presented.

A sample laboratory report written by one of my former students illustrates the points mentioned here. (It is unedited to illustrate a good *student* laboratory report.)

SAMPLE REPORT— INTRINSIC VISCOSITY: CHAIN LINKAGE IN POLYVINYL ALCOHOL

Introduction: Theory and Experimental

The purpose of this experiment was to determine the fraction of head-to-head linkage in a sample of polyvinyl alcohol (PVOH). The 1,2-glycols are quantitatively cleaved by periodic acid; in PVOH, a head-to-head attachment is a 1,2-glycol. By treating a sample of PVOH with periodic acid (or periodate ion), the long-chain polymer is broken down into shorter pieces; the molecular weight is decreased. Thus, to determine the fraction of head-to-head linkages, a determination of molecular weight before and after periodate treatment is required.

Viscosity measurements can be used in the determination of molecular weight; what one must determine is intrinsic viscosity. In this experiment, viscosities of solutions were determined by measuring the time required for a given volume of liquid to flow through a vertical capillary tube under the influence of gravity. Throughout the measurements, temperature was kept constant, because η is dependent on temperature.

$$\eta_{rel} = \frac{\eta}{\eta_0} = \frac{t\rho_0}{t_0\rho}$$

η = viscosity of solution
η_0 = viscosity of H_2O (pure fluid solvent)
ρ_0/ρ = ratio of densities ≈ 1

t = time of flow of solution
t_0 = time of flow of H_2O } experimentally determined

By measuring times of flow, η_{rel} can be calculated directly. One can then calculate $\eta_{sp} = \eta_{rel} - 1$. The intrinsic viscosity $[\eta]$ is the ratio of the specific viscosity to the weight concentration of solute, in the limit of zero concentration

$$[\eta] = \lim_{c \to 0} \frac{\eta_{sp}}{c} = \lim_{c \to 0} \left(\frac{1}{c} \ln \frac{\eta}{\eta_0}\right)$$

By plotting either η_{sp} vs. c or $1/c \ln \eta/\eta_0$ vs. c or both and extrapolating to zero concentration, one can determine the intrinsic viscosity.

Polyvinyl alcohol is a statistically coiled molecule. $[\eta] = KM^a$ (a is a shape factor). For PVOH in aqueous solution at $25°$ C. $[\eta] = 2.0 \times 10^{-4} M^{0.76}$. In our experiment, involving a polydisperse sample of PVOH, the M in this formula is $\bar{M}_v$, the viscosity average molecular weight. When one determines the molecular weight of a polydisperse sample of PVOH, an average is obtained. If a colligative property is used as the basis of the molecular weight determination, a number average $\bar{M}_n$ is obtained for the molecular weight. If one uses light scattering measurements, the larger molecules contribute more than the smaller molecules to the average and one obtains a weight average molecular weight, $\bar{M}_w$. Whereas $\bar{M}_n = \bar{M}_w = \bar{M}_v$ for a monodisperse polymer, $\bar{M}_n < \bar{M}_v$ in a polydisperse sample. For PVOH, $\bar{M}_v/\bar{M}_n = 1.89$.

The purpose of the experiment is to obtain the fraction of attachments that are backward, i.e., head-to-head. The ratio one wants is the ratio of backward monomer units to total monomer units. (The method is predicated on the assumption that only and all 1,2-glycol structures are cleaved.) The ratio is Δ, the increase in number of molecules divided by total number of monomer units.

$$\Delta = 83[(1/\bar{M}_{v'}) - (1/\bar{M}_v)] \qquad\qquad \text{for PVOH}$$
$\bar{M}_v = $ before cleavage
$\bar{M}_{v'} = $ after cleavage

The procedure described in the manual has been followed.

PRIMARY DATA

Temperature of water bath $= 25.4°$ C.
Weight of PVOH $= 4.4035$ grams in 250 ml.
Weight of $KIO_4 = 0.2500$ gram
Viscosity of H_2O

First trial: 1 min. 35.9 sec.
          1 min. 35.5 sec.
          1 min. 35.3 sec.
Second trial: 1 min. 35.5 sec.   Average: 95.6 sec.
          1 min. 35.5 sec.
          1 min. 35.9 sec.

UNCLEAVED POLYMER

4 min. 45.5 sec.
1.3211 gram/100 ml.   4 min. 46.2 sec.   Average: 286.3 sec.
4 min. 47.3 sec.

0.8807 gram/100 ml.	3 min. 18.7 sec. 3 min. 21.2 sec. 3 min. 18.7 sec.	Average: 199.5 sec.
0.70456 gram/100 ml.	2 min. 53.1 sec. 2 min. 52.6 sec. 2 min. 53.7 sec.	Average: 173.1 sec.
0.4404 gram/100 ml.	2 min. 17.0 sec. 2 min. 17.4 sec. 2 min. 18.1 sec.	Average: 137.5 sec.

CLEAVED POLYMER

0.8807 gram/100 ml.	1 min. 48.5 sec. 1 min. 48.5 sec. 1 min. 48.9 sec.	Average: 108.6 sec.
0.7046 gram/100 ml.	1 min. 53.1 sec. 1 min. 52.9 sec. 1 min. 56.3 sec.	Average: 114.1 sec.
0.7926 gram/100 ml.	1 min. 50.9 sec. 1 min. 49.5 sec. 1 min. 50.3 sec.	Average: 110 sec.
0.4404 gram/100 ml.	1 min. 43.3 sec. 1 min. 43.7 sec. 1 min. 44.0 sec.	Average: 103.7 sec.

CALCULATIONS AND DISCUSSION

$t_{H_2O} = 95.6$ sec.

Sample calculation: uncleaved (original) polymer

0.8807 gram/100 ml.   199.5 sec.

$$\eta_{rel} = \frac{199.5}{95.6} = 2.09$$

$$\eta_{sp} = 2.09 - 1 = 1.09$$

$$\frac{\eta_{sp}}{c} = 1.24$$

$$\frac{1}{c}\ln\frac{\eta}{\eta_0} = 0.837$$

RESULTS OF CALCULATIONS

	η_{sp}/c	$(1/c) \ln \eta/\eta_0$	cg polymer/100 ml. soln.
Uncleaved (original) PVOH	1. 1.52	0.831	1.3211
	2. 1.24	0.837	0.8807
	3. 1.15	0.842	0.7046
	4. 1.00	0.827	0.4404
Cleaved polymer	5. 0.1535	0.144	0.8807
	6. 0.195	0.181	0.7926
	7. 0.277	0.253	0.7046
	8. 0.193	0.185	0.4404

The values are then plotted (see graphs.)

For the uncleaved (original) polymer, the intrinsic viscosity $[\eta] = 0.74$. This is the intercept of the η_{sp}/c vs. c graph, extrapolated to zero concentration. This graph shows good linearity. The $(1/c) \ln \eta/\eta_0$ vs. c graph, on the other hand, is not, by our data, strictly linear. Extrapolating to give concentration by use of this graph would be most difficult.

For the cleaved [periodate-treated] polymer the intrinsic viscosity is given as 0.225 in one graph and 0.232 in the other. In the calculations I have used the value 0.23. Two experimental points have been disregarded on the graph because the concentrations corresponding to those two points were the solutions used *last* in the experiment. The relative viscosities of these two solutions were *greater* than that of the more concentrated cleaved polymer solution (which was used *before* these two solutions were). Clearly something is amiss. We noted a few specks of foreign material in the two final solutions. This was probably the cause of the unexpectedly high viscosity registered. Possibly the capillary became semiclogged. Since these two solutions were the final ones used and the bizarre readings are associated with them, we have disregarded these two experimental points.

$$[\eta] = 2.0 \times 10^{-4} M_v^{0.76}$$

$$[\eta] = 0.74 \text{ original (undegraded) polymer}$$

$$0.74 = 2.0 \times 10^{-4} \bar{M}_v^{0.76}$$

$$3700 = \bar{M}_v^{0.76}$$

$$\log 3700 = 0.76 \log \bar{M}_v$$

$$4.7000 = \log \bar{M}_v$$

$$50{,}120 = \bar{M}_v \text{ original}$$

$$1.89 = \frac{50{,}120}{M_n}$$

$$\bar{M}_n = 26{,}500$$

The same procedure is followed for the degraded polymer, using $[\eta] = 0.23$

$$\bar{M}'_v = 10,718$$

$$\bar{M}'_n = 5,670$$

$$\Delta = 83\left[\frac{1}{\bar{M}'_v} - \frac{1}{M_v}\right]$$

$$\Delta = 83\frac{1}{10,700} - \frac{1}{50,120} = 83(0.934 \times 10^{-4} - 0.200 \times 10^{-4}) = 60.9 \times 10^{-4}$$

Thus,

$$\left.\begin{array}{l} \bar{M}_v = 50,120 \\ \bar{M}_n = 26,500 \end{array}\right\} \text{ undegraded (original) polymer}$$

$$\left.\begin{array}{l} \bar{M}'_v = 10,700 \\ \bar{M}'_n = 5,670 \end{array}\right\} \text{ degraded polymer}$$

$$\Delta = 0.00609/1$$

The ratio of backward monomer units to total monomer units is 0.00609/1; there are very few backward units in comparison with total number of units; most units are joined, therefore, head-to-tail. We do not know the polymerization temperature; however, we do know that the frequency of head-to-head linkages increases with increasing polymerization temperature. Since we found the frequency of head-to-head linkages in this sample of PVOH to be relatively small, the polymerization temperature must have been relatively low.

In this experiment, the temperature must be kept constant, because $\eta = f(t)$. We assume $\rho_0/\rho = 1$; for more accurate work, the densities of the solutions must be determined by use of a Westphal balance or pycnometer. Our H_2O bath was 25.4° C. The formula $[\eta] = 2.0 \times 10^{-4}M^{0.76}$ was determined for PVOH in aqueous solution at 25° C. The difference in temperature, of course, led to an error in the calculations.

Thus, viscosity measurements of solutions of high polymers can be used to determine their molecular weights, which can then be used in various ways, such as determining Δ.

II

TREATMENT OF EXPERIMENTAL DATA

The task in a physical chemical experiment is to produce numerical solutions to the problem investigated. The numerical value is the result of the application of arithmetical calculations to the raw data obtained in the experiment. These calculations are based on the formulation of laws of physics and chemistry, most of which are statistical. This means that the laws result from observation of the behavior of systems containing very large numbers of particles and, therefore, represent the average behavior. The averaging process is necessary in the formulation because local fluctuations do exist in the systems observed. For example, if we consider the refractive index of a macroscopic sample, which can be given in terms of a numerical value under a set of conditions, such as sodium, D line, 25° C., 1 atm. it represents the average value, $\bar{n}$, as indicated in Figure 1.

On the other hand, at any specific microscopic location, x, in the sample this may not be the true refractive index since density fluctuations do occur because of random thermal motion of the molecule. In order to describe the behavior of the system some parameter besides the average value often must be supplied. This other parameter may be the average (mean) deviation from the mean value ($\bar{d}$ in Figure 1).

This problem is not restricted to statistical behavior of systems of large numbers of particles and the interpretations of such behavior on the atomic or molecular level. It similarly applies to *nonuniform* macroscopic samples, such as a powder sample in which the grain sizes may vary widely. In the case of vapor sorption on solid powder, the overall effect depends upon the surface area of all the grains regardless of size. This average behavior may be quite different from the local behavior of sorption of vapor on one specified powder particle. Again, in the total description of the system a quantity is necessary that relates the average value obtained to the average local fluctuation. A similar effect to that represented in Figure 1 can be observed, for instance, if the deflections of a torsion balance are plotted on the ordinate against time. The fluctuation of the balance, a macroscopic body, is caused by Brownian motion, but now it is a function of time rather than space.

It is obvious from this discussion that in calculating numerical values of experimental data one always must be concerned about the degree of uncertainty to which this numerical value is subject. The usual laboratory experiment is subject to a number of errors in its execution, and the sum total of these errors represents the uncertainty of the result.

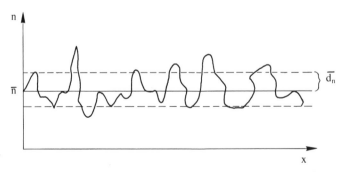

Figure 1   *Local fluctuation in refractive index of a sample as a function of distance.*

Systematic and Random Errors

Analytical errors are usually divided into two types—systematic and random. *Systematic errors* are due to factors operating according to certain laws. Since we know the mode of operation of these laws, they can be corrected. Hence, they are sometimes called correctable or corrigible errors. Examples of systematic error are measurements with an instrument at a temperature other than that for which the instrument is calibrated, unequal buoyancy of air on the unknown and standard weights in using a balance, leakage of gases in a high vacuum apparatus, incomplete experimental conditions (not allowing sufficient time for equilibration in the osmotic pressure experiments, incomplete dehydration of a sample), and consistent operational errors (not leveling the cathetometer in reading mercury column heights). All these may be eliminated or corrected for (at least in principle) if sufficient attention is focused on calibration, controls and blank experiments, and other factors. The most serious problem with systematic errors is that quite often they are not detected and, therefore, they are not corrected for.

Random errors are those beyond the control of the observer. They affect the reproducibility of the results. In contrast, most systematic errors are reproducible since they are errors of method or technique.

Examples of random errors are in reflexes of the observer in timing efflux time in viscosity measurements, in the polarimeter readings in which the observer has to match the intensity of two half-shade fields, and in the density measurements of solids by flotation technique. In each case each measurement represents a fluctuation about a mean and in principle the *precision* may be improved by performing more and more measurements so that the averaging process has more statistical significance.

The key word is the precision of the measurements. This refers strictly to the *reproducibility* of the measurement and, therefore, the uncertainty produced by random errors. In contrast, *accuracy* refers to the total errors, both systematic and random.

The distribution of errors between systematic and random errors is sometimes relative. For example, if we use a set of weights that are defective, the weighing errors are systematic errors. On the other hand, if we consider the defect in the set of weights as being an error in their manufacture, it may be considered a random error by the manufacturer.

We make this distinction between systematic and random errors for operative purposes. Systematic errors, when detected, must be corrected by applying known laws. Random errors, on the other hand, are treated according to the laws of probability.

The first important aspect of this treatment is that the result calculated from a greater number of measurements on the same system has a greater probability of approaching the *true* value, assuming that the systematic errors have been corrected. This, however, does not mean that the probability factor is directly proportional to the number of trials. In fact, the probability is proportional to the square root of

TABLE I. Precision of Measurements

Number of Measurements (n)	Increase in Number of Measurements	Uncertainty Ratio
1	1	1:1
4	3	1:2
9	5	1:3
16	7	1:4
25	9	1:5
36	11	1:6
144	108	1:12

the number of trials. This is a very important point in our quest for the given quantity, since economy of time may be a very significant factor. We conclude from the foregoing that we should take more than one measurement, and if time were not a consideration the more measurements we could take, the better the results would be. When we consider the element of time, however, Table 1 gives us an idea of the diminishing return in the improvement of results with additional measurements.

The uncertainty ratio expresses the improvement of chances (or the diminishing of random errors) that the true value will be obtained when n number of trials are compared to just one trial. It is evident that four measurements will have an uncertainty ratio of 1 to 2 over one measurement, whereas only three additional measurements are required. However, 144 measurements have an uncertainty ratio of 1 to 2 over 36 measurements, but 108 additional measurements are needed. In the latter case one must carefully weigh the justification in expanding such effort for a comparatively small increase in precision. For this reason, ordinary measurements are rarely taken more than 9 or 16 times.

Frequency Distribution; Location and Dispersion Indices

As the number of measurements is increased, more and more of the fluctuation is recorded (as in Fig. 1) and, therefore, a set of measurements has a frequency distribution. In such a distribution the frequency with which a definite value appears in the set of measurements is recorded. This can be represented either in tabular form or more often in graphical form.

The terminology of distribution distinguishes between discrete and continuous distribution. A typical example in physical chemistry of a discrete frequency distribution is a gel filtration experiment. A solution containing one type of polymer molecules but of different molecular weights is put on a column containing gel beads. The lower molecular weight polymers penetrate the pores of the gel beads while the

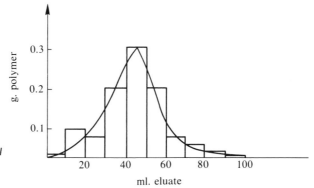

Figure 2 *Histogram of gel filtration experiment.*

larger molecular weight compounds follow a pathway between the beads. The consequence is that first the high molecular weight polymers appear in the eluate and those of smaller molecular weights appear in later elution volume. Thus, the elution volume is inversely related to the molecular weight. The experimenter collects 10 ml. portions of the eluate and determines the polymer content. The results are given in Figure 2. Since only every 10 ml. of eluate is collected and analyzed, the data have to be presented in block diagram form. Each block represents a *class*, which has both upper and lower limits. The first box represents the eluate from 0 to 9 ml., the second from 9 to 19, and so forth. The frequency (concentration in this case) of the polymer molecules is recorded for each box. For such block-area type distribution the term histogram is often used. It is a discrete distribution since the volumes of eluates collected are discrete quantities. The smooth curve fitted over the histogram in Figure 2 is the *frequency curve*. A histogram approaches a frequency curve as the size of the *class* is reduced to zero and the number of classes in a set of measurements goes to infinity. The fitted frequency curve is, then, a continuous frequency distribution. A similarly discrete distribution is obtained, for example, if one takes 20 measurements on the optical rotation of a sample, as in Figure 3. The broken line is the frequency polygon and the fitted smooth curve again is the frequency curve that would be approached if the size of the classes could be reduced to zero. (In practice, the optical rotation should be obtained to more than just two significant figures.)

The main distinction between a discrete and a continuous frequency distribution is that the discrete distribution has a unit sum, whereas the continuous distribution

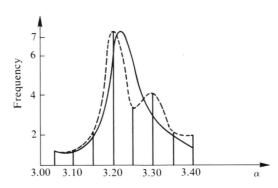

Figure 3 *Optical rotation of a sugar solution.*

has a unit area. Therefore, the latter cannot have ordinate values greater than one.

Once a frequency distribution graph is presented it is of primary interest to design numerical descriptions of the distribution. Two main indices serve this purpose: the location of the center of the distribution and the spread of the dispersion. The former is the location index and the latter the dispersion index.

Location Indices. The most important location index is the *mean* or arithmetic average. Referring to Figure 2, assume that $f_1, f_2, \ldots, f_n$ are the frequencies of the various classes and $x_1, x_2, \ldots, x_n$ are mid-values of the variable (eluate volume). The weighted mean, $\bar{x}$, of this distribution is

$$\sum_{i=1}^{n} f_i x_i \bigg/ \sum_{i=1}^{n} f_i = \bar{x} \tag{1}$$

where n is the total number of observations. This is the mean of a discrete (experimental) distribution.

The hypothetical or mathematical mean, $\bar{\mu}$, is defined by a relation similar to equation 1 under the condition that $n \to \infty$. Since the hypothetical or mathematical mean refers to a frequency curve (continuous distribution), the summation in equation 1 is replaced by an integral. Only $\bar{x}$ can be determined experimentally, but the value of $\bar{\mu}$ is approached when n, the number of observations, increases as was illustrated in Table 1.

The calculation of $\bar{x}$ is usually laborious even with aids such as desk calculators, and it can be simplified by selecting a *working mean* or *assumed mean*, w, which is not far from the mean, $\bar{x}$. Thus

$$\bar{x} = w + \bar{\Delta} \tag{2}$$

and if the selection of w is right, $\bar{\Delta}$ is a small number. An example will illustrate this technique. The data in Figure 2 can be represented in tabular form as in Table 2. If we calculate the *mean* according to equation 1, we have to tabulate the $f_i x_i$ listed in the third column. The mean is found to be 44.10. If, on the other hand, a quick inspection of the second column indicates that the mean may be approximately 45, we can select this as a working mean, w. The Δ_i are then listed in the fourth column, and the $\Delta_i f_i$ products are listed in the fifth column. Since $\sum \Delta_i f_i / \sum f_i = \bar{\Delta}$, the $\bar{x}$ can be calculated from equation 2.

$$\bar{x} = 45 - 0.9 = 44.1$$

TABLE 2. Calculation of Mean

x_i	f_i	$f_i x_i$	Δ_i	$\Delta_i f_i$
5	0.01	0.05	−40	−0.4
15	0.10	1.5	−30	−3.0
25	0.06	1.5	−20	−1.2
35	0.20	7.0	−10	−2.0
45	0.30	13.5	0	0
55	0.20	11.0	+10	+2.0
65	0.06	3.9	+20	+1.2
75	0.04	3.0	+30	+1.2
85	0.02	1.7	+40	+0.8
95	0.01	0.95	+50	+0.5
Σ	1.0	44.10		−0.9

The use of a working mean is especially useful with a large number of observations, since the amount of arithmetic is reduced and thereby the likelihood of computational error is also reduced.

Another location index is the *median*. This is the middle measurement of the set when they are arranged in increasing order. If there are an odd number of observations, $2n + 1$, the median is the $(n + 1)^{th}$ value. If the number of observations is even, $2n$, the median is the average of the nth and $(n + 1)^{th}$ observation. For example, in Table 2 there were an even number of measurements (ten); the *median* is 50 (i.e., the average of 45 and 55).

The third location index is the *mode* or the most probable value. This is the value of the abscissa where the distribution curve has its maximum. In the case of the data in Figure 2 or Table 2, the mode is 45. Thus, if, for example, the molecular weight of the polymer, M.W., is related to the elution volume by $V = 1 \times 10^6/\text{M.W.}$, the mean molecular weight of the sample would be 22,700, the median 20,000, and the mode 22,200.

Among the three location indices, the mean is the most frequently used. The median and the mode are given together with the mean, especially when the frequency distribution is strongly skewed (asymmetrical).

Dispersion Indices. Beyond the location indices, mean, median, and mode, the dispersion or scatter about the mean, median, or mode is a very important characteristic. We shall deal with the dispersion problem in two parts. First, we will discuss the general indices of any kind of dispersion, and then we will classify special types of dispersion and characterize the dispersion indices in these special cases.

In general, the three most important dispersion indices of any distribution are the range, the mean deviation, and the standard deviation. The *range* simply indicates the limits of the dispersion. In the case used in Figure 2 and Table 2, the range is 0 to 100 ml.—that is, the smallest and the greatest value of the variable. Obviously, as the number of measurements increases, the range usually increases; therefore, it is useful to indicate the number of trials whenever the range of the dispersion is specified.

The *mean* or *average deviation* can be visualized by referring to Figure 1. In a set of data $x_1, x_2, \ldots, x_n$, in which the mean is indicated by $\bar{x}$, the individual deviations from the mean are indicated by $x_1 - \bar{x}, x_2 - \bar{x}, \ldots, x_n - \bar{x}$. Let us designate these deviations by $d_1, d_2, \ldots, d_n$. Some of these differences are positive and some of them are negative, but the sum of these differences is zero.

$$d_1 + d_2 + \cdots d_n = 0$$

The mean deviation, $\bar{d}$, is defined as

$$\frac{1}{n} \sum_{i=1}^{n} |d_i| = \bar{d} \tag{3}$$

that is, the sum of the absolute values of the deviations (disregarding signs) divided by the number of observations.

The most important measure of dispersion is the standard deviation, σ. This is defined in terms of the squares of the individual deviations.

$$\sigma^2 = (d_1^2 + d_2^2 + \cdots d_n^2)/(n - 1) \tag{4}$$

Hence

$$\sigma = \left[\frac{1}{(n-1)}\sum_{i=1}^{n}d_i^{\,2}\right]^{1/2} = \left[\frac{1}{(n-1)}\sum_{i=1}^{n}(x_i - \bar{x})^2\right]^{1/2} \qquad (5)$$

The use of $(n-1)$ instead of n in equations 4 and 5 is known as Bessel's correction, and it is important if n is not large.

If each observation, x_i, has a frequency of f_i, the standard deviation has the form

$$\sigma = \left[\sum_{i=1}^{n}f_i(x_i - \bar{x})^2 \Big/ \left(\sum_{i=1}^{n}f_i\right)\right]^{1/2} = \left[\sum_{i=1}^{n}f_i\, d_i^{\,2} \Big/ \left(\sum_{i=1}^{n}f_i\right)\right]^{1/2} \qquad (6)$$

The σ^2 is known as the *variance*, and the standard deviation, σ, is the root mean square deviation of the data measured from the mean.

Similarly, the variance in the mean can also be obtained from an equation analogous to equation 4.

$$\sigma_{\text{mean}}^2 = [1/n(n-1)] \sum_i d_i^{\,2} \qquad (7)$$

The two dispersion indices, the mean deviation and the standard deviation, are used frequently in scientific literature. However, statisticians prefer the standard deviation to the mean deviation as a dispersion index. The reason is that if, for example, the whole set of data is divided into subsets and, for economy, the dispersion of a small subset is calculated as an index for the total set, the mean deviation values of subsets, d_j's, will show a rather larger scatter about the mean deviation of the total set, $\bar{d}$. The scatter of the standard deviation values of subsets, σ_j, about the standard deviation of the whole set, σ, is much less. Conversely, a larger number of observations are needed to achieve the same precision index if the mean deviation rather than the standard deviation is used as an index.

The evaluation of the standard deviation of the set of data presented in Table 2 and Figure 2 is given in Table 3. (The mean, 44.1, is calculated in Table 2.) According to equation 6, σ is determined by calculating the values in the third, fourth, and fifth columns from those in the first and second columns and using the sums of the pertinent columns.

TABLE 3. Calculations of Standard Deviation

x_i	f_i	$(x_i - \bar{x})$	$(x_i - \bar{x})^2$	$f_i(x_i - \bar{x})^2$	Δ_i	Δ_i^2	$f_i\,\Delta_i^2$
5	0.01	−39.1	1528.8	15.28	−40	1600	16
15	0.10	−29.1	846.8	84.68	−30	900	90
25	0.06	−19.1	364.8	21.88	−20	400	24
35	0.20	−9.1	82.8	16.25	−10	100	20
45	0.30	+0.9	0.8	0.24	0	0	0
55	0.20	+10.9	118.8	23.76	+10	100	20
65	0.06	+20.9	436.8	26.20	+20	400	24
75	0.04	+30.9	954.8	38.19	+30	900	36
85	0.02	+40.9	1672.8	33.45	+40	1600	32
95	0.01	+50.9	2590.8	25.90	+50	2500	25
Σ	1.0	251.80	8598.1	286.28	+10	8500	287.0

$\sigma = (286.28/1)^{1/2}$ 　　　　　　　　　$\sigma^2 = 287.0 - (45 - 44.1)^2 = 286.2$

$\sigma = 16.9$ 　　　　　　　　　　　　　　$\sigma = 16.9$

Considerable economy in arithmetic is achieved if we again use a working mean, w, which is selected by visual inspection and which is not far from the actual mean, $\bar{x}$, of the set. Again, as in equation 2

$$\Delta_i = x_i - w$$

Since the quantity needed to calculate σ can be written in terms of working mean, w, as

$$\sum f_i(x_i - \bar{x})^2 = \sum f_i[(x_i - w) + (w - \bar{x})]^2 \tag{8}$$

by expanding the square term and dividing both sides by $\sum f_i$, we obtain

$$\sigma^2 = \frac{\sum f_i(x_i - \bar{x})^2}{\sum f_i} = \frac{\sum f_i(x_i - w)^2}{\sum f_i} + 2(w - \bar{x})(\bar{x} - w) + (w - \bar{x})^2$$

$$= \frac{\sum f_i(x_i - w)^2}{\sum f_i} - (w - \bar{x})^2 = \frac{\sum f_i \Delta_i^2}{\sum f_i} - (w - \bar{x})^2 \tag{9}$$

Such a calculation of σ is given in the sixth, seventh, and eighth columns in Table 3.

TYPES OF FREQUENCY DISTRIBUTIONS

Beyond the location and dispersion indices, a frequency distribution is characterized by the shape of the frequency curve. There are an infinite variety of shapes of distribution curves. However, most of them approach closely one of the three types of frequency distributions that were derived mathematically by using the theory of probability—binomial, Poisson, and Gaussian or normal distributions.

The *binomial distribution* originates from a probability situation in which there are only two possible outcomes of a trial, called the Bernoulli trials. Tossing a coin is one example, the two possible outcomes being heads or tails.

In most physical and chemical measurements similar situations may exist. There are two possibilities that a certain measurement will or will not have a definite value, x; these two possibilities are called success or failure. The x is within the domain of the total number of trials, n, which determines the range of distribution. If the probability for success is p and for failure is q, the total probability of having value x in n trials is

$$B(x; n, p) = \binom{n}{x} p^x q^{n-x} \tag{10}$$

In equation 10, $B(x; n, p)$ is the probability given by the Bernoulli or binomial expression of having x successes in n trials. This is called a binomial expression of distribution because the total probability of successes and failures must be one

$$(p + q) = 1 \tag{11}$$

$$1 = (p + q)^n = p^n + \binom{n}{1} p^{n-1} q + \binom{n}{2} p^{n-2} q^2 + \cdots + \binom{n}{x} p^x q^{n-x} + \cdots + q^n \tag{12}$$

The right side of equation 12 is the Newtonian binomial expansion and the symbols

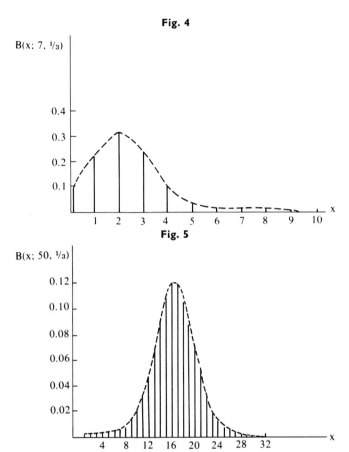

Figures 4 and 5 *Binomial probability distributions with a one-third probability for success but a different number of trials (n).*

of $\binom{n}{x}$, and so forth, are called the binomial coefficients. These are the number of combinations and, therefore, they are

$$\binom{n}{x} = \frac{n!}{x!\,(n-x)!} \tag{13}$$

In Figures 4 and 5 two binomial distributions are given with the same probability of success, $p = 0.33$, but with a different number of trials. It can be seen that the binomial distribution is *asymmetrical* except when the probability of a success has a value of 1/2, as in tossing a coin. However, the asymmetry decreases as the number of trials increases.

Furthermore, the mean (average) of a binomial distribution is

$$\bar{x} = np \tag{14}$$

and the standard deviation is

$$\sigma = (npx)^{1/2} \tag{15}$$

The most important thing about binomial distribution is that it is a discrete function. The variable x can have only positive integer values. This is important to remember because the binomial distribution can be looked upon as the general case of the two most often used error functions, i.e., the Poisson and Gaussian distributions.

The *Poisson distribution* is a special case of the binomial distribution and it occurs when n becomes very large and p, the probability of the event, very small, but in such a fashion that the mean, np, is appreciable. For example,

$$np \ll (n)^{1/2} \tag{16}$$

Under such conditions, the Poisson probability of x successes in n trials, with p individual probabilities, $p(x; n, p)$, is

$$P(x; np) = \frac{np^x e^{-np}}{x!} \tag{17}$$

Equation 17 is the Poisson distribution equivalent to the binomial distribution formula given in equation 10.

There are many cases in physical and biological sciences in which the Poisson distribution is applicable also for errors. The important feature to remember is that the Poisson distribution, as well as its general case, the binomial distribution, is discrete and not continuous.

In contrast, the most often used error curve, the *normal or Gaussian distribution*, is a continuous function. It can be looked upon as a special limiting case of the binomial distribution when n becomes very large and p is also large, and $np \gg 1$. For example, if the probability of an event is $p = 1/2$, the approximation of the distribution by a normal or Gaussian distribution is fairly good as long as $np > 5$ and $nq > 5$.

The Gaussian (normal) probability of x successes in n trials is $G(z, h)$.

$$G(z, h) = \frac{h}{\sqrt{\pi}} e^{-h^2 z^2} \tag{18}$$

where $h = 1/(2npq)^{1/2}$ and $z = (x - np)$.

Since in a binomial distribution for which the Gaussian type is a special case the mean, $\bar{x}$, is equal to np and the standard deviation, σ is $(npq)^{1/2}$, we can rewrite equation 18 in these terms

$$G = \frac{1}{\sigma(2\pi)^{1/2}} e^{-(x-\bar{x})^2/2\sigma^2} \tag{19}$$

This then is the normal distribution curve, and the area under the curve is equal to unity. The most important aspect of the Gaussian distribution is that *it is continuous and x can take any value.* Such a distribution is given in Figure 6 with different standard deviations.

An experimental histogram more often than not can be fitted with a normal distribution curve, as in Figure 7. In such a case one has to be aware that although the histogram is discontinuous, the normal Gaussian distribution is continuous, but the former may closely approximate the normal distribution.

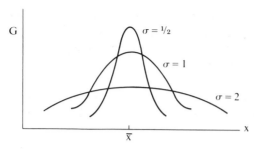

G

$\sigma = \frac{1}{2}$

$\sigma = 1$

$\sigma = 2$

x

$\bar{x}$

Figure 6   *Normal error curves.*

Among the dispersion indices (equations 3, 5, and 6) we used the mean, $\bar{d}$, and standard, σ, deviations. For normal distribution curves, the relationship between these two dispersion indices is

$$\bar{d} = \sigma/(1/2\pi)^{1/2} \cong 4/5\sigma \tag{20}$$

Therefore, the $\bar{d}/\sigma$ ratio is sometimes used to estimate whether a distribution departs substantially from the normal distribution.

In most normal distribution curves the quantity plotted on the abscissa is not the variable x, but the deviation from the mean $x - \bar{x} = d$. Therefore, the normal distribution curve has its maximum at $d = 0$.

Finally, because a Gaussian distribution fitted to the experimental histogram is only an approximation, we may have errors in both the location and the dispersion index (mean and standard deviation) because of inadequate sampling. This problem can be understood if we refer again to the nature of the Gaussian (normal) distribution, which is continuous, and its range is, therefore, between $+\infty$ and $-\infty$.

If we select a random number of n observations from an infinite population, it is likely that the mean of this sample will not be the same as the mean of the whole population. If n is large, the two means may not differ very much; but if n is small, the difference may be appreciable.

The *means* of different samples of the total population have a normal distribution, and the mean of the distribution is the mean of the total population. Furthermore, the standard deviation (error) in the mean, σ_m, of a sample of n observations is given by

$$\sigma_m = \sigma/(n^{1/2}) \tag{21}$$

where σ is the standard deviation of the whole population. Therefore, it is useful in dealing with any set of data to find the mean, $\bar{x}$, first and then to write it as $\bar{x} \pm \sigma/n^{1/2}$.

Similarly, the standard deviation of a set of data may differ from the standard deviation of the total population. Therefore, when the standard deviation of a finite

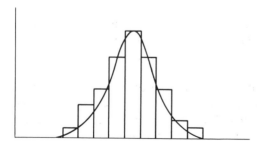

Figure 7   *Histogram   fitted   with Gaussian distribution curve.*

set is given, the *standard error* of the standard deviation also has to be given. Since the latter is $\sigma/(2n)^{1/2}$, the standard deviation of a set is written as

$$\sigma_s = \sigma\left[1 \pm \frac{1}{(2n)^{1/2}}\right] \tag{22}$$

In scientific literature another index, the *probable error*, is also often used, although its use as an index is discouraged by statisticians. The probable error, PE, is defined as the size of an error such that half of all observations taken will be within the range $\bar{x} \pm$ PE. For normal distributions the probable error is

$$PE = 0.674\sigma \tag{23}$$

This is true regardless of whether we are talking about the probable error of a single observation or the probable error of the mean, in which case equation 23 becomes

$$PE_m = 0.674\sigma_m \tag{24}$$

From this discussion it is clear that various error indices appear in the literature. Therefore, it is very important that the student mention *explicitly* which error index is used in his calculations.

Significant Figures

At this point it is advisable to review the subject of significant figures. A digit is significant when it may be considered to be of importance, no matter how small, in a given computation. The exact dividing line between significant and nonsignificant figures is a very difficult one to draw. For this reason, on the basis of experience, experimenters arbitrarily set the line of demarcation. Although this may differ slightly for different groups of experimenters, the variation in the final result is of small consequence. We will consider a figure or digit not significant when it is in doubt by a value of 50 or more as indicated by its mean deviation. Thus, in the quantity 25.36 ± 0.81, the digit 6 is not significant and should be dropped from the number. It is now 25.4 ± 0.8.

Precision measures should never be represented by more than two significant figures and sometimes even by only one.

In performing calculations, first determine the number of significant figures in the quantities involved. Never add nonsignificant digits to or subtract them from significant ones. In multiplication and division, remember that the number of significant figures in the result cannot exceed the number of figures in the term with the least number of significant figures. It cannot be sufficiently emphasized that applying these rules not only greatly shortens the time for computations but also is the only correct procedure. Failure to apply these rules leads to false results. The following short problem illustrates the use of significant figures as described.

In the determination of the density of a liquid, an empty pycnometer is weighed and filled with water and the liquid at 25° C. The following data are obtained:

Weight of empty pycnometer = 5.9342 ± 0.0005 gram
Weight of pycnometer filled with water = 15.8974 ± 0.0005 gram
Weight of pycnometer filled with liquid = 16.2365 ± 0.0005 gram
Density of water at 25° C. = 0.997044 ± 0.000002 gram/cc.

Calculations

$$
\begin{array}{ll}
\quad 15.8974 & \quad 16.2375 \\
-5.9342 & -5.9342 \\
\hline
\quad 9.9632 \text{ grams water} & \quad 10.3033 \text{ grams liquid}
\end{array}
$$

$$
\frac{10.3033}{9.9632/0.9970} = 1.0310 \text{ grams/cc.} \quad \text{density of liquid}
$$

In this example the estimates of error in weights may come from the analysis of multiple weight measurements on the analytical balance. On the other hand, estimates of error such as here are frequently attached to a single measurement on the basis of intuition and prior experience with the analytical balance. The estimates of error in density are taken from the *Handbook of Chemistry and Physics*.

Propagation of Errors

In the previous sections we have seen that when the physical chemical quantity desired is measured directly, we can calculate the errors and, hence, the precision of the measurements by giving location and dispersion indices. The most frequently used location index is the *mean*, $\bar{x}$, and the most frequently used dispersion index is the mean (average) deviation, $\bar{d}$.

Often, however, the task of a physical chemist is to determine a quantity that cannot be measured directly and must be calculated from a number of directly measured quantities. The measured quantities are usually averaged and the arithmetic mean is used in the final computation. The precision of the final result depends upon the precision (dispersion) indices of the measured quantities. Thus, the errors of the individual measurements are propagated in the final result.

Let us assume that the calculated quantity, Y, is a function of a number of measured quantities, x_1, x_2, x_3, and so forth.

$$
Y = f(x_1, x_2, x_3, \ldots) \tag{25}
$$

The uncertainty in Y is the result of the errors in x_1, x_2, x_3, and so forth. The uncertainty in x_1 causes

$$
-\delta Y_1 = \left(\frac{\partial Y}{\partial x_1}\right)\delta x_1 \tag{26}
$$

In equation 26 the δx_1 can be the mean deviation, standard deviation, or any other precision (dispersion) index for the arithmethic mean of the experimental quantity. The total uncertainty in Y, δY, is *the most probable error in* Y.

$$\delta Y = [(\delta Y_1)^2 + (\delta Y_2)^2 + (\delta Y_3)^2 + \cdots]^{1/2} \tag{27}$$

Combining equations 26 and 27 we obtain

$$\delta Y = \left[\left(\frac{\partial Y}{\partial x_1}\right)^2 (\delta x_1)^2 + \left(\frac{\partial Y}{\partial x_2}\right)^2 (\delta x_2)^2 + \cdots\right]^{1/2} = \left[\sum_{J=1}^{J}\left(\frac{\partial Y}{\partial x_J}\delta x_J\right)^2\right]^{1/2} \tag{28}$$

If, on the other hand, the maximum propagated error, δY_m, rather than the most probable error is desired, a simple summation formula is used.

$$\delta Y_m = \delta Y_1 + \delta Y_2 + \delta Y_3 + \cdots = \sum_{j=1}^{J}\delta Y_J \tag{29}$$

All values in the partials and the δ's of δY_J (equation 26) are considered to be positive; hence, the maximum propagated error is also positive.

In general, if the computed quantity, Y, is a function of three or more experimentally measured quantities, the maximum propagated error overstates the error in the Y and it is better to use the most probable propagated error as an index.

For example, one way to determine the viscosity coefficient of a liquid, η, would be to use the directly measured quantities, i.e., the efflux time of the liquid, t, and a number of instrumental parameters, such as the volume of liquid, V, the length, l, and the radius, R, of the capillary tube, and the hydrostatic pressure, ΔP. In general, we can write that

$$\eta = f(V, l, R, \Delta P, t) \tag{30}$$

and the mean deviation of the viscosity coefficient is

$$d_\eta = \left[\sum_{j=1}^{J}\left(\frac{\partial \eta}{\partial x_j}\right)^2 \delta x_j^{\,2}\right]^{1/2} \tag{31}$$

for J independent variables.

The relationship between the viscosity coefficient and the directly measure quantities is given by Poiseuille's equation

$$\eta = \frac{\pi \Delta P R^4 t}{8lV} \tag{32}$$

In order to determine the mean (average) deviation of the viscosity coefficient, $d_\eta = |\eta_i - \bar{\eta}|$, the mean deviations of the individual variables ($\delta \Delta P$, δR, δl, δV, and δt, commonly designated as δx_j in equation 28) must be known. Then, with the aid of differential calculus, we can write

$$\delta \eta_i = \left(\frac{\partial \eta}{\partial t}\right)_{\Delta P, R, l, V} \delta t + \left(\frac{\partial \eta}{\partial \Delta P}\right)_{R, l, V, t} \delta \Delta P$$

$$+ \left(\frac{\partial \eta}{\partial R}\right)_{\Delta P, t, l, V} \delta R + \left(\frac{\partial \eta}{\partial V}\right)_{R, l, \Delta P, t} \delta V + \left(\frac{\partial \eta}{\partial l}\right)_{R, \Delta P, V, t} \delta l \tag{33}$$

and

$$d_\eta = \left[\left(\frac{\partial \eta}{\partial t}\right)^2 \delta t^2 + \left(\frac{\partial \eta}{\partial R}\right)^2 \delta R^2 + \left(\frac{\partial \eta}{\partial V}\right)^2 \delta V^2 + \left(\frac{\partial \eta}{\partial l}\right)^2 \delta l^2\right]^{1/2} \quad (34)$$

Using the preceding example and Poiseuille's equation (32), equation 33 is rewritten as follows:

$$\delta\eta = \frac{\pi R^4 \Delta P}{8lV} \delta t + \frac{\pi R^4 t}{8lV} \delta \Delta P + \frac{\pi \Delta P t R^3}{2lV} \delta R - \frac{\pi \Delta P R^4 t}{8V^2 l} \delta V - \frac{\pi \Delta P R^4 t}{8 l^2 V} \delta l \quad (35)$$

Let us assume that a student wishes to determine the viscosity coefficient of an unknown liquid by using Poiseuille's equation, and the following mean values and mean deviations have been established in the individual measurements

x_j	$\bar{x}_j$	δx_j
P	0.010	± 0.0050 atm.
R	0.052	± 0.001 cm.
l	10.10	± 0.01 cm.
V	5.856	± 0.001 cm.^{3}
t	100.00	± 0.15 sec.

Substituting the mean values, $\bar{x}_j$, into Poiseuille's equation (32), we obtain an average (mean) viscosity coefficient

$$\bar{\eta} = 4.849 \times 10^{-8} \text{ atm. sec.} = 4.913 \times 10^{-2} \text{ poise}$$

Substituting now the pertinent values into equation 35,

$$\delta_\eta = 4.85 \times 10^{-10} \times 0.15 + 4.85 \times 10^{-6} \times 0.005 + 3.73 \times 10^{-6} \times 0.001$$
$$- 8.28 \times 10^{-9} \times 0.001 - 4.80 \times 10^{-9} \times 0.01$$

we see that the first, the fourth, and the fifth terms on the right hand side are small and can be ignored, especially since they almost cancel each other. Hence, for equation 34 we use only two terms

$$d_\eta = \left[\left(\frac{\partial \eta}{\partial \Delta P}\right)^2 (\delta \Delta P)^2 + \left(\frac{\partial \eta}{\partial R}\right)^2 (\delta R^2)\right]^{1/2}$$

$$d_\eta = [4.85^2 \times 5^2 \times 10^{-18} + 3.73^2 \times 10^{-18}]^{1/2} = 2.45 \times 10^{-8} \text{ atm. sec.} =$$

$$2.48 \times 10^{-2} \text{ poise}$$

The mean deviation of the viscosity coefficient is, therefore, approximately 50 percent of the mean viscosity coefficient. This would make the calculated value, η, imprecise.

The preceding example was selected to demonstrate that in spite of the precision of the individual measurements, the propagated error can make the calculated quantity meaningless, especially when some of the variables appear on a power in the function, such as R in equation 32. This example is also instructive when another problem faces the experimenter; the converse problem is how precisely each measurement should be carried out in order to have a certain degree of precision in the final result.

We have seen that the propagated error in the preceding problem is mainly due to two terms containing $\delta \Delta P$ and δR. In order that the precision of the final result should be in the magnitude of $\bar{d}_\eta \sim 10^{-4}$ poise, the precision of the ΔP and R measurements should be improved a hundredfold, that is, $\delta \Delta P \sim 0.00005$ atm. and $\delta R \sim 0.00001$. The latter especially is an almost impossible task. Surveying the coefficients of δR and $\delta \Delta P$ in equation 35, however, we see that $(\delta \eta / \partial R)$ and $(\partial \eta / \partial \Delta P)$ contain R^3 and R^4, respectively. Furthermore, these coefficients are squared in equation 35, which yields the mean deviation of the viscosity coefficient. If one used a wider capillary tube, for example, one with a radius of 0.1 cm. (twice the radius of the one in the preceding problem), the R^4 would be 10^{-4} cm. instead of 6.25×10^{-6}, and the R^3 would be 10^{-3} instead of 1.25×10^{-4} as in the preceding example. Therefore, approximately a hundredfold improvement in the precision of the final result can be effected by simply using the larger capillary tube.

In this discussion we concentrated on one dispersion index, the mean deviation. Similar consideration could be given to the standard deviation, σ, of the final result. However, since the standard deviation of the final result may place too much weight on large deviations in some measurements (as compared to the mean deviation) and since its calculation involves more computation, we shall not use this index with reference to propagated error. The interested student may refer to texts given in the references.

Curve Fitting by Least Squares

Some physical chemical measurements are made to evaluate a fundamental constant or a calibration constant of an instrument. An example of the former is the determination of the absorption (absorptivity) coefficient, ϵ, of a compound when the absorbency (optical density, A) is measured in a cell of d pathlength as a function of concentration of the compound, c, and $[A = \epsilon dc]$. An example of the latter is the calibration of the light-scattering instrument with benzene, the scattering intensity of which has been established accurately. Here, $R_{90}^{\text{benzene}} = kw$, where R_{90}^{benzene} is the Rayleigh ratio $(I/I_0)r^2$ of benzene at 90 degrees (taken from the literature), w is the instrument response (scattered light intensity of benzene at 90 degrees), and k is the instrument calibration constant.

If there are n constants to be determined from n pairs of measurements, n simultaneous equations could be solved to obtain these constants *exactly*. This, however, does not allow the determination of precision of the constant. But, if more measurements are available than the number of constants to be determined, the errors in the measured quantities do allow estimation of the precision, although they may not yield an exact solution.

If the data is presented in graphic form, the usual process is to fit the best curve through the experimental points. To rely on the eye for such a fit may introduce large errors. The eye has the tendency to give undue weight to the first and last experimental points in trying to fit a curve, which is wrong because most errors occur in the extremes of the range.

Fitting such a set of data with the least squares techniques simply involves the assumption that the best fit is attained when the sum of the squares of the deviations from a certain function is minimum. This assumption is correct only if the distribution is normal (Gaussian). Because most random errors follow Gaussian distribution, this technique is applicable.

The method of least squares is used under two conditions: (1) When the functional relationship between the independent and dependent variables is known, least square fitting gives us the best constants of such a function; (2) when the form of the function is not known, the least square method allows us to choose between two possible functional relationships.

Most experimental data have simple functional forms or can be manipulated in such a way as to assume simple forms. These may be linear

$$y = a + bx \qquad (36)$$

or parabolic

$$y = a + bx + cx^2 \qquad (37)$$

or power series.

$$y = a + bx + cx^2 + dx^3 + \cdots \qquad (38)$$

Let us assume that our data follows a linear relationship, for example, the osmotic pressure of polymer solution.

$$\pi/c = RT/M + Bc$$
$$y = a + bx$$

The first question to be dealt with is the direction in which the deviation should be counted. In the ideal case, random errors are present in both the dependent variable, y (osmotic pressure), and the independent variable, x (concentration). The deviations, therefore, should be perpendicular to the straight line, as in Figure 8A, assuming that the variances in x and y are the same. This would involve lengthy arithmetic in evaluating the constants a and b of equation 36. We usually select deviations in y only, since the deviations in the dependent variables are almost always greater than those in the independent variable, x (concentration), since the values of x can be selected with negligible errors. The deviation can then be written as

$$\delta y_i = y_i - y_0 = y_i - (a + bx_i) \qquad (39)$$

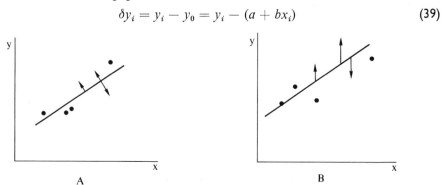

Figure 8 *Fitting curves by least squares method.*

The δy_i values are represented by the arrows in Figure 8B, and all of them are equally weighted. For n measurements we get n deviations of δy_i. According to the principles of least squares, we are looking for values of a and b that will make the sum of the square of the deviation a minimum. Therefore,

$$\sum_{i=1}^{n} (\delta y_i)^2 = \sum_{i=1}^{n} [y_i - (a + bx)]^2 \tag{40}$$

$$\frac{\partial [\sum (\delta y_i)^2]}{\partial a} = 2na - 2 \sum y_i + 2b \sum x_i = 0 \tag{41}$$

$$\frac{\partial [\sum (\delta y_i)^2]}{\partial b} = 2b \sum x_i^2 + 2a \sum x_i - 2 \sum (x_i y_i) = 0 \tag{42}$$

Solving equations 41 and 42 simultaneously, we obtain

$$a = \frac{\sum x_i^2 \sum y_i - \sum x_i \sum (x_i y_i)}{n \sum x_i^2 - (\sum x_i)^2} = \frac{\bar{y} \sum x_i^2 - \bar{x} \sum (x_i y_i)}{\sum (x_i - \bar{x})^2} \tag{43}$$

$$b = \frac{n \sum (x_i y_i) - \sum x_i \sum y_i}{n \sum x_i^2 - \sum (x_i)^2} = \frac{\sum (x_i - \bar{x})(y_i - \bar{y})}{\sum (x_i - \bar{x})^2} \tag{44}$$

An example for evaluation is given in Table 4.

TABLE 4

$y_i(\pi/c)$	$x_i(c)$	$(x_i - \bar{x})$	$(x_i - \bar{x})^2$	x_i^2	$(y_i - \bar{y})$	$(x_i - \bar{x})(y_i - \bar{y})$	$x_i y_i$
0.126	1.0	−2.25	5.06	1.00	−0.0284	0.0639	0.126
0.140	1.5	−1.75	3.06	2.25	−0.0144	0.0252	0.210
0.138	2.0	−1.25	1.56	4.00	−0.0164	0.0205	0.276
0.147	2.5	−0.75	0.56	6.25	−0.0074	0.0056	0.368
0.156	3.0	−0.25	0.06	9.00	0.0016	−0.0004	0.468
0.154	3.5	+0.25	0.06	12.25	−0.0004	−0.0001	0.539
0.160	4.0	+0.75	0.56	16.00	0.0056	0.0420	0.640
0.172	4.5	+1.25	1.56	20.25	0.0176	0.0220	0.774
0.173	5.0	+1.75	3.06	25.00	0.0186	0.0325	0.865
0.178	5.5	+2.25	5.06	30.25	0.0236	0.0530	0.979
Σ 1.544	32.5	0	20.63	126.25		0.2265	5.244

$\bar{y} = 0.1544$

$\bar{x} = 3.25$

$$a = \frac{\bar{y} \sum x_i^2 - \bar{x} \sum (x_i y_i)}{\sum (x_i - \bar{x})^2} = \frac{0.1544 \times 126.25 - 3.25 \times 5.244}{20.63} = 0.119$$

$$b = \frac{\sum (x_i - \bar{x})(y_i - \bar{y})}{\sum (x_i - \bar{x})^2} = \frac{0.2265}{20.63} = 0.011$$

From the data given here, the osmotic pressure gradient as a function of concentration is described by the equation

$$\pi/c = 0.119 + 0.011c$$

Parabolic, hyperbolic, power series, or any other relationship can be handled in a similar manner and the constants of the analytical expressions are obtained.

In estimating the errors in the calculated quantities a and b, one uses the mathematical convenience of assuming that the error is solely in the dependent variable Y and that all experimental data points are equally reliable. With these assumptions, the variance of the slope, b, is

$$\sigma_b^2 = \frac{n\sigma_x^2}{n\sum x_i^2 - (\sum x_i)^2} \tag{45}$$

and the variance of the intercept, a, is

$$\sigma_a^2 = \frac{\sigma_Y^2 \sum x_i^2}{n\sum x_i^2 - (\sum x_i)^2} \tag{46}$$

Although the least square fitting cannot indicate the best functional relationship between variables, it can be used to allow us to choose between two possible functional forms. Let us assume that we may have two possible functional forms—parabolic, α, and power series, β. For each value of the independent variable x_i we now have two dependent variables computed from functions α and β, namely, $y_{i\alpha}$ and $y_{i\beta}$. We also have the experimental values of y_i corresponding to x_i. Let us designate the constants of the respective functions c_α and c_β. Then,

$$\Omega_\alpha = \frac{\sum (y_i - y_{i\alpha})^2}{n - c_\alpha} \qquad \Omega_\beta = \frac{\sum (y_i - y_{i\beta})^2}{n - c_\beta}$$

The form of the function that has the smallest Ω is the one that best fits the data obtained.

A particular and important case is that in which the functional form is a power series and the experimenter wants to know the order of the power series needed to describe the relationship. Especially this can be done by quick inspection if the differences between the independent variables are constant. Under such conditions a table of differences may be set up as in Table 5.

TABLE 5. Table of Differences

x	y	Δy	$\Delta^2 y$	$\Delta^3 y$	$\Delta^4 y$
x_1	y_1				
		Δy_1			
$x_2 = x_1 + \Delta x$	y_2		$\Delta^2 y_1$		
		Δy_2		$\Delta^3 y_1$	
$x_3 = x_1 + 2\Delta x$	y_3		$\Delta^2 y_2$		$\Delta^4 y_1$
		Δy_3		$\Delta^3 y_2$	.
$x_4 = x_1 + 3\Delta x$	y_4		$\Delta^2 y_3$	.	.
		Δy_4		.	.
$x_5 = x_1 + 4\Delta x$	y_5	.	.	.	.
.	.	.	.	.	.
.	.			.	.

In such a table the first order differences are $y_i = y_{i+1} - y_i$, the second order differences are $\Delta^2 y_i = \Delta y_{i+1} - \Delta y_i$, and so forth.

By inspecting such a table of differences one can see at what order, n, the differences become constant (or within the standard deviation of their differences). A power series function of such order, n^{th}, is then required to describe the experimental data.

Rejection of Data

In all the preceding discussions we assumed that all experimental data are equally weighted when location and dispersion indices are calculated. In many experiments individual data are obtained that are in discord with the rest of the data. An objective experimenter needs criteria by which such discordant data may be rejected since if there is something obviously wrong with a measurement we do not want to include that in the averaging process.

Data may be rejected if the observer *knows* that something definitely went wrong during the experiment; for example, a leak developed in a vacuum system. In the absence of such precise knowledge of experimental mistakes, *discordant data may be rejected if it deviates from the mean by more than four times the average deviation.*

Graphic Methods

The data obtained by the student can be presented frequently in graphic form, which is more instructive than tabulated data. For instance, trends in the function, such as minima and maxima, or deviation from linearity may be spotted and also the total data and its significance may be grasped at once. Besides the visual presentation some manipulations, such as differentiation (by taking slopes) or integration (by measuring areas), may be accomplished with relative ease. Make certain that the axes of your graph are labeled.

In any graphic presentation be sure that you plot your independent variable on the abscissa and the dependent variable on the ordinate.

The experimental points may be presented as small circles and the function as a smooth unbroken line. The scales of abscissa and ordinate should be chosen so that the graph paper will cover the whole range of experimental data. (The tendency has been observed in the reports of inexperienced students that they compressed the scale of one of the axes. This sometimes results in an undesirable smoothing out of curves, making a zero slope of a straight line where none exists.)

If graph paper were large enough, there would be no problem with the precision of the plot, but to meet these demands, in most cases the graph paper would have to be too large and clumsy. For standard size co-ordinate paper the precision cannot be much greater than 0.5 percent. The result of plotting any data having a precision better than 0.5 percent would be that the final result could not be better than 0.5 percent regardless of the precision of the data.

To overcome this difficulty, a residual plot is made. The residual plot consists of plotting the deviations of the observed data from the best representative line on an enlarged scale. If from the direct plot in the case of a linear graph we determine the equation $y = ax + b$ with its constants, we can compute the values of y corresponding to the observed values of x. The differences between the computed values of y and the observed values are called residuals. If now we plot the residuals as ordinates against the corresponding values of x as abscissas, and draw the best representative line, we can obtain the equation of this correction line, $y' = a'x' + b'$. Applying these corrections to the originally determined equation, our corrected equation becomes

$$y = (a + a')x + (b + b')$$

For satisfactory results, the residuals should be plotted to a considerably enlarged scale. It is important to realize that this procedure can in no way improve the precision of the original data, and once the procedure has corrected the errors of plotting to the degree of precision of the data, further application of this method is meaningless.

For graphical differentiation, the slope of a curve has to be obtained. Drawing slopes visually usually involves large errors. A stepwise approach to the problem minimizes these errors (Fig. 9).

Let us assume that we want to obtain the slope at x_i. Selecting a random point as the center, A, of a circle, we draw two intersections, B and C, with radius R_1, and another two intersections, D and E, with radius R_2. Connecting B to C and D to E,

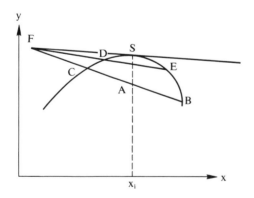

Figure 9 *Drawing the slope of a curve*

we have two chords that intersect at point F. We now connect point F to S and the slope is obtained.

When a definite integral is desired, graphic integration may be the best way to obtain the desired value. The area under a curve (function) between two limits may be obtained by using a planimeter or simply by counting squares of the graph paper. Instead of this tedious counting, the area may be cut out and its weight may be obtained on an analytical balance. A certain square area (1 cm.^{2}) of the same paper is also weighed. From the two weights, the area under the curve may be calculated.

REFERENCES

1. L. G. Parratt. Probability and Experimental Errors in Science. Wiley & Sons, Inc.. New York, 1961.
2. A. G. Worthing and J. Geffner. Treatment of Experimental Data. Wiley & Sons, Inc., New York, 1943.
3. J. Topping. Errors of Observation and Their Treatment. Reinhold Publishing Corp., New York, 1960.
4. P. G. Guest. Numerical Methods of Curve-Fitting. Cambridge University Press, London, 1961.
5. E. B. Wilson, Jr. An Introduction to Scientific Research. McGraw-Hill Book Co., New York, 1952.
6. T. C. Fry. Probability and Its Engineering Uses. 2nd Edition, O. Van Nostrand Co., Inc., Princeton, N.J., 1965.

III

GASES

1

SECOND VIRIAL COEFFICIENT AND PRESSURE-VOLUME-TEMPERATURE DATA OF GASES

The simple kinetic theory of gases provides a good account of the nearly ideal behavior of gases. It assumes that: (1) the volume of gas is negligible compared to the available volume of the container and (2) all collisions between molecules and between molecules and the container are elastic collisions.

The ideal gas law is a limiting law, which real gases approach when the two conditions are approximated at low pressures and high temperatures. However, deviations from ideality can be observed even at moderate pressures (especially with polar molecules). A large number of empirical pressure-volume-temperature relationships have been proposed for different gases over various temperature and pressure ranges. The most general empirical form of the equation that expresses the behavior of gases is a power series:

$$PV = A + BP + CP^2 + DP^3 + \cdots \tag{1}$$

A is the first virial coefficient, B, the second, C, the third, and so on. It is obvious that at very low pressures this equation reduces to the ideal gas law and $A = RT$ for one mole of a gas. Through some intermediate ranges of pressure and temperature, the behavior of real gases can be expressed by using only two terms of equation 1. The second virial coefficient can then be obtained easily, and some interpretation of it can be derived from theories.

The well known van der Waals equation explains the deviation from ideality in terms of excluded volumes and inelastic collisions or intermolecular interactions.

$$\left(P + \frac{an^2}{V^2}\right)(V - nb) = nRT \tag{2}$$

The excluded volume correction term, b, is related to the molecular diameter if the

37

molecules are visualized as hard spheres or rapidly rotating hard bodies sweeping out a spherical space.

$$b = 4N[\tfrac{4}{3}\pi(\sigma/2)^3] \tag{3}$$

where σ is the molecular diameter and N is Avogadro's number. The term correcting for the intermolecular interaction is proportional to the square of the concentration $(n/V)^2$, but the proportionality constant, a, is not defined in physical terms. Rather, the best a and b values are selected that when employed in equation 2 will reproduce the empirical pressure-volume-temperature data.

The van der Waals constants can be expressed in terms of critical constants

$$b = \tfrac{1}{3}V_c \tag{4}$$

$$a = 3P_cV_c{}^2 \tag{5}$$

by applying equation 2 to the critical isotherm and to the inflection point within it.

The empirical second virial coefficient, B, can also be obtained in terms of van der Waals' constants, a and b.

For one mole of gas the rearrangement of equation 2 takes the form

$$PV = RT + Pb - a/V + ab/V^2 \tag{6}$$

For large volumes the ab/V^2 term can be neglected. At low pressures the V in the correction term can be approximated by the ideal gas law

$$PV = RT + P(b - a/RT) \tag{7}$$

Hence, the second virial coefficient for one mole of gas is

$$B = b - a/RT \tag{8}$$

or

$$B = nb - \frac{n^2a}{RT} \tag{9}$$

for n moles of gas.

Equations 8 and 9 clearly show the temperature dependence of the second virial coefficient.

Another physical interpretation of B is obtained from the *virial theorem*, which Clausius originally developed by considering the interaction between gas molecules and the average force exerted on one molecule.

Clausius argued that although the space co-ordinates of a single molecule are time dependent, when a system with a large number of molecules is investigated, the net change for the sum of the space co-ordinate squares of all molecules must be zero. If this sum changed with time, we would observe random expansions and contractions of the system.

Two kinds of forces operate in the system. One originates from collisions of the molecules with the container, and the total contribution due to this effect is $3PV$. The second force is the result of intermolecular collision. Clausius expressed it in terms of the interaction energy, ϵ, operating between a pair of molecules separated at r distance. Rayleigh introduced into the Clausius expression the concept that the concentration of the molecules separated from a reference molecule at r distance can be expressed in terms of the Maxwell-Boltzmann distribution function. The combined equation, known as the virial of Clausius in Rayleigh's extension, is

$$PV = RT + P2\pi N \int_0^\infty [1 - e^{-\epsilon(r)/kT}]r^2 \, dr \qquad (10)$$

A comparison of equations 10 and 1 gives

$$B = 2\pi N \int_0^\infty [1 - e^{-\epsilon(r)/kT}]r^2 \, dr \qquad (11)$$

The temperature dependence of the second virial coefficient is clearly indicated in equation 11.

The potential energy, ϵ, is usually the difference between energies due to attraction and repulsion, each being a function of intermolecular separation, although they may depend differently on separation. The Lennard-Jones potential assumes a force field of $\lambda/r^n - \mu/r^m$, where n and m can be different integers, usually 12 and 6.

Other theoretical derivations, such as statistical mechanical considerations, give the second virial coefficient as

$$\left(\frac{\partial P/\rho kT}{\partial \rho}\right)_{T,\rho=0} = B \qquad (12)$$

where ρ is the density of the gas.

Experimental

Wearing safety glasses throughout this experiment is mandatory.

To obtain P-V-T data for gases, a simple and relatively safe apparatus was designed (Fig. 1-1). The principal components of the apparatus are a gas buret (A), an adjustable mercury column (B), and a Kammerlingh-Onnes type manometer (C).

The gas buret consists of five or six spherical bulbs of increasing volume that are connected by short capillaries. A marking on each capillary defines the boundaries of each volume. The gas buret is jacketed, and a thermostat can be provided by circulating liquid through the jacket from a bath. The gas buret is calibrated by the instructor, before the apparatus is assembled, by filling the gas buret with mercury. A stopcock is attached temporarily to the lower end of the buret. The mercury contained in bulbs 1, 2, 3, and so forth, as defined by the capillary markings, is allowed to flow into preweighed beakers. The volumes of the gas buret can be obtained from the weight of mercury from each bulb and the density of mercury at the temperature measured. It is practical to calibrate at a minimum of two different temperatures to establish whether the volume expansion coefficient of the gas buret actually corresponds to the volume expansion coefficient of Pyrex glass. Such a calibration is given in Table 1-1.

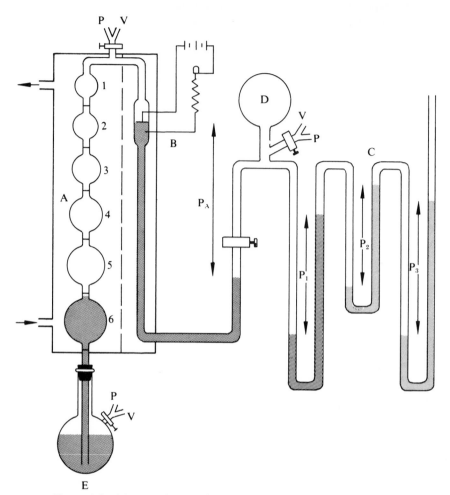

Figure 1-1   *Schematic drawing of the pressure-volume-temperature apparatus.*

Once the calibration is accomplished, the stopcock is removed, and the gas buret is lowered into the mercury reservoir bulb, *E*, which can be connected to either vacuum or compressed air. By manipulating the pressure on the side arm of the mercury reservoir bulb, the mercury can be raised or lowered to the desired height in the gas buret.

The second part that must be prepared before the apparatus is assembled is the manometer, *C*. For high pressures, a battery of 10 to 20 connected U tubes can be constructed. For the present experiment, the pressure range of 200 to 1600 mm. of mercury can be achieved with three interconnected U tubes, each about 1 meter high. Each U tube is filled to less than half its height with mercury. Air is trapped between the columns of mercury; the air used is at normal atmospheric pressure or is compressed to 1.5 to 2 atm. Difference in pressure of the trapped air in the two arms of a tube

TABLE I-I. Calibration of a Gas Buret

Bulb No.	T = 299.7° K Weight of Mercury (Grams)	Volume (cc.)	T = 345.1° K Weight of Mercury (Grams)	Volume (cc.)
1	34.2738	2.53	34.5012	2.57
2	262.10	19.37	259.41	19.33
3	500.09	36.96	496.13	36.97
4	736.39	54.42	730.86	54.46
5	1382.44	102.16	1379.16	102.27
6	2422.30	179.03	2403.53	179.11
Total		394.47		394.71

$$\Delta V = \text{difference in volume} = \begin{array}{r} 394.71 \\ -394.47 \\ \hline 0.24 \text{ cc.} \end{array}$$

Cubical expansion coefficient of Pyrex glass is
0.099×10^{-4} cc./degree.

Theoretical $\Delta V = 0.099 \times 10^{-4} \times 45.4 \times 394.47 = 0.18$ cc.

$$\frac{0.06}{394.71} = 0.016 \text{ percent error.}$$

does not influence the measurements. The pressure is measured by computing the sum of the differences of the mercury heights in the U tubes. One end of the manometer is open to atmospheric pressure. The reading of the actual atmospheric pressure can be obtained on a separate barometer. This atmospheric pressure must be added to the pressure measured in the series of U tubes. The manometer, C, is connected through a 1 liter ballast bulb, D, to the adjustable mercury column, B. (To prevent accidental leaks from shooting large amounts of mercury in the U tubes and consequently breaking the apparatus, each U tube in B and C should be constructed so that its lower end narrows into a capillary; see Figure 1-2.)

The adjustable mercury column, B, which is in a U tube, maintains a reference boundary for the different volumes. Two tungsten needles are sealed in the top of the column. The needles are connected to a small signal lamp operating on 110 line voltage. When the mercury column reaches the top of the second tungsten needle, the electrical circuit is closed and the

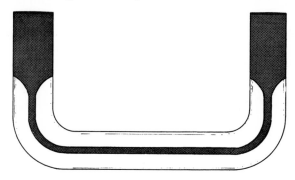

Figure I-2 *Detailed construction of a U tube.*

light bulb is on. The adjustable mercury column is connected with a capillary to the gas buret. In this connecting capillary there is also a three-way stopcock, which can connect the gas buret to either vacuum or pressure or can seal it off from the environment. The interconnecting capillary (between the gas buret and the adjustable mercury column) is also thermostated or at least well insulated with asbestos tape.

The pressure changes caused in the gas buret by compressing or expanding the gas under investigation are balanced by applications of the appropriate pressure in the ballast bulb. A balance of pressures is achieved when the mercury column in the gas buret reaches the desired height and at the same time the height of mercury in the adjustable column is raised to the reference level (the second needle tip). The actual pressure is then read from the manometer, C. The absolute pressure of the gas is obtained as follows:

$$P_T = P_1 + P_2 + P_3 + P_B + P_A \tag{13}$$

where P_1, P_2, and P_3 are the pressures obtained in the three columns of the manometer, P_B is the barometric pressure, and P_A is the difference in the height of mercury between the two arms of the adjustable column. If the level in the adjustable column at the reference needle exceeds that in the other arm, the P_A must be subtracted from the sum of the other pressures; if the level is lower than in the other arm, the P_A must be added to the sum of pressures.

For the experiment, the student should first calculate the dead volume between the last marking in the gas buret and the height of mercury adjusted to the standard reference level. Air can be used for this purpose. The mercury in the gas buret is raised to the third marking while the stopcock on the gas buret is open to the atmosphere. The gas buret is thermostated at 25° C. The mercury in the adjustable column, B, is raised to the reference level, the stopcock on the gas buret is closed, and the pressure reading corresponding to atmospheric pressure is read from the manometer, C.

In the next step, the mercury in the adjustable column, B, is lowered *first* so that the mercury is not siphoned over into the gas buret in the following operation. The mercury in the gas buret is now lowered to the fourth mark. The pressures are balanced again and read off in the manometer. It is important to remember that every time the mercury level in the gas buret has to be lowered, the mercury level in the adjustable column should be lowered first. The same procedure is repeated when the mercury level in the gas buret is lowered to the fifth mark. Boyle's law is applied, and the dead volume is calculated.

$$P_1(V_1 + V_2 + V_3 + V_d) = P_2(V_1 + V_2 + V_3 + V_4 + V_d) \tag{14}$$

and

$$P_1(V_1 + V_2 + V_3 + V_d) = P_3(V_1 + V_2 + V_3 + V_4 + V_5 + V_d)$$

where V_1, V_2, V_3, V_4, and V_5 are the calibrated volumes of the bulbs 1, 2, 3, 4, and 5. The average V_d, or dead volume, is calculated from these equations.

In this experiment, the second virial coefficient of carbon dioxide gas will be determined. The first isotherm is obtained at 50° C. Evacuate the gas buret again while the stopcock connecting B and C is closed. Lower the mercury level in the gas buret by applying vacuum on the reservoir bulb. After five minutes' evacuation, raise the mercury level in the gas buret to fill bulbs 5 and 6. Add carbon dioxide to the gas buret, and open the stopcock separating B and C. When a pressure of about 1.5 atm. is reached, turn the three-way stopcock to isolate the system. By applying vacuum on the mercury reservoir, lower the mercury level in the gas buret to the lowest mark. Apply the appropriate pressure on the ballast bulb and read the pressure on the manometer.

Assume that at this low pressure the ideal gas law is applicable and calculate the number of moles of gas present. By raising the level of mercury in the gas buret to each successive mark and obtaining the pressure, the P-V isotherm at 50° C. is established. Using the same quantity of gas, lower the temperature and obtain P-V data for three more isotherms—20, 30, and 40° C.

Plot your data. By using least square fitting, obtain for each isotherm the second virial coefficient of equation 1.

Establish the temperature dependence of the second virial coefficient.

Fit your P-V-T data to the van der Waals equation 2 and calculate the empirical van der Waals constants. From equation 3 obtain the molecular diameter. Does this diameter correspond to those given in handbooks and literature from gas viscosity and x-ray diffraction measurements?

Obtain the critical constants from equations 4 and 5 and compare them with those reported in the literature.

Evaluate the isothermal coefficient of compressibility of carbon dioxide.

$$\beta = -\frac{1}{V_0}\left(\frac{\partial P}{\partial V}\right)_T$$

Material and Equipment. One cylinder of compressed carbon dioxide gas; an adjustable thermostat of about 20 liter capacity with heater, stirrer, and thermoregulator; circulating pump; jacketed gas buret; Kammerlingh-Onnes manometer; barometer; about 600 cc. of mercury.

REFERENCES

1. E. A. Moelwyn-Hughes. Physical Chemistry. 2nd Revised Edition, Pergamon Press, Inc., New York, 1964.

2. J. R. Partington. An Advanced Treatise on Physical Chemistry. Vol. I, Section VII, Longmans, Green & Co., Ltd., London, 1949.
3. J. H. Hildebrand. An Introduction to Kinetic Theory. Reinhold Publishing Corp., New York, 1963.
4. T. L. Hill. An Introduction to Statistical Thermodynamics. Addison-Wesley Publishing Co., Inc., Reading, Mass., 1960, Chapter 15.

2

CRITICAL CONSTANTS; LAW
OF RECTILINEAR DIAMETER

When a liquid in equilibrium with its vapor is sealed in a glass tube and the tube is gradually heated, a change in the volume of the liquid occurs, as is observable by a change in the height of the meniscus. At a critical temperature the meniscus suddenly disappears and only one transparent phase is present. Upon cooling the system, a cloud precedes the reappearance of the meniscus at the same critical temperature. This description of the critical state is still the most useful operative description, although it was published first in 1822 by Cagniard de la Tour.[5]

The critical constants refer to the critical isotherm in Figure 2–1. The critical isotherm is below the coaxial rectangular hyperbolas expected to describe gases that obey Boyle's law and is above the curves that show discontinuity by having a constant pressure portion of the isotherm. The critical isotherm has an inflexion point, and the transition from gas to liquid and vice versa occurs without the coexistence of two phases.

The criteria for the critical state according to Figure 2–1 and incorporated into all equations of state are

$$\left.\begin{array}{c} \left(\dfrac{\partial P}{\partial V}\right)_T = 0 \\[2mm] \left(\dfrac{\partial^2 P}{\partial V^2}\right)_T = 0 \end{array}\right\} \quad T = T_c \tag{1}$$

Through the conditions in equation 1 the constants of the different equations of state can be related to the critical temperature, T_c, critical pressure, P_c, and critical volume, V_c. For example, over the range in which the van der Waals equation is

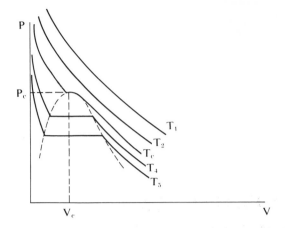

Figure 2–1   *Isotherms of a gas: T_c P_c, and V_c are the critical constants.*

applicable, the relations will be

$$a = \tfrac{1}{3}V_c \tag{2}$$

$$b = 3P_cV_c^2 \tag{3}$$

$$R = \frac{8P_cV_c}{3T_c} \tag{4}$$

We shall return to the significance of such relationships, especially that implied by equation 4. First, however, we must determine whether there are techniques for obtaining critical constants that are simpler and easier than the direct technique, which consists of running compressibility isotherms until the critical isotherm is obtained.

It is obvious that the simple setup of Cagniard de la Tour can yield the critical temperature, and if the tube is connected to a manometer, one can also obtain the critical pressure at the point at which the meniscus disappears or reappears. It is much more difficult, however, to obtain the critical volume.

Critical volume or critical density data can be obtained from the so-called law of rectilinear diameter of Cailletet and Mathias. This law states that the average of the orthobaric densities of liquid and its saturated vapor is a linear function of the temperature.

$$\tfrac{1}{2}(\rho_l + \rho_g) = a - bT \tag{5}$$

where a and b are constants and ρ_l and ρ_g are the densities of liquid and its saturated vapor, respectively. If it is assumed that such a relationship holds from the critical point down to absolute zero, the constants can be evaluated in terms of physical properties of the material.

The limiting conditions then will be (a) if $T = 0$, $\rho_g = 0$, $\rho_s = \rho_l$, then $a = \tfrac{1}{2}(\rho_s)$ (In writing this, one assumes that the density of the liquid subcooled to absolute zero is the density of the solid at absolute zero.)   (b) if $T = T_c$, $\rho_l = \rho_g = \rho_c$, then $\rho_c = a - bT_c$. Using the relationships of the limiting conditions, we can rewrite equation 5 as

$$\rho_g + \rho_l = \rho_s - (\rho_s - 2\rho_c)\frac{T}{T_c} \tag{6}$$

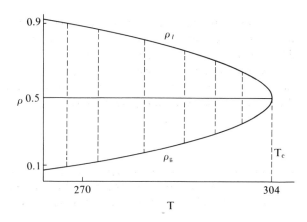

Figure 2–2 *The Cailletet-Mathias curve for carbon dioxide.*

In Figure 2–2 a plot of the law of rectilinear diameter is given. Thus, if the rectilinear diameter $(\rho_g + \rho_l)/2$ is plotted against temperature, ρ_c can be obtained by extrapolating the straight line to the critical temperature, T_c.

The task is then to obtain the orthobaric densities, ρ_l and ρ_g, at different temperatures. If a sealed tube under observation at a constant temperature contains w grams of pure substance and if the volume occupied by the liquid in the tube is V_l and the corresponding gas volume is V_g, the total volume is

$$V_T = V_g + V_l \tag{7}$$

and

$$w = V_g \rho_g + V_l \rho_l \tag{8}$$

In equation 8 two unknowns appear—ρ_g and ρ_l; therefore, at least two sealed tubes that contain different total weights of substances w_1 and w_2 must be measured.

Solving two equations of the nature of equation 8 simultaneously, we can obtain ρ_l and ρ_g at different constant temperatures. If equation 6 is divided by ρ_c and the proportionality that exists between densities and concentrations is used,

$$\frac{c_g + c_l}{c_c} \cong \frac{\rho_g + \rho_l}{\rho_c} = \frac{\rho_s}{\rho_c} - \left(\frac{\rho_s}{\rho_c} - 2\right)\frac{T}{T_c} \tag{9}$$

The left side of equation 9 is the mean of the molecular concentration of the liquid and vapor phases. As the temperature increases, this mean concentration (or average density) always decreases. The straight line in Figure 2–2 joining the bisections (the so-called rectilinear diameter) has a negative slope. This can occur only if

$$\frac{\rho_s}{\rho_c} > 2 \tag{10}$$

The ratio of limiting density to critical density (ρ_s/ρ_c) can be evaluated from the slope and the intercepts of the Cailletet-Mathias curves (equation 9). The average value for many liquids is about 4, although considerable variation occurs. For example, ρ_s/ρ_c for helium is 2, for neon, 3.51, and for xenon, 4.34; for many nonpolar diatomic molecules it averages 3.67.

The variation of ρ_s/ρ_c is a puzzling theoretical problem and illustrates the difficulty of finding a suitable theory for liquid structure.

Many empirical relationships have been proposed, and all have some similarity to equations 6 and 9, which may work for certain homologous series but require different constants for different substances.

At the same time, however, the ratios of critical constants also vary from substance to substance and they are by no means the value predicted, for example, by the van der Waals equation (equation 4)

$$\frac{RT_c}{P_c V_c} = \frac{8}{3} = 2.666 \tag{11}$$

Even monatomic gases have higher values than 2.666; e.g., helium is 3.08 and xenon is 3.50. For polar molecules the $RT_c/P_c V_c$ ratio seems to be high; for example, for water it is 4.39 and for methanol it is 4.56.

Experimental

Wearing safety glasses throughout this experiment is mandatory.

The critical temperature and critical density of benzene will be determined. Benzene is selected because its critical pressure is about 48 atm., and therefore in obtaining the orthobaric densities below the critical temperature, only pressures up to 20 to 25 atm. will be reached. This will minimize the possibility of explosion of the pycnometer tube under high pressures. Other liquids that could be used under the experimental conditions to be described are n-butane, isobutane, n-pentane, isopentane, n-hexane, and cyclohexane.

For each experiment five pycnometer tubes will be used. The tubes are Pyrex glass tubes with a 2 mm. wall thickness, an outside diameter of 10 or 12 mm., a varying length of 5 to 20 cm., and are closed at one end (Fig. 2–3). It is important that these tubes be properly annealed and that no residual strains be left in them because in withstanding pressures up to 50 atm. the wall thickness is of less importance than local weaknesses due to strain. The four largest tubes should have a fine reference mark of India ink on them. (Do not use file scratches because they might weaken the wall.) The

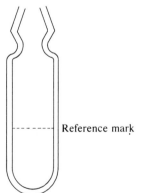
Reference mark

Figure 2–3 *Pycnometer.*

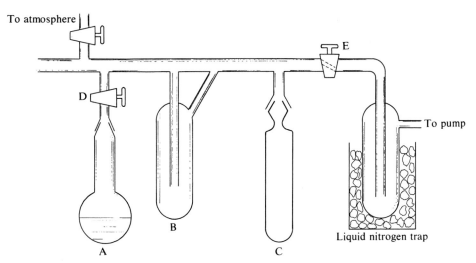

To atmosphere

E

D

To pump

B

C

Liquid nitrogen trap

A

Figure 2–4 *Vacuum distillation apparatus.*

bottom of the tube can also be used as a reference mark, although one must consider the parallax.

Benzene should be vacuum distilled directly into each pycnometer tube in an apparatus similar to that given in Figure 2–4. The clean pycnometer tube, C, should be cemented to a vacuum apparatus with wax or de Khotinsky cement. The whole apparatus should be evacuated while the benzene reservoir, A, is frozen in a liquid nitrogen bath. After five to ten minutes of pumping, close stopcock D (silicone grease) and warm the benzene reservoir to room temperature. Place a liquid nitrogen bath around trap B and distill benzene from A to B by opening stopcock D. After 10 to 15 ml. of benzene has been condensed in B, close stopcock D. Remove the liquid nitrogen bath from B and place another liquid nitrogen bath around the pycnometer C. Isolate the vacuum pump by closing stopcock E. By allowing B to warm, distill enough benzene into the pycnometer to fill it to approximately half its volume. To stop the distillation, place the liquid nitrogen bath around B again. Open stopcock E and vacuum seal the pycnometer. (Be sure that the seal is annealed so that no residual strain is left in the glass.) After the sealed end cools to room temperature, remove the pycnometer from the liquid nitrogen bath and let it come to room temperature.

Similarly fill the other four pycnometers.

To save time, the instructor may carry out this procedure; the students are given the sealed samples. Another way of saving time is to provide triply distilled benzene to the students and to allow them to fill the clean pycnometer tubes by pipetting the sample. The benzene is then frozen in a liquid nitrogen bath and the pycnometers are sealed with a torch *under vacuum*.

Place the smallest pycnometer (5 cm.) in a conventional melting point apparatus or in a beaker filled with silicone oil that can be heated to 350° C. Be certain that the pycnometer tube is vertical and fully submerged in the silicone oil. In the past we have used a sulfuric acid bath, but silicone oil is safer.

Place the melting point apparatus in a container with glass walls and place the container in a hood. Warm the melting point apparatus with the sealed pycnometer in it, and observe the temperature at which the meniscus disappears. Cool the tube and observe the critical temperature at which the meniscus reappears. If the two temperatures are different, repeat the observations until they are the same.

Placing the melting point apparatus in a container and the container in a hood are important precautions; if elevated temperature and pressure cause the pycnometer tube to explode because of weakness in the glass, the silicone oil (or sulfuric acid) will not spill and serious injury can be averted. (Pycnometer tubes have never exploded during the six years that this experiment has been carried out by about 100 students in my laboratory, nor have any other accidents occurred with the procedure. Care must be taken also in the cooling process. The hood should not be open until the pycnometer has cooled to room temperature.)

After the critical temperature is obtained, completely submerge the remaining four pycnometer tubes in a vertical position in silicone or paraffin oil at room temperature. Be sure that the reference mark of the pycnometer tube is clearly visible.

After four to five minutes' equilibration time, turn the pycnometer tubes upside down, shake them, and place them back in their original position. After another four to five minutes has elapsed and the liquid benzene has drained from the side of the tube, take a cathetometer reading of the position of the *meniscus* and the *reference mark* of *each* tube.

Repeat the same procedure eight more times at temperatures ranging from 25 to 150° C., but without inverting, shaking or, in general, removing the tubes from the oil bath.

At the end of the experiment, cool the pycnometer tubes to room temperature, wash them with detergent and distilled water, and wipe them clean; do not wipe off the reference mark.

Weigh each tube with an analytical balance, and then partially submerge each pycnometer tube in a liquid nitrogen bath, and make a small hole in the tip of the pycnometer with a hand blast lamp or a microtorch.

Remove most of the benzene by shaking it out of the tube. Remove the last traces by gentle heating and evacuation with a vacuum pump with a liquid nitrogen cooled trap.

Reweigh the pycnometers, filled first with air and then with freshly boiled distilled water. For the distilled water procedure, place the empty pycnometer tube in a large test tube containing distilled water. Stopper the test tube and connect it to a vacuum line. Pump out the air. The distilled water replaces the air in the pycnometer, filling it completely. An alternative technique may be to fill the pycnometer tubes with water by using a syringe and hypodermic needle. Knowing the densities of water and air at the weighing temperature, one can calculate the total volume of each pycnometer tube and the weight of each benzene sample in the pycnometer tubes.

After this has been accomplished, break off the top of the pycnometer by making a scratch mark with a file. Establish a calibration curve for each pycnometer tube: obtain the weight of the dry pycnometer after its top has been removed. Fill the tube with freshly distilled water to the *approximate* heights at which the menisci of the liquid benzene were observed. Four or five such heights are sufficient to establish a calibration curve.

At each meniscus height take a cathetometer reading and immediately afterward weigh the pycnometer tube.

Calculate the volume corresponding to each meniscus height and construct a calibration curve for each tube by plotting meniscus height against volume.

The task is to calculate ρ_g and ρ_l values at each temperature using equation 8. The data obtained for two pycnometer tubes are used in the following way:

$$w_1 = V_{g_1}\rho_g + V_{l_1}\rho_l$$

$$w_2 = V_{g_2}\rho_g + V_{l_2}\rho_l$$

where 1 and 2 refer to pycnometer tubes 1 and 2. The weight of the substance in each pycnometer tube has been obtained (w_1 and w_2). The volume of the liquid, V_{l_1} and V_{l_2}, is read from the calibration curves since the position of the meniscus in relation to the reference mark has been determined. From these volumes and from the total volume of the tubes determined earlier, V_{g_1} and V_{g_2} are calculated. Thus, ρ_l and ρ_g are obtained from the measurement on pycnometer tubes 1 and 2. Calculate the ρ_l and ρ_g values from the data on pycnometer tubes 1 and 3, 2 and 3, 1 and 4, 2 and 4, and 3 and 4. Average these values.

From the average values of ρ_l and ρ_g at each temperature, construct a plot of rectilinear diameter vs. temperature. Extrapolate your line to the critical temperature and obtain the critical density of benzene.

Compare your data with the values reported in the literature. Calculate the mean error or mean deviation from the average ρ_l and ρ_g values obtained using different pairs of pycnometer tubes. Estimate the maximum random errors by taking into account the errors in cathetometer readings and analytical balance weighing. Is the mean error within the maximum random error?

Calculate the *systematic* errors introduced by not correcting for the following:

1. The thermal expansion of the glass since the calibration curves were obtained at room temperature (the thermal expansion coefficient of Pyrex glass is 9.9×10^{-6} cc. [cc. degree]$^{-1}$).

2. The dilation of the glass under pressure (the pressure expansion coefficient of Pyrex glass is approximately 1×10^{-7} cc. [cc. atm.]$^{-1}$).

3. The different curvatures of the meniscus of benzene and of water (the calibration liquid). This correction actually should be made empirically by determining the curvatures at the glass surface for both liquids and calculating the volume occupied by the two liquids above the meniscus.

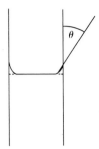

Figure 2-5 *Contact angle of the meniscus.*

For estimation of this systematic error, assume that at 30° C. the weight of water above the meniscus in a 12 mm. O.D. pycnometer tube is 0.05 gram. The contact angle of the meniscus, θ, is related to the surface tension, γ, and the density, ρ, and the ratio of two contact angles can be given as

$$\frac{\cos \theta_{benzene}}{\cos \theta_{water}} = \frac{\rho_{benzene}\gamma_{water}}{\rho_{water}\gamma_{benzene}}$$

From these data calculate the systematic errors introduced by not correcting for the difference in contact angle.

Do any of these systematic errors (under 1, 2, and 3) fall within the maximum random error and therefore need no correction?

Are there any other systematic errors not mentioned here that have to be taken into account?

Material and Equipment. Triply distilled thiophene-free benzene; five Pyrex pycnometers of 2 mm. wall thickness and 12 mm. O.D. (5, 7, 12, 15, and 20 cm. long); vacuum line for vacuum distillation; sealing wax or de Khotinsky cement; liquid nitrogen bath; torch; silicone oil bath; two thermometers, 100 to 400° and 0 to 200° C.; heater; stirrer; (paraffin oil bath); cathetometer.

REFERENCES

1. J. R. Partington. An Advanced Treatise on Physical Chemistry. Vol. I, Longmans Green & Co., Ltd., London, 1949, Chapter VII.
2. E. A. Moelwyn-Hughes. Physical Chemistry. 2nd Revised Edition, Pergamon Press, Inc., New York, 1964, Chapter XIV.
3. M. Cailletet and M. E. Mathias. Compt. Rend., **102**, 1202 (1886).
4. E. L. Quinn and G. Wernimont. J. Am. Chem. Soc., **51**, 2002 (1929).
5. J. S. Rowlinson. Nature, **224**, 541 (1969).
6. A. N. Campbell and R. M. Chattersee. Can. J. Chem., **46**, 575 (1968) and **47**, 3893 (1969).

3

VISCOSITY OF GASES

The viscosity of a gas is a transport phenomenon in which the properties under investigation are not evenly distributed in the system. To treat a system mathematically in which pressure, concentration, velocity, and other gradients exist is a difficult problem that is beyond the scope of this presentation. However, one may approximate the real situation by assuming an equilibrium condition and superimposing upon it a small perturbance, which will not appreciably alter properties that follow a Maxwell-Boltzmann distribution. This provides a number of useful expressions for the rate of transport of the perturbation relating it to the mean free path.

For example, the viscosity of gas is imagined as a consequence of the flow of layers past each other (Fig. 3–1). Adjacent layers are separated from each other by a distance, λ, which is the mean free path. The layers, which are parallel to the xy plane, move in the x direction. There is a velocity gradient du/dz along the z axis, so that if the middle layer in Figure 3–1 has a velocity of u, the layer above it has a velocity of $u + \lambda(du/dz)$ and the layer below it has a velocity of $u - \lambda(du/dz)$.

We consider that each layer has an area of A cm.^{2}. The concentration in the system is n^* molecules per cc., each having a mass of m, and the average velocity is $\bar{u}$ with which the molecules move in the z direction. The rate of transport of the momenta of the gas molecules from the upper layer to the middle layer is $n^*A\bar{u}[mu + m\lambda(du/dz)]$ and that from below is $n^*A\bar{u}[mu - m\lambda(du/dz)]$. The net rate of transport of momentum in the z direction is

$$\Gamma = -2n^*A\bar{u}\lambda m(du/dz) \tag{1}$$

and is the force, F, acting on the molecules in the direction of the x axis and perpendicular to the yz plane according to Newton's second law of motion. If the velocity gradient is positive, the force is negative and acts counter to the direction of motion.

53

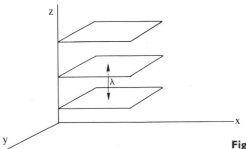

Figure 3–1 *Transport phenomena of gases.*

In a Newtonian flow the drag one layer exerts on another is proportional to the velocity gradient of the layers and to the area.

$$F = -\eta A \frac{du}{dz} \qquad (2)$$

The viscosity coefficient, η, is then obtained from the combination of equations 1 and 2.

$$\eta = 2mn^*\bar{u}\lambda \qquad (3)$$

When the average velocity in space, $\bar{c}$, is substituted for the average velocity in one direction, $\bar{u}$,

$$\bar{u} = \tfrac{1}{4}\bar{c} \qquad (4)$$

$$\eta = \tfrac{1}{2}mn^*\bar{c}\lambda \qquad (5)$$

If the molecular diameters, σ, are the quantities to be determined from viscosity measurements, $\bar{c}$ and λ can be substituted from

$$\bar{c} = \left(\frac{\pi kT}{2m}\right)^{1/2} \qquad (6)$$

and

$$\lambda = \frac{1}{\sqrt{2}\,\pi\sigma^2 n^*} \qquad (7)$$

Hence, equation 5 becomes

$$\eta = \frac{1}{\pi\sigma^2}\left(\frac{kTm}{\pi}\right)^{1/2} \qquad (8)$$

It is pertinent to note that the result of this derivation (equation 8) differs by a factor of $\tfrac{2}{3}$ from the one that appeared in Maxwell's original paper (see Moelwyn-Hughes).

In equation 8 a relationship between molecular diameter and viscosity coefficients is given. The limitation of this equation should be understood clearly. A relationship was obtained first between the viscosity coefficient and the mean free path. That is, our model of layers was set up in such a manner that each molecular collision occurred between molecules only and therefore carried a transport of momentum. In most experimental setups, however, the viscosity coefficient of a gas is obtained by timing a laminar flow through a capillary tube. If the gas is too dilute (i.e., low pressures) the mean free path may be of the same order as the dimension of the capillary, and the frequency of collision with the walls may be of considerable importance.

Under such conditions, the Newtonian flow profile corresponding to equation 2 is not tenable. This profile would indicate that the first layer near the capillary wall has zero velocity. Each successive layer proceeding toward the center of the capillary tube has higher velocity and the central core layer has the maximum velocity. When the collisions with the wall are appreciable in addition to intermolecular collisions, the first layer of gas near the wall is not immobile but has a finite velocity because only half the molecules in this first layer have collided with the wall; the other half collided with molecules in the neighboring layers at some distance, λ, from the wall. The average velocity of this layer is approximately $4\lambda/r$, where r is the radius of the capillary.

Since this first layer slips along the wall, the correction to be made is known as slip correction. The apparent viscosity, η_{app}, obtained in the experiment is then related to the true viscosity coefficient, η:

$$\eta = \eta_{app}\left(1 + \frac{4\lambda}{r}\right) = \eta_{app}\left(1 + \frac{4}{\pi\sigma^2 n^* r}\right) \tag{9}$$

and

$$\eta_{app} = \frac{n^* r}{4 + \pi\sigma^2 n^* r} \cdot \left(\frac{kTm}{\pi}\right)^{1/2} \tag{10}$$

Thus, molecular diameters can be obtained with the aid of equation 10 if the radius of the capillary is known.

In both the kinetic theory of gases and the correction for the excluded volume in the van der Waals equations the molecules are visualized as hard spheres or rapidly rotating hard bodies sweeping out a spherical space in any given instant (rotational motions being much faster than translational motion). Thus it is of interest to compare molecular diameters obtained from viscosity coefficients (equation 10) to that obtainable from the van der Waals volume correction term, b,

$$b = 4N[\tfrac{4}{3}\pi(\sigma/2)^3] \tag{11}$$

This comparison points out the limitations in the interpretations of both the equation of state and the transport phenomena.

The viscosity coefficient, η, of a gas is best measured by observing the efflux time of a laminar flow through a capillary. Poiseuille derived the following relationship for these flows. The frictional drag due to a Newtonian flow is given by equation 2. In terms of the capillary dimensions of l, length, and r, radius, of the layer in question, the area of the central layer is $2\pi r l$. Thus equation 2 becomes

$$F = -\eta(2\pi r l)\left(\frac{du}{dr}\right) \tag{12}$$

The hydrostatic pressure, ΔP, drives the central core layer in opposition to the viscous drag, and for steady flow the two are equal.

$$-\eta(2\pi r l)\frac{du}{dr} = \Delta P(\pi r^2) \tag{13}$$

If we rearrange equation 13 and integrate between the boundary conditions so that $r = R$ (the radius of the capillary) and $u = 0$,

$$\int_{u=0}^{u=u} du = -\frac{\Delta P}{2\eta l}\int_{r=R}^{r=r} r\, dr \tag{14}$$

$$u = \frac{\Delta P}{4\eta l}(R^2 - r^2) \tag{15}$$

Thus the velocity of each layer of r, radius, is obtained. In practice, however, the efflux time, t, of a certain volume, V, of a gas is measured. The volume rate of flow is obtained by integrating the product of the cross sectional areas of cylindrical segments and the velocity flow of segments:

$$\frac{V}{t} = \int_0^R (2\pi r)u\, dr = \frac{\pi \Delta P R^4}{8\eta l} \tag{16}$$

Thus

$$\eta = \frac{\pi \Delta P R^4 t}{8lV} \tag{17}$$

Equation 17 is the well known Poiseuille equation and can be applied to gases if ΔP is small compared to the atmospheric pressure. Equation 17 can be written in the abbreviated form

$$\eta = Kt \tag{18}$$

where K is the instrument constant containing the geometrics of the viscometer, such as the length of the capillary tube, l, its radius, R, the total volume, V, and the hydrostatic pressure.

Equation 18 can also be used in a more general case when the compressibility of gas is also taken into account. Under this condition, however, the proportionality constant, K, includes factors besides that of the geometry of the instrument. Since these other factors (such as compressibility) are not easily evaluated in each case, the instrument constant is best obtained by calibration with a gas having an accurately known viscosity coefficient. Dry air can be used for this purpose. The following empirical equation gives the viscosity coefficient of dry air as a function of temperature.

$$\eta = \frac{(145.8 \times 10^{-7})T^{3/2}}{T + 110.4} \tag{19}$$

Experimental

The experimental apparatus shown in Figure 3–2 consists of a capillary tube attached to a container acting as a mercury reservoir. The container has two marks, a and b, which define the viscometer volume, V. When gas is admitted into the viscometer the mercury level is depressed below mark a. When the stopcock to the capillary tube is opened the gas is allowed to escape

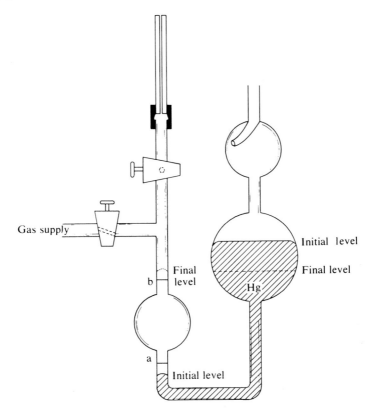

Figure 3–2 *Gas viscometer.*

and the mercury level rises. The efflux time is the time it takes the mercury level to move from mark *a* to mark *b*.

Mercury is poisonous. Do not inhale mercury vapor by pipetting. If accidental spills occur, collect the mercury droplets into a pool by wiping the working area with a wet cloth. The mercury pool should be collected through a capillary into a suction flask connected to an aspirator.

Before beginning the experiment, measure the length of the capillary. Obtain the radius of the capillary. Connect one end of the capillary to a vacuum pump or an aspirator. Dip the other end into a small mercury pool. When a small amount of mercury is sucked into the capillary, stop the vacuum and quickly turn the capillary to a horizontal position. Measure the length of the mercury column. Allow the mercury from the capillary to run into a preweighed beaker and weigh it. From the weight and density of the mercury, calculate its volume. From the known length of this volume of mercury in the capillary, calculate the the radius of the capillary. Carry out this measurement twice more. Average the radii of the capillary thus obtained.

The viscometer is calibrated with compressed air that passes through a drying tube containing Drierite. With both stopcocks open, allow dry air to flow through the capillary. The pressure under which the air flows should be regulated so that it is just above the atmospheric pressure. Close the stopcock leading to the capillary. Watch the mercury meniscus as it moves slowly downward. Increase the air pressure slightly to depress the meniscus of the mercury below mark *a*. Close the stopcock leading to the gas supply. Shut off the gas supply. Open the stopcock to the capillary and measure with a stopwatch the time it takes the mercury meniscus to move from mark *a* to *b*. Repeat the experiment three or four times until reproducible results are obtained. Note the temperature of the experiment.

Repeat the same measurements with helium, carbon dioxide, and oxygen. Each time a new gas is used, flush the viscometer with the new gas.

From the average efflux time of air and from the viscosity coefficient calculated from equation 19, obtain the instrument constant.

Using this instrument constant, calculate the apparent viscosity coefficients, η_{app}, of helium, carbon dioxide, and oxygen from equation 18.

Using the slip correction, calculate from equation 10 the corresponding molecular diameters. In order to use equation 10 assume that the gases are ideal gases; the concentration n^*(molecules/cc.) can be obtained from the ideal gas law.

Also calculate the molecular diameters from the corresponding van der Waals constants (equation 11) and compare the results.

Are the two values of molecular diameters in agreement? If not, is the disagreement random or systematic? What conclusion do you draw from these results?

Material and Equipment. Gas viscometer; 100 cc. of mercury; cylinders of compressed air, helium, carbon dioxide, and oxygen; stopwatch; 0 to 30° C. thermometer; millimeter scale.

REFERENCES

1. E. A. Moelwyn-Hughes. Physical Chemistry. 2nd Revised Edition, Pergamon Press, Inc., New York, 1964, Chapter II.
2. J. R. Partington. An Advanced Treatise on Physical Chemistry. Vol. I, Section VII, Longmans Green & Co., Ltd., London, 1949.

4

THE JOULE-THOMSON
COEFFICIENT OF REAL GASES

The definition of an ideal gas can be phrased in different terms. In the well known equation of state

$$PV = nRT \tag{1}$$

the relationship between pressure, P, volume, V, and absolute temperature, T, is established. The gas constant, R, is 0.082 atm. 1 mole^{-1} degree^{-1} and n is the number of moles of ideal gas in the system. The thermodynamic criterion for ideality, on the other hand, states that the internal energy, E, of the system is a function of the absolute temperature alone.

Therefore,

$$\left(\frac{\partial E}{\partial V}\right)_T = \left(\frac{\partial E}{\partial P}\right)_T = 0 \tag{2}$$

in an isothermal process the internal energy of a system will remain constant. It is easy to show that the same criterion also applies to the enthalpy, H, of an ideal gas.

$$H = E + PV \tag{3}$$

$$\left(\frac{\partial H}{\partial V}\right)_T = \left(\frac{\partial E}{\partial V}\right)_T + \left(\frac{\partial (PV)}{\partial V}\right)_T \tag{4}$$

According to equations 1 and 2, both the first and second terms on the right side of equation 4 are zero. Therefore,

$$\left(\frac{\partial H}{\partial V}\right)_T = 0 = \left(\frac{\partial H}{\partial P}\right)_T \tag{5}$$

In the case of real gases both equations, 2 and 5, differ from zero.

59

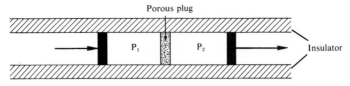

Figure 4–1   *The throttling process through a porous plug.*

The Joule-Thomson experiment, in which a gas is allowed to pass from a high pressure area to a low pressure area, measures the temperature change that accompanies such a throttling process. If the system is well insulated, and no heat can enter or leave the system ($q = 0$), the process is adiabatic. A pressure change will occur in the system, though, and so we cannot assume, a priori, that ΔH is also zero.

A schematic diagram of the Joule-Thomson experiment is given in Figure 4–1. The work done on the gas by the first piston in Figure 4–1 is

$$W_1 = P_1 V_1 \tag{6}$$

and the work done by the gas on the second piston is

$$W_2 = P_2 V_2 \tag{7}$$

The change in the internal energy of the system, ΔE, is therefore equal to

$$\Delta E = E_2 - E_1 = \Delta q - (P_2 V_2 - P_1 V_1) \tag{8}$$

But since the process is adiabatic, $\Delta q = 0$, by rearranging equation 8, we obtain

$$E_2 + P_2 V_2 = E_1 + P_1 V_1 \tag{9}$$

and

$$H_2 = H_1 \tag{10}$$

Thus the throttling process is not only adiabatic but also isoenthalpic (has constant enthalpy).

Therefore, we may write the experimental quantity measured in the Joule-Thomson experiment as

$$\left(\frac{\partial T}{\partial P}\right)_H = \mu_{\text{JT}} \tag{11}$$

and call it the Joule-Thomson coefficient, μ_{JT}.

It is easy to show that the Joule-Thomson coefficient is zero for ideal gases:
For an isoenthalpic process $dH = 0$ and

$$0 = dH = \left(\frac{\partial H}{\partial P}\right)_T dP + \left(\frac{\partial H}{\partial T}\right)_P dT \tag{12}$$

Rearrangement and statement of the constant enthalpy process produces equation 13.

$$\mu_{\text{JT}} = \left(\frac{\partial T}{\partial P}\right)_H = -\left(\frac{\partial H}{\partial P}\right)_T \bigg/ \left(\frac{\partial H}{\partial T}\right)_P = -\frac{1}{C_P}\left(\frac{\partial H}{\partial P}\right)_T \tag{13}$$

Since $(\partial H/\partial P)_T = 0$ for ideal gases (equation 5), the Joule-Thomson coefficient is zero for ideal gases.

This is not the case for real gases. The coefficient $(\partial H/\partial P)_T$ can be obtained from the thermodynamic relationship.

$$dH = T\,dS - P\,dV + P\,dV + V\,dP = T\,ds + V\,dP \tag{14}$$

Dividing by dP and maintaining constant T, we obtain

$$\left(\frac{\partial H}{\partial P}\right)_T = T\left(\frac{\partial S}{\partial P}\right)_T + V \tag{15}$$

However,

$$dG = V\,dP - S\,dT \tag{16}$$

and two relationships can be written

$$\left(\frac{\partial G}{\partial P}\right)_T = V \tag{17}$$

and

$$\left(\frac{\partial G}{\partial T}\right)_P = -S \tag{18}$$

Differentiating equation 17 with respect to T and equation 18 with respect to P, we obtain the equality

$$\left(\frac{\partial V}{\partial T}\right)_P = \frac{\partial^2 G}{\partial P\,\partial T} = -\left(\frac{\partial S}{\partial P}\right)_T \tag{19}$$

When equation 19 is substituted into equation 15 the following expression is obtained:

$$\left(\frac{\partial H}{\partial P}\right)_T = -T\left(\frac{\partial V}{\partial T}\right)_P + V \tag{20}$$

and the Joule-Thomson coefficient becomes

$$\left(\frac{\partial T}{\partial P}\right)_H = \mu_{JT} = \frac{1}{C_P}\left[T\left(\frac{\partial V}{\partial T}\right)_P - V\right] \tag{21}$$

If $T(\partial V/\partial T)_P > V$, then μ_{JT} will be positive; if $T(\partial V/\partial T)_P < V$, then μ_{JT} will be negative.

A positive μ_{JT} means that upon expansion of a gas the temperature decreases; we observe this phenomenon with most gases near room temperature. A negative μ_{JT} indicates that the temperature of the gas will increase upon expansion.

Whether a gas has a positive or negative μ_{JT} depends upon the working temperature. When $T(\partial V/\partial T)_P = V$, the Joule-Thomson coefficient becomes zero for real gases. This temperature is called the inversion temperature.

The range in which a gas has a positive Joule-Thomson coefficient is useful for gas liquefaction purposes. The Linde process is an example in which, by repeated gas expansion and compression cycles, the temperature of the gas is lowered to a point at which the last compression liquefies the gas.

From known equations of state one may predict the sign and the magnitude of the Joule-Thomson coefficient. For example, when the equation of state is a virial expression:

$$PV = A + BP + \cdots = nRT + BP + \cdots \tag{22}$$

$$\left(\frac{\partial V}{\partial T}\right)_P = \frac{nR}{P} + \left(\frac{\partial B}{\partial T}\right)_P + = V + \left(\frac{\partial B}{\partial T}\right)_P \tag{23}$$

Thus the μ_{JT} can be predicted from the temperature dependence of the second virial coefficient $(\partial B/\partial T)_P$.

The van der Waals equation of state can be used for the same purpose.

$$\left(P + \frac{an^2}{V^2}\right)(V - nb) = nRT \tag{24}$$

If we neglect the small term of n^3ab/V^2 for the moment and write P/nRT for $1/V$

$$PV = nRT - \frac{an^2P}{nRT} + nbP \tag{25}$$

$$\left(\frac{\partial V}{\partial T}\right)_P = \frac{nR}{P} + \frac{na}{RT^2} \tag{26}$$

Equation 25 can be rearranged as

$$V - nb = T\left(\frac{nR}{P} - \frac{na}{RT^2}\right) \tag{27}$$

and from this

$$(nR/P) = \frac{V - nb}{T} + \frac{na}{RT^2} \tag{28}$$

Substituting equation 28 into 26 we obtain

$$\left(\frac{\partial V}{\partial T}\right)_P = \frac{V - nb}{T} + \frac{2na}{RT^2} \tag{29}$$

Hence, the Joule-Thomson coefficient (equation 21) in terms of the van der Waals constant is expressed as

$$\mu_{\mathrm{JT}} = \frac{(2na/RT) - nb}{C_P} \tag{30}$$

Thus, the predicted inversion temperature is

$$T_i = \frac{2a}{Rb} \tag{31}$$

This prediction is not very good because, for one thing, the van der Waals constants vary with temperature and pressure. In addition, neglecting the term n^3ab/V^2 in the derivation (see equation 25) makes the prediction (equation 31) applicable only at low pressures.

If the term n^3ab/V^2 is also included (in the derivation), the condition for inversion temperature will be (still using $V = nRT/P$)

$$\frac{2a}{RT_i} - b - \frac{3abP}{R^2T_i^2} = 0 \tag{32}$$

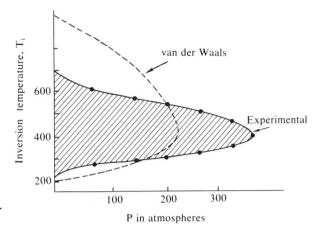

Figure 4-2 *Inversion tem-*
perature of nitrogen vs. pressure.

Multiplying by T_i^2 and dividing by b, we can write equation 32 in the quadratic form

$$T_i^2 - \frac{2a}{Rb} T_i + \frac{3a}{R^2} = 0 \tag{33}$$

Therefore, for every pressure, two inversion temperature values are predicted. A comparison of experimental data and the prediction by equation 33 is given in Figure 4-2.

Data such as those given in Figure 4-2 are of considerable industrial importance. The Joule-Thomson coefficients of a gas are positive within the shaded area and negative outside it. Therefore, for all temperature and pressure combinations within the curve, the rapid expansion of a gas will result in cooling.

Experimental

Two possible experimental setups are given here; the choice depends upon the accuracy desired and the expense involved. The first setup requires the direct measurement of temperatures on two sides of the porous plug as given in Figure 4-3. It is the more accurate but also the more expensive method. A porous plug made of asbestos pressed between two perforated porcelain disks or of a fritted glass plate is used. There is a calibrated platinum resistance thermometer on each side of the porous plug. The whole apparatus is well insulated to prevent heat loss.

Pressure is applied on one side of the porous plug by letting the gas flow from a cylinder of carbon dioxide, and the pressure is monitored by an open arm manometer. Open the valve of the cylinder very slowly not only to prevent the mercury from being blown from the manometer but, more important, to prevent the porous plug from being cooled below its steady state temperature. Readings are taken when a steady pressure of about 1.5 atm. (about 400 torrs on the open arm of the manometer) is reached and

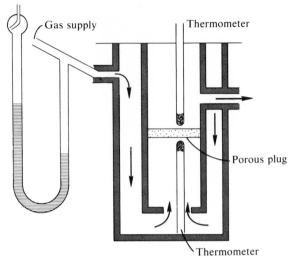

Figure 4–3   *The Joule-Thomson apparatus.*

a steady temperature difference is observed. This is usually achieved in 20 to 30 minutes.

The second setup is very similar except that the temperature is measured on only one side (exit side) of the porous plug (Fig. 4–4). On the other side is a thermostat into which a long copper coil is submerged (100 feet of $\frac{1}{4}$ inch O.D.). The assumption is that the gas passing through this long coil attains the temperature of the thermostat, so that its temperature can be obtained by simply measuring the temperature of the bath.

As in the previous method, a carbon dioxide pressure of about 1.5 atm. is applied by opening the cylinder valve very slowly, with caution so that the mercury is not blown from the manometer and the porous plug is not cooled below its steady state value. Steady pressure and steady temperature are reached in 20 to 30 minutes.

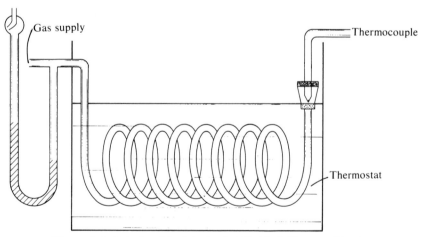

Figure 4–4   *Experimental setup of a simplified Joule-Thomson apparatus.*

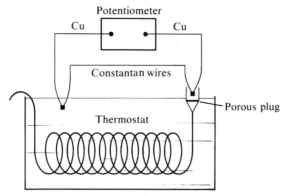

Figure 4-5 *The thermocouple circuit.*

While waiting for the system to reach a steady state the student may calibrate the thermocouples. The thermocouple used is a constantan (copper-nickel alloy) which in connection with copper wire generates an average thermoelectric potential of 0.03 millivolt per degree between $-200°$ C. and room temperature. The potential is generated as a result of an electron transfer that takes place when two dissimilar metals are placed in contact with each other. This transfer creates an electric double layer at the junction, and its potential is measured with reference to a standard potential. The standard potential is the thermoelectric potential generated by the same two wires at a standard temperature, usually $0°$ C. The potential thus is measured between the *ends* of the copper wires—one coming from the sample and the other from the reference ice water temperature bath. The constantan wires complete the circuit (see Fig. 4-5).

The advantage of connecting the copper wires to the potentiometer is that the binding posts or junctions of the potentiometer are usually also made of copper; hence, no thermoelectric potential will be generated at the sites of the connection. The constantan–copper thermocouple is useful between -200 and $+300°$ C., and the *Handbook of Chemistry and Physics,* the *International Critical Table,* and other reference sources provide tables to convert the potentials measured into temperatures.

The finest wires should be selected. A sensitive galvanometer (5×10^{-4} μa mm.$^{-1}$ sensitivity) should be used. The thermocouple leads are connected directly across the terminals of the galvanometer. One of the thermocouples, embedded in a small amount of wax, is placed in the thermostat. The other thermocouple is placed in a beaker that contains water about $1°$ C. warmer than the thermostat. This thermocouple is taped to the bulb of a Beckman thermometer. The temperature difference between the bath and the water in the beaker is noted and the corresponding galvanometric reading is taken. By adding small amounts of hot and cold water one can change the temperature of the beaker, and a calibrated curve can be established by plotting galvanometric readings vs. temperature difference. Four to five points should be obtained for both positive and negative ΔT, with the thermostat as the reference temperature. A more detailed account of the potentiometer, galvanometer, and other electronic components may be found in Part XIV on electronics.

After the calibration curve is obtained, dry the thermocouple used in the beaker. Insert the dry thermocouple just above the porous plug, and after steady state is reached, note the galvanometric reading.

Measure $(\Delta P/\Delta T)_H$ twice more with carbon dioxide and then twice with helium.

Calculate the Joule-Thomson coefficients for carbon dioxide and helium. Calculate the predicted values using the van der Waals constants and equation 30.

To what do you attribute the difference between the μ_{JT} of carbon dioxide and of helium? To what degree is the van der Waals equation applicable to the Joule-Thomson coefficient? What would be the inversion temperatures of helium and carbon dioxide under the experimental conditions available to you?

Material and Equipment. Cylinders of compressed helium and carbon dioxide; porous plug made of asbestos pressed between two perforated porcelain disks or of a fritted glass plate; open arm (mercury) manometer; *either* (1) a well insulated apparatus divided into two parts by a porous plug and (2) two calibrated platinum resistance thermometers, Muller resistance bridge, *or* (1) 30 liter tank of thermostat with heater, stirrer, and thermoregulator, (2) 100 feet of copper coil $\frac{1}{4}$ inch O.D., (3) constantan thermocouple, (4) galvanometer of $5 \times 10^{-4} \mu a$ mm.$^{-1}$ sensitivity, and (5) Beckmann differential thermometer, range 5° C., readable to 0.001°C.

REFERENCE

1. S. Glasstone. Thermodynamics for Chemists. D. Van Nostrand Co., Inc., Princeton, N.J., 1949.

IV

THERMODYNAMICS

5

HEAT CAPACITY RATIOS OF GASES

Heat capacity of a gas is defined as the heat absorbed by one mole of a gas per degree rise in the temperature. Heat capacity measurements can be performed either at constant volume or at constant pressure. If the work performed is work of expansion only (no electrical or other work involved), this will be zero for a constant volume process. The first law of thermodynamics under this condition is written as

$$\Delta E = q \tag{1}$$

where ΔE is the change in internal energy and q is the heat absorbed. The definition of heat capacity at constant volume will then be

$$C_V = \frac{dq_V}{dT} = \left(\frac{\partial E}{\partial T}\right)_V \tag{2}$$

the dq being an inexact differential. By definition, the enthalpy change is

$$dH = dE + P\,dV + V\,dP \tag{3}$$

For a constant pressure process the last term on the right side of equation 3 becomes zero.

Again, if no work other than work of expansion is involved,

$$dE + P\,dV = dE + dw = dq \tag{4}$$

and

$$dq_P = dH \tag{5}$$

for a constant pressure process. Hence,

$$C_P = \frac{dq_P}{dT} = \left(\frac{\partial H}{\partial T}\right)_P \tag{6}$$

The temperature derivative of equation 3 can be written as

$$\frac{dH}{dT} = \frac{dE}{dT} + \frac{d(PV)}{dT} \tag{7}$$

For one mole of an ideal gas the last term of equation 7 becomes

$$\frac{d(PV)}{dT} = \frac{d(RT)}{dT} = R \tag{8}$$

Combining equations 2, 6, 7, and 8, we obtain the following equation for an *ideal gas*:

$$C_P = C_V + R \tag{9}$$

These equations can be combined because for ideal gases

$$(\partial E/\partial T)_V = (\partial E/\partial T)_P$$

and

$$(\partial H/\partial T)_V = (\partial H/\partial T)_P$$

For reversible adiabatic expansion of an ideal gas, if it is assumed that C_V is not a function of temperature,

$$\frac{C_V}{R} \ln \frac{T_2}{T_1} = -\ln \frac{V_2}{V_1} \tag{10}$$

where the expansion proceeds from a volume, V_1, at a temperature, T_1, to a new volume, V_2, and temperature, T_2.

For an ideal gas, the ratio of the temperatures is given by equation 11:

$$\frac{T_2}{T_1} = \frac{P_2 V_2}{P_1 V_1} \tag{11}$$

Substituting equation 11 into 10, combining the V_2/V_1 terms, and using equation 9, we obtain

$$\ln \frac{P_2}{P_1} = -\frac{C_P}{C_V} \ln \frac{V_2}{V_1} \tag{12}$$

According to equation 12, the ratio of heat capacities can be obtained if the pressure and volume changes in an adiabatic expansion can be measured.

Pressure changes are easy to measure in an adiabatic expansion, but the experimental setup for measuring volume changes in a system having a *frictionless* piston is quite difficult.

It is easier to measure the volume ratios in equation 12 by following the adiabatic step (I) with an additional constant volume process (II) in which the gas is allowed to warm to the original temperature.

$$P_1 V_1 T_1 \rightarrow P_2 V_2 T_2 \tag{I}$$
$$P_2 V_2 T_2 \rightarrow P_3 V_2 T_1 \tag{II}$$

From the second step it follows that for one mole of ideal gas

$$\frac{T_2}{T_1} = \frac{P_2}{P_3} \tag{13}$$

and this combined with the first step as in equation 11 yields

$$\frac{V_2}{V_1} = \frac{P_1}{P_3} \tag{14}$$

Therefore, equation 12 becomes

$$\frac{C_P}{C_V} = \frac{\ln P_1 - \ln P_2}{\ln P_1 - \ln P_3} \tag{15}$$

Thus the ratio of the heat capacities can be obtained by simply observing the pressure changes in the two steps: (I) a rapid adiabatic expansion of the gas and (II) the thermal reequilibration of the gas with the thermostat bath.

In the derivation of equation 15, it was assumed that the gas behaves ideally. This assumption should be valid for a monatomic rare gas, such as helium. To test this, the following comparison can be made: At room temperature only translational, rotational, and vibrational contributions make up the heat capacity term. For a monatomic gas only the translational contribution exists. Therefore, the value of C_V for helium should be $3/2R$ and the ratio of heat capacities should be $C_P/C_V = 5/3$.

In the case of a diatomic or polyatomic molecule, the heat capacity at constant volume, C_V, can be calculated if the vibrational frequencies of all normal modes of vibrations are known. The total C_V calculated will be the sum of three contributions:

$$C_V$$

Translational	$3/2R$
Rotational	$3/2R$ (or R for linear molecules)
Vibrational	$Rx^2 e^x/(e^x - 1)^2$ for each normal mode

and $\Delta E = h\nu$ where $x = \Delta E/kT$, ν is the frequency of the vibrations, h is Planck's constant, and k is the Boltzmann constant.

Thus, the C_V can be calculated from spectral data. Again the assumption of an ideal gas can be used

$$C_P = C_V + R$$

and the theoretical C_P/C_V can be calculated.

Experimental

The following experiment involves a very simple apparatus, and the results obtained with it are only approximate heat capacity ratios. Nevertheless, the values can be compared to calculated values.

The apparatus consists of a large vessel with a capacity of 25 to 30 liters (Fig. 5–1). The room as a whole is used as a thermostat if the temperature fluctuation can be kept within a degree during the laboratory period. The vessel is connected to a gas source (helium and carbon dioxide tanks). The three outlets of the vessel go: (a) to the gas source, (b) to the atmosphere, and (c) to an open manometer. The pressure hoses should have screw clamps

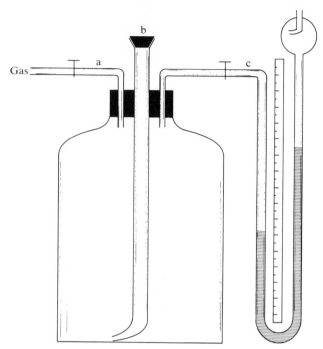

Figure 5–1   *The Clement and Desormes apparatus for measuring heat capacity ratios.*

on them so that the gas flow can be regulated. The hose leading to the atmosphere is closed by a stopper that can be removed and reinserted quickly.
The manometer should be about 1 meter long, and dibutyl phthalate (density = 1.046 grams/cc.) or Octoil (density = 0.875 gram/cc.) can be used as
a manometric fluid. The densities are those at 25° C.

The first experiment should be done with helium gas. Since helium is
less dense than air, care must be taken in filling the vessel with helium so
that no air remains to influence the results. Before the vessel is swept with
helium, close the screw clamp on the hose leading to the manometer. Remove
the stopcock on hose *b* and allow helium to sweep the vessel for 10 to 15
minutes at a pressure of about 1.5 atm. The rate of sweep should be about
10 liters/minute so that about 5 volumes of gas is swept out. (Note that
the helium, since it is lighter than the air, should enter the top of the vessel
and the air should exit through the hose reaching to the bottom of the vessel.
The rate of sweep can be measured roughly by leading the swept-out gases
through the hose into a water tank and collecting them in an inverted beaker
held under the water.)

Reduce the flow rate about twentyfold by adjusting the screw clamp on
hose *a*. While hose *b* is still open to the atmosphere, open the screw clamp
on hose *c*, but *very slowly* so that the manometric fluid is not blown out of the
manometer. (A safety bulb should be attached to the open arm of the
manometer; see Figure 5–1.) With the flow rate maintained at this reduced

level, insert the stopper very slowly into hose b, keeping an eye on the manometer and taking care not to cause a rapid rise. Stop the gas flow by closing the screw clamp on hose a when the pressure reaches about 80 cm. in the manometer. Allow the gas to equilibrate for about ten minutes and take the reading of the initial equilibrium pressure, P_1.

Remove the stopper from hose b with a small sweeping motion and then reinsert it *quickly*, making certain that it is fitted tightly. Allow the gas in the vessel to reach thermal equilibrium. During this time (10 to 15 minutes) the pressure will increase because of warming of the gas to room temperature. When the final equilibrium is reached, the pressure will become steady. Record this pressure, P_3. Also record the barometric pressure, P_2, from a barometer, and note the room temperature.

Carry out this procedure twice more with helium, without the flushing process. Initial manometric fluid pressures of 50 to 80 cm. can be used.

Then connect a carbon dioxide gas tank to hose b (the one that reaches to the bottom of the vessel). Use hose a as the outlet to the atmosphere, to be closed with a stopper. Since carbon dioxide is denser than helium and air, it will be introduced at the bottom of the vessel and the residual helium will be swept out through the top. Otherwise the carbon dioxide experiment is the same as the helium experiment, and it also should be done three times.

In calculating the heat capacity ratios, convert all your pressure readings into torrs (mm. of mercury), using the manometric liquid densities. Be sure that the converted readings of the open arm manometer are added to the barometric pressure in calculating P_1 and P_3; the difference measured in the open arm manometer is the difference between the pressure in the vessel and the atmospheric pressure. The P_2 is the barometric pressure because we assume that when the stopcock on the vessel is opened, adiabatic expansion causes the pressure inside to drop to the atmospheric pressure.

Use equation 15 to calculate the ratios of heat capacities for helium and carbon dioxide.

Calculate the theoretical values of C_V for helium and carbon dioxide, using the spectroscopic data for carbon dioxide as $\nu_1 = 1340$ cm^{-1}, $\nu_2 = 667.3$ cm.$^{-1}$, and $\nu_3 = 2349.3$ cm.$^{-1}$ (carbon dioxide is linear).[2]

Assume that equation 9 is operative for these gases within the pressure range of the experiment and compute the experimental C_V values. Is the deviation from the calculated value within your experimental error? Is there any systematic deviation? If there is, to what do you attribute it?

Material and Equipment. Cylinders of compressed helium and carbon dioxide; large vessel of 25 to 30 liter capacity; open manometer containing Octoil or dibutyl phthalate; stopper with three holes; glass tubing with stopcocks or rubber tubing with screw clamps; barometer; thermometer.

REFERENCES

1. J. R. Partington. An Advanced Treatise of Physical Chemistry. Vol. I, Longmans Green & Co., Ltd., London, 1949, pp. 792–848.
2. G. Herzberg. Molecular Spectra II—Infrared and Raman Spectra of Polyatomic Molecules. D. Van Nostrand Co., Inc., Princeton, N.J., 1956, p. 272.

6

HEAT OF MIXING*

When two liquid components are mixed, the change in the enthalpy during the process can be given by

$$\Delta H_M = H - (n_1 H_1 + n_2 H_2) \tag{1}$$

where ΔH_M is the heat of mixing, H is the total enthalpy of the solution, H_1 and H_2 are the molar enthalpies of pure components 1 and 2, respectively, and n_1 and n_2 are the number of moles of components 1 and 2.

The heat of mixing as given in equation 1 is an extensive property that depends on the size of the sample. The corresponding intensive property is defined as the heat of mixing per mole of solution, Δh_M,

$$\Delta h_M = \frac{\Delta H_M}{n_1 + n_2} = \frac{H}{n_1 + n_2} - (x_1 H_1 + x_2 H_2) \tag{2}$$

where x_1 and x_2 are the mole fractions of components 1 and 2.

$$x_1 = \frac{n_1}{n_1 + n_2} \tag{3}$$

$$x_2 = \frac{n_2}{n_1 + n_2} \tag{4}$$

The heat of mixing usually refers to an isothermal isobaric process, although we will later see that the actual experiment is performed somewhat differently. The ΔH_M and Δh_M are called the integral heats of mixing.

If equation 1 is differentiated with respect to n_2 while assuming constant T, P, and n_1, then

$$\left[\frac{\partial(\Delta H_M)}{\partial n_2}\right]_{T,P,n_1} = \left(\frac{\partial H}{\partial n_2}\right)_{T,P,n_2} - H_2 = \bar{H}_2 - H_2 \tag{5}$$

75

The partial or differential molar heat of solution of component 2 is

$$\left(\frac{\partial H}{\partial n_2}\right)_{T,P,n_1} = \bar{H}_2$$

The physical interpretation of $\bar{H}_2$ is the amount of heat evolved or absorbed if one mole of component 2 is dissolved in a very large quantity of solution at a specified mole fraction, x_2, or molality, m_2, so that the actual concentration is not appreciably changed.

Equation 5 can be rewritten by adding $\bar{H}_2^0$ to and subtracting it from the quantity on the right side of the equation.

$$\left[\frac{\partial(\Delta H_M)}{\partial n_2}\right]_{T,P,n_1} = (\bar{H}_2 - H_2) = (\bar{H}_2 - \bar{H}_2^0) - (H_2 - \bar{H}_2^0) = \bar{L}_2 - L_2 \quad (6)$$

$\bar{L}_2$ is the relative partial molar heat content of component 2 in a solution of specified concentration, and L_2 is the relative molar heat content of the pure component 2. In both cases, the reference state is $\bar{H}_2^0$, the partial molar heat content in infinitely dilute solution when $n_2 \to 0$.

Similarly, differentiating equation 1 with respect to n_1 gives

$$\left[\frac{\partial(\Delta H_M)}{\partial n_1}\right]_{T,P,n_2} = \bar{H}_1 - H_1 \quad (7)$$

Since, however, the molar heat content of component 1 in an infinitely dilute solution ($n_2 \to 0$) is the same as the partial molar heat content

$$H_1 = \bar{H}_1^0 \quad (8)$$

$$\left[\frac{\partial(\Delta H_M)}{\partial n_1}\right]_{T,P,n_2} = \bar{L}_1 \quad (9)$$

In a perfect solution or mixture, there is no heat of mixing. The partial molar heat contents and their variations with composition, therefore, indicate the extent of deviation from an ideal solution. This is similar to the deviation encountered in gases in which the ideal gas is a limiting condition. In an analogous manner an ideal (perfect) mixture or solution is a limiting condition that is approached as the solution is made infinitely dilute.

Although the integral heats of mixing are experimentally obtained in calorimetric studies, it is useful to evaluate the partial molar quantities. These can be determined as follows: Using the integral heat of mixing per mole of solution as defined in equation 2, we can write:

$$\Delta H_M = (n_1 + n_2)\Delta h_M \quad (10)$$

Differentiation with respect to n_2 gives

$$\left[\frac{\partial(\Delta H_M)}{\partial n_2}\right]_{T,P,n_1} = \Delta h_M + (n_1 + n_2)\left[\frac{\partial(\Delta h_M)}{\partial n_2}\right]_{T,P,n_1} \quad (11)$$

Differentiation of the mole fraction x_1 with respect to n_2 gives

$$\left(\frac{dx_1}{dn_2}\right)_{n_1} = -\frac{n_1}{(n_1 + n_2)^2} = -\frac{x_1}{(n_1 + n_2)} \quad (12)$$

Also

$$\left[\frac{\partial(\Delta h_M)}{\partial n_2}\right]_{T,P,n_1} = \left[\frac{\partial(\Delta h_M)}{\partial x_1}\right]_{T,P,n_1}\left(\frac{\partial x_1}{\partial n_2}\right)_{T,P,n_1} \tag{13}$$

Substituting equations 12 and 13 into equation 11, we obtain

$$\left[\frac{\partial(\Delta H_M)}{\partial n_2}\right]_{T,P,n_1} = \Delta h_M - x_1\left[\frac{\partial(\Delta h_M)}{\partial x_1}\right]_{T,P,n_1} \tag{14}$$

Therefore, if the integral heat of mixing per mole of solution (Δh_M) vs. the mole fraction (x_1) is plotted, the differential heat of mixing can be obtained from equation 14 at any concentration. The definition of this differential heat of mixing is given by equation 5.

Analyzing equation 14, we find that there are two conditions under which the second term of the right-hand side becomes zero:

(1) when $x_1 \to 0$ and

(2) when $\dfrac{\partial(\Delta h_M)}{\partial x_1} = 0$

The first condition indicates that the partial or differential heat of mixing of component 2 equals the integral heat of mixing per mole of solution in an infinitely dilute solution (from the point of view of component 1).

The second condition implies that, if there is an extremum in the Δh_M vs. mole fraction curve, at that specified concentration the partial or differential heat of mixing of component 2 becomes the integral heat of mixing per mole of solution.

When the Δh_M shows a maximum in a binary solution, a maximum in the partial molar enthalpy of mixing is also evident, although at a slightly different concentration. A maximum in the partial molar enthalpy of mixing usually indicates a strong association of the two components of the mixture at the specified concentration. This association can also be interpreted as compound formation.

During the previous discussion we assumed that the experimental conditions are such that the heat of mixing is obtained in an isothermal process. However, in order for isothermal conditions to exist, the heat evolved in the system during mixing must be removed as fast as it is produced. If the heat capacity of the system is known, it is much more convenient to carry out the mixing process adiabatically in an insulated (Dewar) flask and to calculate the heat of mixing from the change in the temperature.

Since ΔH is a thermodynamic property that depends only on the initial and final states and thus is independent of the path, the isothermal process can be visualized to proceed in two steps.

Step I

$$C(T_1) + nDMSO(T_1) + mH_2O(T_1) \to C(T_2) + nDMSO \cdot mH_2O(T_2) \tag{15}$$

Step II

$$C(T_2) + nDMSO \cdot mH_2O(T_2) \to C(T_1) + nDMSO \cdot mH_2O(T_1) \tag{16}$$

In equations 15 and 16 dimethyl sulfoxide (DMSO) and water are used as components 1 and 2, respectively. The same applies to other binary systems.

The first step is the adiabatic mixing process in which one allows the temperature to increase from T_1 to T_2. The C in the equations refers to the calorimeter (i.e., the inside walls of the Dewar flask, the sample holder, the stirrer, and the thermometer), the temperature of which also changes in the first step of the actual mixing process. The calorimeter is covered at the top to allow thermal insulation, but this covering is usually not a hermetic seal and it generally allows pressure equilibration. Therefore, the process can be considered a constant pressure process. In essence, the second step returns the mixture to the original temperature by removing the generated heat from the system. We can write the total enthalpy change for the two processes as follows:

$$\Delta H = \Delta H_I + \Delta H_{II} \tag{17}$$

But the first step is an adiabatic mixing; therefore, by definition

$$\Delta H_I = q_P = 0 \tag{18}$$

and

$$\Delta H = \Delta H_{II} \tag{19}$$

The measurement of the change in enthalpy in the second step is of great importance. It is not convenient to measure the heat removed from a system in an experiment; it is easier to use electrical work and to determine the heat necessary to produce a temperature increase equal to that observed in the first step. The enthalpy change obtained from electrical heating, ΔH_h, is equal but opposite in sign to ΔH_{II}, which is the value that would have been obtained if the system had been cooled. Therefore,

$$-\Delta H_{II} = \Delta H_h = -w_{el} = \int_0^t Ei\,dt \tag{20}$$

In equation 20, w_{el}, the electrical work, is negative, because work is done *on* the system. E is the potential (measured in volts), i, the current measured in amperes, and t, the time of heating in seconds.

Up to now we have assumed that the calorimeter is a perfectly insulated system and that no heat escapes. Obviously this is not the case. However, even under laboratory conditions, extraneous heat effects can be minimized. The stirrer's motion produces heat. On the other hand, there is a heat leakage because of imperfect insulation. The rate of stirring can usually be adjusted so that the two effects cancel each other out, and thus one obtains a constant temperature.

Heat loss may cause errors in the second step, however. In DMSO-water (DMF-water, and so forth) systems large amounts of heat are produced in the mixing process. The temperature rise in the adiabatic step is fast, and therefore not much heat is lost during the short interval of the experiment. When the same temperature rise is produced by means of electrical work, it takes considerably longer. During such a long heating period, losses due to heat leakage may be considerable. For this reason it is preferable to determine the heat capacity of the system (calorimeter + mixture) by measuring the electrical work necessary to increase the temperature of the system by only *one degree* at T_1 and at T_2.

$$C_P = \frac{-w_{el}}{\Delta T} = \frac{1}{\Delta T} \int_0^t Ei\,dt \tag{21}$$

As a first approximation, the average value of the heat capacity determinations made at the two temperatures for each mixture can be used for the heat capacity that would be operative over the whole temperature range $T_1 \rightarrow T_2$. Thus

$$\Delta H_{II} = C_P(T_2 - T_1) \tag{22}$$

Experimental

The calorimeter is constructed from a Dewar flask (A) capable of holding 1 liter of solution (Fig. 6–1). It is insulated from the environment by a cork. There are four holes through the cork so that the sample tube (B), thermometer (C), stirrer (D), and heater (E) may be placed in the calorimeter. A precision calibrated thermometer with a range of 20 to 60° C. is used and the temperature readings are estimated to 0.01° C.

The heater is made of Nichrome wire of appropriate resistivity. A Bakelite tube serves as the support for the heater, and grooves are provided for the winding of the resistance wire. The wire should be selected to provide approximately 15 watts of D.C. power operating on 5 volts. A voltmeter, an ammeter, and a variable D.C. power supply complete the circuit as shown in Figure 6–2. A more detailed account of the electronic components of the apparatus may be found in Part XIV of this book.

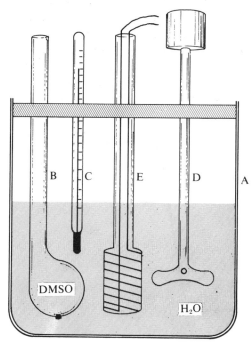

Figure 6–1 *Calorimeter.*

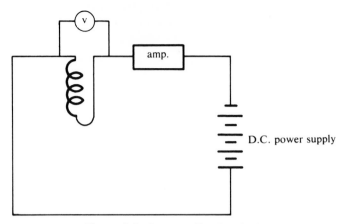

Figure 6–2 *Diagram of heating circuit.*

Portions of water and dimethyl sulfoxide (DMSO) are carefully weighed. (Dimethylformamide [DMF], acetone, dioxane, and so forth can be substituted for DMSO and for water.) Eight mixtures should be measured covering the range of 0.1 to 0.95 mole fraction of DMSO. The total weight of the two components for each mixture should be about 400 grams. (Although the weights of the mixtures need not be identical, the weights of the components in each mixture should be known accurately.)

Transfer the first portion of water to the calorimeter, *A*, and the first portion of DMSO to the sample tube, *B* (Fig. 6–1). The sample tube has a silicon gel (silicon grease) plug at the bottom, which keeps the DMSO from mixing with the water. Turn on the motor of the stirrer and, with the aid of a rheostat, establish a slow rate of stirring so that a constant temperature is maintained for three to five minutes (bottom part of Fig. 6–3). Apply pressure on the top of the sample tube, *B*, to remove the silicone gel plug and permit mixing. Record the temperature rise every two to three seconds

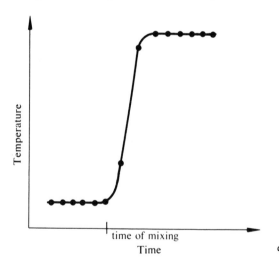

Figure 6–3 *Plot   of   the adiabatic step of mixing.*

until a maximum is reached, and then record the temperature for another two to three minutes. (An alternative procedure for mixing may be to simply pour the DMSO through a funnel into a calorimeter instead of employing a sample tube. If this procedure is adopted it is necessary to ensure that the two components (DMSO and water) are at exactly the same temperature before mixing.)

After the temperature has leveled off, turn on the power to the heater and simultaneously start the stopwatch. Note the time required for the temperature to rise one degree. Note the current and the voltage readings and repeat the procedure to obtain two heat capacity measurements at the high temperature.

Remove the mixture from the Dewar flask, put it in an Erlenmeyer flask, and thermostat it at 25° C or at the temperature of the mixing. Clean and dry the calorimeter and repeat the whole procedure with seven other portions of components 1 and 2.

After obtaining the temperature changes in the adiabatic mixing of eight different concentrations of mixtures, determine the heat capacities at the lower temperature as follows:

Place 400 grams of the first mixture, cooled to room temperature (the temperature at which the mixing was done), in the calorimeter and allow it to come into thermal equilibrium with the calorimeter. With a slow stirring rate, monitor the temperature as a function of time. After establishing a constant temperature plateau for two to three minutes, set the D.C. power supply to about 5 volts. Turn on the power and simultaneously start a stopwatch. Observe the rise in temperature and the readings on the voltmeter and ammeter. When a 1° C. temperature rise is reached, stop the watch and shut off the power. When the interval of time that elapsed, the voltage, and the current are known, the electrical work can be obtained. (Note that the electrical work is obtained in joules when E is given to volts, i, in amperes, and t, in seconds. To convert this to calories, divide by 4.184.) The heat capacity is then obtained from equation 21.

Repeat the heat capacity measurement twice at room temperature in order to check the reproducibility. If the reproducibility is poor, perform additional heat capacity determinations, since the final results depend to a large extent on the reliability of these measurements.

Repeat the process with the other seven mixtures. Average the heat capacities obtained at the high and low temperatures for each mixture. Using this average heat capacity and the temperature rise obtained in the mixing process, calculate the integral heat of mixing of each mixture (equation 22).

Calculate the integral heat of mixing per mole of solution at each concentration.

Plot Δh_M (integral heat of mixing per mole of solution) vs. mole fraction of water (or component 1).

By drawing tangents, obtain

$$\left[\frac{\partial(\Delta h_M)}{\partial x_1}\right]_{T,P,n_1}$$

and, using equation 14, obtain the differential heat of mixing of DMSO (or DMF, acetone, dioxane, and so forth)

$$\left[\frac{\partial(\Delta H_M)}{\partial n_2}\right]_{T,P,n_1}$$

as a function of concentration. Plot the latter function.

Is there any evidence for compound formation in the DMSO-water mixture (or DMF-water, acetone-water, dioxane-water, and so forth)?

In what respect is the curve of the integral heat of mixing vs. mole fraction different from the curve of differential heat of mixing vs. mole fraction.

Material and Equipment. Dimethyl sulfoxide (or dimethylform-amide, acetone, dioxane, and so forth); 1 liter Dewar flask; precision thermometer graduated to 0.01° C., range 20 to 60° C.; stirrer with rheostat; sample tube, maximum capacity of 200 ml., with silicon gel plug; Nichrome wire; D.C. power supply; ammeter; voltmeter; stopwatch; Erlenmeyer flask.

REFERENCES

1. S. Glasstone. Thermodynamics for Chemists. D. Van Nostrand, Co., Inc., Princeton, N.J., 1949, Chapter 18.
2. I. Prigogine, A. Bellemans and V. Mathot. The Molecular Theory of Solutions. Interscience Publishers, Inc., New York, 1957.
3. G. N. Lewis and M. Randall. Thermodynamics (revised by K. S. Pitzer and L. Brewer). 2nd Edition, McGraw-Hill Book Co., New York, 1961.
4. J. M. Cowie and P. M. Toporowski. Can. J. Chem., **39**, 2240 (1961).
5. J. M. Sturtevant. Calorimetry. In Weissberger, A. (ed.). Techniques of Organic Chemistry. 3rd Edition, Vol. I, Part I, Interscience Publishers, Inc., New York, 1959.
6. M. J. Blandamer et al. Trans. Faraday Soc., **65**, 2633 (1969).
7. T. Murakami, S. Murakami and R. Fujishiro. Bull. Chem. Soc. Japan, **42**, 35 (1969).

7

HEAT OF COMBUSTION

In a combustion reaction the test material is burned in pure oxygen atmosphere. In the simplest form of the reaction, when the test material is an organic compound containing only carbon, hydrogen, and oxygen, the final products are carbon dioxide and water. Thus

$$C_aH_bO_c(s) + (a + b/4 - c/2)O_2(g) \rightleftharpoons aCO_2(g) + (b/2)H_2O(l) \qquad (1)$$

is the stoichiometric form of the reaction.

When we refer to the enthalpy, ΔH, or energy, ΔE, of a reaction we have isothermal conditions in mind in the sense that both the initial state (reactants) and the final state (products) are at the same temperature.

Combustion reactions, on the other hand, are usually performed under adiabatic conditions in a bomb calorimeter.

$$C_aH_bO_c(T_0) + (c + b/4 - c/2)O_2(T_0) \rightleftharpoons aCO_2(T_1) + (b/2)H_2O(T_1) \qquad (2)$$

and the final state is at a higher temperature than the initial state. In order to be able to refer to isothermal conditions, a second process must be performed to cool the products to the original temperature.

$$aCO_2(T_1) + (b/2)H_2O(T_1) \rightleftharpoons aCO_2(T_0) + (b/2)H_2O(T_0) \qquad (3)$$

Thus the heat (ΔH) or energy (ΔE) of the reaction can be looked upon as the sum of two terms associated with reactions 2 and 3.

The combustion reaction in a bomb calorimeter is performed at constant volume; hence ΔE is the thermodynamic property that can be obtained directly:

$$\Delta E = \Delta E_a + \Delta E_c \qquad (4)$$

where ΔE_a is the internal energy change in the adiabatic process (reaction 2), and ΔE_c is the energy change in the cooling process (equation 3).

83

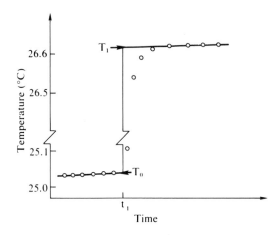

Figure 7-1 *Plot showing the extrapolation procedure necessary to obtain* ΔT *in the adiabatic calorimeter.*

According to the first law of thermodynamics

$$\Delta E = q - w \tag{5}$$

If the work, w, is restricted to work of expansion in a constant volume process, $w = P\,dV = 0$. Since in the adiabatic process $q = 0$,

$$\Delta E_a = 0 \quad \text{and} \quad \Delta E = \Delta E_c \tag{6}$$

Thus, in order to obtain the internal energy change of the combustion reaction, ΔE_c must be calculated.

$$\Delta E_c = [aC_V(CO_2) + b/2\ C_V(H_2O) + C_V(S)](T_0 - T_1) \tag{7}$$

In equation 7 the C_V's are the heat capacities at constant volume; $T_0 - T_1$ is the temperature change observed in the adiabatic step (reaction 2); S refers to the system, that is, to the calorimeter and its parts, such as the bomb lining and the sample crucible.

Calculation of ΔE requires two steps. First, the test material is combusted in a bomb and the accompanying $(T_1 - T_0)$ is obtained. In the laboratory experiment described later, the temperature rise is monitored as a function of time, Figure 7-1.

The temperature scale in Figure 7-1 is interrupted so that an expanded scale can be presented without excessively large graph paper. The slow rise in the temperature-time curve at the first five minutes is due to the heat generated by the stirrer of the calorimeter. At time t_1 the bomb is ignited, and the sudden rise in temperature, due to the heat of combustion is monitored. The curve tapers off eventually into another slow gradual rise that is again due to stirring. The two temperature rises caused by stirring should be linear in time and each can be extrapolated back to the time of ignition. Thus, the ΔT due to combustion is obtained.

In the second step, the sum of the heat capacities in equation 7 must be obtained. A standard substance (benzoic acid) the heat of combustion of which is known accurately is burned under conditions identical to those used for the test material, and the temperature rise due to its heat of combustion is obtained.

$$\frac{\Delta E_{\text{benzoic acid}}}{\Delta T} = C \tag{8}$$

The C is assumed to be equivalent to the bracketed term in equation 7. Admittedly, this introduces some error since the burning of benzoic acid may not produce the same number of moles of carbon dioxide and water as burning of the test sample. These errors are only minor, however, because usually only 1 to 1.5 grams of test material is used, whereas the weight of the calorimeter is about 2500 grams; therefore, the $C_V(S)$ in the bracketed term of equation 7 is dominant and outweighs small differences in the a and $b/2$ coefficients between test material and benzoic acid. These minor errors are acceptable in the laboratory experiment, but the student should be cognizant of how such errors can be corrected for high precision work, and the applicable corrections are given at the end of the experimental description.

Now that the heat capacity or the energy equivalent, C, has been obtained, the ΔE of the reaction can be calculated from equation 7.

Furthermore, if the ΔH of combustion is the desired thermodynamic property, it can be obtained from equation 9

$$\Delta H = \Delta E + (\Delta n)RT \qquad (9)$$

where Δn is the change in the number of moles of gas in the system as a result of combustion.

The ΔH obtained from this calculation refers to an initial and final state, which may differ from compound to compound. To make the heats of combustion of different materials comparable, they have to refer to a standard state. The heat of combustion in the standard state refers to heat obtained by burning one mole of test substance in its standard state and temperature (25° C.) in a sufficient amount of oxygen at 1 atm., with the formation of carbon dioxide gas and liquid water, also at 1 atm., the reaction taking place without the production of any external work.

Since the experimental conditions are not such that oxygen, carbon dioxide, and the other substances are under 1 atm. pressure, corrections must be made in order to be able to refer to the standard state.

In the present laboratory experiment these corrections will not be applied because the calculations are tedious and, without a computer, time-consuming. Furthermore, the errors introduced by not making corrections are tolerable for low precision work. Corrections are necessary only for high precision work, such as that reported in the literature.

Justification for not making corrections to the standard state is based on two facts: (1) The enthalpy changes due to phase transitions are small compared to those due to chemical changes, and (2) the enthalpy changes due to changes in pressure are also small (i.e., the compressibilities of condensed phases are small, and if the ideality of gases is invoked the enthalpy changes due to pressure changes at constant temperature are also small).

Experimental

The essential components of a bomb calorimeter are: (1) the bomb, which contains the sample and the oxygen, (2) an electrical ignition system, (3) the calorimeter, in which a known weight of water surrounds the bomb,

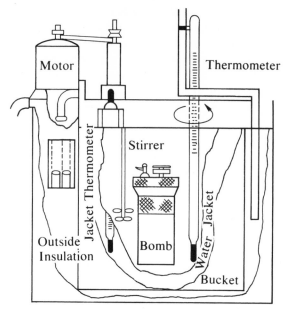

Figure 7-2 *Cross section of an adiabatic calorimeter (Parr No. 1230).*

and which includes a precision thermometer graduated to 0.01° C. and a
stirrer, and (4) an insulating wall or adiabatic jacket surrounding the
calorimeter (Sturtevant; Reilley and Rae; Hubbard, Scott and Waddington).
Figures 7-2 and 7-3 show some calorimetric designs.

The heat of combustion of anthracene (phenanthrene, naphthalene,
glucose, sucrose, galactose, or other test material) is determined in this

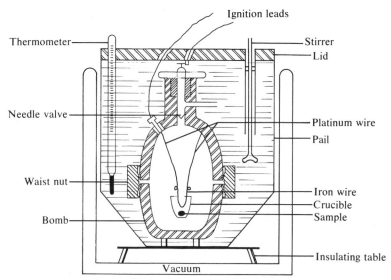

Figure 7-3 *Detailed cross section of a general purpose bomb calorimeter.*

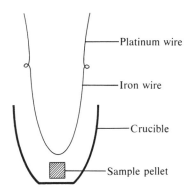

Platinum wire

Iron wire

Crucible

Figure 7–4 *Positioning of the sample by the first method.*

Sample pellet

experiment. First, however, the heat capacity of the calorimeter must be determined with the aid of benzoic acid, the heat of combustion of which has been established ($\Delta E = -6316$ cal./gram).

1. The sample should be large enough to provide 6 to 8 kcal. for a bomb calorimeter that has a volume of 300 to 400 ml. Approximately 1 gram of benzoic acid is pressed into a pellet in a die under pressure of 1 to 2000 psi by using a hydraulic press. (This must be done because loose powders usually do not burn completely. Even with pressed pellets the student must check at the end of the experiment to see whether there are any signs of incomplete combustion, such as soot or unburned sample. If there are, the experiment must be repeated, possibly at a higher oxygen pressure.)

There are two techniques by which the sample may be ignited.

(A) In one technique the sample pellet is placed in the crucible. The two platinum electrodes inside the bomb calorimeter are connected to a piece of special iron wire of about 0.005 mm. in diameter (Fig. 7–4). The iron wire passes slightly above the sample pellet but does not touch it. On ignition, the iron wire burns and its burning pieces fall on the sample, starting the combustion reaction. If this method of ignition is used, the weight of the iron wire as well as the weight of the sample pellet must be obtained within ±0.5 mg. by direct weighing.

(B) The benzoic acid (and other sample) pellets may be ignited by an iron wire that is fused into the pellet. A piece of iron wire free of bends and kinks is cut to the desired length to connect the electrodes inside the bomb calorimeter. After cutting, the wire is weighed. Next the wire is heated with a dry cell battery of 1.5 volts. While it is hot it is pushed through the sample (Fig. 7–5). The pellet is positioned so that it is in the middle of the wire and above the sample pan (crucible). Under this condition the burning iron wire will ignite the sample pellet directly, and the sample will fall into the sample pan (crucible) and continue to burn there. If this second technique of ignition is used, the pellet is weighed with the fused wire in it, and the weight of the pellet is obtained by determining the difference between this weight and the original weight of the wire. The sample should be handled very carefully after weighing.

Choice of method of ignition depends on the preference of the experimenter. In our laboratory the first technique has been used with good results.

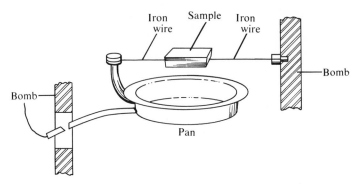

Figure 7–5 *Positioning of the sample by the second method.*

2. After the sample (benzoic acid) has been prepared, the bomb is inspected. Be sure that the bomb is dry and clean (that no iron globs remain from previous experiments, and so forth).

Install the sample pellet in the crucible and insert the iron wire into the nuts that lead to the ignition system. Add about 1 ml. of distilled water to ensure saturation of the oxygen in the bomb with water vapor. Assemble the two parts of the bomb and tighten the waist nut with a wrench. Be sure that the waist nut is correctly aligned with the threads on the bomb; otherwise the threads may be ruined.

Atmospheric nitrogen must be removed from the bomb by flushing with oxygen. First open the needle valve of the bomb about one turn and connect the bomb to an oxygen supply. Slowly fill the bomb with 20 to 30 atm. oxygen (300 to 400 psi), and then release the pressure. During this process most of the atmospheric nitrogen will be flushed out. Refill the bomb with oxygen to about 30 atm. pressure and close the needle valve of the bomb, tightening it with the hand only. Disconnect the oxygen supply and place the bomb containing the sample and oxygen in a water bath to check it for leaks. Most leaks occur around the waist nut, and if bubbles form more rapidly than one per five to ten seconds the waist nut needs to be tightened.

Dry the bomb. Connect the ignition wires to the appropriate places on the bomb (on the side and at the top in an Emerson type calorimeter, or both at the top if the Parr 1230 or a similar commercial calorimeter is used). Make certain that no shorting occurs in the ignition wires.

3. Place the bomb in the pail of the calorimeter. Position the pail in the center of the calorimeter; it should stand on an insulating table made of wood or cork. Fill the pail with accurately weighed water (~2000 grams) at 25° C. This can be accomplished by filling a 2 liter volumetric flask with cold and warm distilled water while monitoring the temperature so that the final temperature is within ±0.2 per cent of 25° C. The weight may be obtained either from the known density of water at 25° C. or by direct weighing of the volumetric flask (empty and filled).

Pour the contents of the flask into the pail carefully to avoid splashing, and thereby loss of water, and allow the flask to drain for a minute.

Assemble the top of the calorimeter, positioning the stirrer and the thermometer inside the calorimeter and tightening them in place. If commercial apparatus is used the precision mercury thermometer supplied is already clamped to the top of the calorimeter. This is usually a thermometer divided into 0.01° with which estimate can be made to 0.001° C. If the calorimeter is homemade, a Beckmann differential thermometer, set at about 24.5° as the lower limit, may be used. The 5° C. range of the Beckmann thermometer, which can be read to 0.001° C., is sufficient to cover the expected temperature rise due to the heat of combustion.

To set the Beckmann differential thermometer at a 24.5° C. low level, mercury must be either added to or taken from the mercury column. This is accomplished by immersing the Beckmann thermometer in a 40° C. water bath and allowing the mercury in the column to make contact with the mercury reservoir. Allow the water bath to cool to 39.5° C. (as measured by another thermometer), and then suddenly invert the Beckmann thermometer; break the mercury column by gently hitting the head of the Beckmann thermometer against your palm.

Check the setting of the Beckmann differential thermometer by immersing it in a 25° C. bath and recording the corresponding temperature (i.e., 25° C. = 0.438° on the differential Beckmann thermometer). If the setting of the mercury level is either too high or too low, repeat the preceding procedure until the correct setting is achieved.

Secure the lid of the calorimeter firmly, with both the thermometer and the stirrer in position.

4. For most work it is sufficient to use an insulating air jacket around the calorimeter to prevent or rather minimize heat leaks. (If a commercial adiabatic calorimeter is used, the jacket surrounding the calorimeter is filled with water, the temperature of which is constantly adjusted so that it is the same as that of the calorimeter itself. In the Parr 1230 adiabatic calorimeter this is accomplished by adding hot and cold water from an external supply. This produces a rapid and uniform heating in the jacket that is controlled by the temperature rise in the calorimeter when the bomb is fired. In some commercial instruments the heating is actuated by a tapping key that starts the heating circuit. In other instruments, such as the Gallenkamp bomb calorimeter, the adiabatic operation is furnished by an automatic thermistor-actuated control circuit.)

5. For the actual run, plug in the stirrer and start a *steady and smooth* stirring of the water in the calorimeter.

Take temperature readings every 30 seconds, recording both time and temperature. Estimate the temperature to 0.001° C. After observing and recording the steady temperature rise for about five minutes, fire the bomb. *Ignition is accomplished by turning the ignition switch on and then immediately off.* The electrical energy coming from a 110 volt 60 cycle supply is sufficient to burn the iron wire fuse and thereby start the reaction.

Time-temperature readings every 30 seconds are continued. The temperature of the calorimeter will start to rise about 15 to 20 seconds after ignition. This may continue for 10 to 20 minutes, after which the temperature will level off or, more frequently, continue increasing, slightly and steadily, because of the action of the stirrer. Monitor the time-temperature data for another ten minutes in order to provide a good base for extrapolation.

6. At the end of the experiment disassemble the calorimeter. Remove the bomb. *Release the pressure slowly by turning the needle valve.* Then open the bomb carefully by turning the waist ring nut with a wrench. Inspect the bomb for any sign of incomplete combustion, such as soot or unburned sample. If such signs are present, the next experiment must be done with somewhat less sample and somewhat higher oxygen pressure.

Remove the unburned parts of the iron wire fuse and *weigh them carefully.* (The globs are usually iron oxide and can be ignored unless they cannot be crushed to powder; in that case they are molten unoxidized iron.) The weight of the unburned iron wire is subtracted from the original weight of the wire to obtain the net weight of the iron that burned.

Clean all parts of the calorimeter.

Repeat the experiment with another sample of benzoic acid and then with two samples of the test material, anthracene (or phenanthrene, sucrose, or some other substance).

In calculating the heat capacity of the calorimeter (also called energy equivalent of the calorimeter) from equation 8, make certain that you use the contributions of both the benzoic acid and the burned iron wire. The following values should be used.

$$\Delta E_{\text{benzoic acid}} = -6316 \text{ cal./gram}$$

$$\Delta E_{\text{iron}} = -1600 \text{ cal./gram}$$

After calculating the heat capacity of the calorimeter, calculate the average heat of combustion of the test material (anthracene, phenanthrene, sucrose). *Be sure that in this calculation you also take into account the amount of heat contributed by the burned iron wire.*

Calculate the standard heat of combustion of your sample from tables of standard heats of formation. Compare your experimental value with that calculated from the tables.

CORRECTIONS

Three corrections are necessary in calculating heat of combustion with high precision: (1) correcting for nitric acid formed during the reaction; (2) correcting for the unequal moles of carbon dioxide and water formed during the combustion of benzoic acid and that of the test material; and (3) correcting for standard conditions.

1. Nitric acid is formed in the bomb calorimeter because of contaminating nitrogen in the bomb. The presence of contaminating nitrogen may be due to insufficient flushing of the ambient atmosphere from the bomb or to use of unpurified commercial oxygen. During the combustion nitrogen

dioxide is formed and it combines with water to yield nitric acid. The heat of formation of nitric acid must be subtracted from the total heat evolved. Usually at the end of the experiment the bomb is rinsed with distilled water and the collected solution is titrated with sodium hydroxide to obtain the amount of nitric acid formed.

For high precision work, then, the correction is calculated by using 227 calories per gram of nitric acid formed.

In most experiments the amount of nitric acid formed from atmospheric impurities is very small ($\sim$3 $\times$ 10^{-4} moles). Furthermore, when benzoic acid (or other substance) is used to obtain the heat capacity of the calorimeter, presumably the same atmospheric nitrogen impurities are present as with the unknown sample. Therefore, the heat capacity calculated from the heat of combustion of benzoic acid and from the temperature rise is not the true heat capacity but includes the nitric acid correction term.

$$\frac{\Delta E_b}{\Delta T} + \frac{\Delta E_{Fe}}{\Delta T} + \frac{\Delta E_{HNO_3}}{\Delta T} = C_V$$

where ΔE_b is the energy of the combustion of benzoic acid, ΔE_{Fe} is the energy of combustion of the iron wire, C_V is the heat capacity of the bomb, ΔE_{HNO_3} is the energy of formation of the nitric acid, and ΔT is the temperature rise observed. The ΔE_{HNO_3} is small and can be neglected as an approximation. Better, if the ΔT values observed with the benzoic acid and with the test material are approximately the same, and the nitrogen impurities in the oxygen atmosphere are also the same in the two subsequent experiments, the $\Delta E_{HNO_3}/\Delta T$ can be included in the heat capacity term.

$$\frac{\Delta E_b}{\Delta T} + \frac{\Delta E_{Fe}}{\Delta T} = C_V' = C_V - \frac{\Delta E_{HNO_3}}{\Delta T}$$

Thus, for medium precision work the nitric acid content at the end of the combustion does not have to be determined.

2. The second correction corrects for the unequal quantities of carbon dioxide, water, and so forth present in the combustion of the standard and of the test material.

We separate the heat capacity into two terms: A, the heat capacity of the calorimeter without the sample and iron wire, and B, the heat capacity of the combustion sample and iron wire.

For a sample with the molecular formula $C_aH_bO_c$, according to Washburn the B is given by equation 10

$$B = 0.158m_{Fe} + (m/M)[(1.77 + 0.0112P)a + 7.74b + 2.5c]$$

$$- 34n_{HNO_3} \text{ cal./degree C. (10)}$$

In equation 10, m_{Fe} is the mass of iron wire burned to ferric oxide. P is the final pressure of the bomb in atmospheres. m is the mass of the sample in grams, and M is its molecular weight. The coefficients a, b, and c refer to the molecular formula of the sample, and n_{HNO_3} is the number of moles of nitric acid formed during the combustion.

Thus the ΔE of combustion obtained is

$$-\Delta E = \Delta T(A + B)M/m \qquad (11)$$

In order to calculate the A of equation 11, benzoic acid is used for calibration.

$$\Delta E_{\text{benzoic acid}} = \Delta T(A + B)M/m \qquad (12)$$

The $\Delta E_{\text{benzoic acid}}$, the molecular weight, M, and the mass, m, of the benzoic acid are known. The ΔT is obtained in the calibration run. The B is calculated from equation 10. Thus the A calculated from equation 12 is used in equation 11 to obtain ΔE for the sample.

3. The third correction brings the enthalpy of combination into the standard state, i.e.,

$$C_a H_b O_c(s, 1 \text{ atm.}) + (a + b/4 - c/2)O_2(g, 1 \text{ atm.})$$
$$\rightleftharpoons aCO_2(g, 1 \text{ atm.}) + (b/2)H_2O(l, 1 \text{ atm.})$$

This hypothetical process can be broken up into a number of steps. Washburn described these steps first and a more recent step by step procedure is given by Hubbard, Scott and Waddington.

Step 1: n_{O_2} moles of oxygen at T_{standard} and 1 atm. is compressed into a bomb that contains n moles of substance and n_w moles of water. The initial pressure of the oxygen in the bomb is P atm. at T_s.

Step 2: the combustion is carried out and $-\Delta E$ is calculated at T_s. The final pressure in the bomb is $(P_2 + P_w)$ atm. at T_s, with P_w being the partial pressure of the water.

Each step has a ΔE term associated with it.

The final

$$\Delta E = \Delta E_{\text{bomb}} + \sum \Delta E_{\text{corr}} \qquad (13)$$

where ΔE_{bomb} is the internal energy change obtained in the particular state in which the experiment was performed and $\sum \Delta E_{\text{corr}}$ is the sum of $\Delta E - s$ in each of the correction steps.

These calculations are tedious. Dr. Randolph C. Wilhoit, Thermodynamics Research Center, Department of Chemistry, Texas A & M University, College Station, Texas 77843, has written a program in Fortran IV language to compute these correction terms. In an article in *Journal of Chemical Education* he graciously offered to supply this program to anyone interested.

Material and Equipment. Benzoic acid; anthracene (phenanthrene, naphthalene, glucose, or other substance); die; hydraulic press; bomb calorimeter; iron ignition wire, 0.005 mm. diameter; cylinder of compressed oxygen.

REFERENCES

1. J. M. Sturtevant. Calorimetry. In Weissberger, A. (ed.). Physical Methods of Organic Chemistry. 3rd Edition, Vol. I, Part I, Interscience Publishers, Inc., New York, 1959.
2. J. Reilley and W. N. Rae. Physico-Chemical Methods. 5th Edition, Vol. I, Methuen & Co., London, 1959, p. 509.
3. E. W. Washburn. J. Res. Natl. Bur. Standards, **10,** 525 (1933).
4. W. N. Hubbard, D. W. Scott and G. Waddington. In Rossini, F. D. (ed.). Experimental Thermochemistry. Vol. I, Interscience Publishers, Inc., New York, 1956.
5. R. C. Wilhoit. J. Chem. Ed., **44,** A571, A629, A685, and A853 (1967).

8

TEMPERATURE DEPENDENCE
OF VAPOR PRESSURE;
HEAT OF VAPORIZATION*

A closed, two-phase system, liquid and its vapor, is in thermal equilibrium when no heat flows across the phase boundary. The necessary condition for thermal equilibrium is that both phases have the same temperature. This can be recognized intuitively. However, two other conditions are necessary for the existence of thermodynamic equilibrium.

The Gibbs free energy is the function of composition and two external variables

$$G = f(T, P, n_i) \tag{I}$$

and

$$dG = \left(\frac{\partial G}{\partial T}\right)_{P, n_i} dT + \left(\frac{\partial G}{\partial P}\right)_{T, n_i} dP + \Sigma \left(\frac{\partial G}{\partial n_i}\right)_{P, T} dn_i \tag{2}$$

where the partial molar free energy $(\partial G/\partial n_i)$ is also called the chemical potential μ. The criterion for total equilibrium is that

$$dG = 0 \tag{3}$$

but for a two-phase system

$$dG = dG_\alpha + dG_\beta \tag{4}$$

where α and β indicate the two phases.

Considering the *thermal* equilibrium first, the equilibrium requirement is that

$$\left(\frac{\partial G}{\partial T}\right)_{P, n_{i\alpha}} dT_\alpha = -\left(\frac{\partial G}{\partial T}\right)_{P, n_{i\beta}} dT_\beta \tag{5}$$

94

and since

$$dT_\alpha = dT_\beta \tag{6}$$

$$\left(\frac{\partial G}{\partial T}\right)_{P,n_{i\alpha}} = -\left(\frac{\partial G}{\partial T}\right)_{P,n_{i\beta}} \tag{7}$$

For *hydrostatic* equilibrium, the two phases must be at the same pressure, unless they are separated by a rigid boundary. This can be shown as follows.

Suppose there is a volume variation dV_α in phase α, and a variation dV_β in phase β. The changes in the free energies of the two phases for an isothermal process, are given by

$$dG_\alpha = -P_\alpha \, dV_\alpha \tag{8}$$

$$dG_\beta = -P_\beta \, dV_\beta \tag{9}$$

The criterion for equilibrium is given in equations 3 and 4.

The combination of equations 3, 4, 8, and 9 gives

$$-P_\alpha \, dV_\alpha - P_\beta \, dV_\beta = 0 \tag{10}$$

But $dV_\alpha = -dV_\beta$, since the total volume is constant. Thus

$$(P_\alpha - P_\beta) \, dV = 0 \tag{11}$$

and we must have

$$P_\alpha = P_\beta \tag{12}$$

Finally, any substance that can pass freely between the two phases must have the same *chemical potential* in both phases. Suppose dn_i moles of substance i pass from phase α to phase β. At constant temperature and pressure, the change in the free energy is

$$dG_\alpha = \left(\frac{\partial G}{\partial n_i}\right)_{P,T,n_j} dn_{i\alpha} = \mu_{i\alpha} \, dn_{i\alpha} \tag{13}$$

$$dG = \left(\frac{\partial G}{\partial n_i}\right)_{P,T,n_j} dn_{i\beta} = \mu_{i\beta} \, dn_{i\beta} \tag{14}$$

According to equation 4 we must have

$$\mu_{i\alpha} \, dn_{i\alpha} + \mu_{i\beta} \, dn_{i\beta} = 0 \tag{15}$$

Since the total number of moles of i is constant, $dn_{i\alpha} = -dn_{i\beta}$,

$$(\mu_{i\alpha} - \mu_{i\beta}) \, dn_{i\beta} = 0 \tag{16}$$

and we must have

$$\mu_{i\alpha} = \mu_{i\beta} \tag{17}$$

Considering two different equilibrium states of a two-phase system at different temperatures and pressures, we can write the following conditions:

Equilibrium 1: $T; P; \mu_{i\alpha} = \mu_{i\beta}$ (18)

Equilibrium 2: $T + dT; P + dP; \mu_{i\alpha} + d\mu_{i\alpha} = \mu_{i\beta} + d\mu_{i\beta}$ (19)

Comparing conditions in equations 18 and 19, we find

$$d\mu_{i\alpha} = d\mu_{i\beta} \tag{20}$$

But the chemical potential, μ, is the partial molar free energy and since the change in free energy in terms of P and T is given by

$$\overline{dG} = \bar{V}\,dP - \bar{S}\,dT \tag{21}$$

it follows from equations 20 and 21 that

$$\bar{V}_\alpha\,dP - \bar{S}_\alpha\,dT = \bar{V}_\beta\,dP - \bar{S}_\beta\,dT \tag{22}$$

Rearranging, we obtain

$$\frac{dP}{dT} = \frac{\Delta\bar{S}}{\Delta\bar{V}} \tag{23}$$

If, as will be examined in the present experiment, the system contains only a single component, the partial molar entropy will be simply the molar entropy.

The entropy of transformation at constant T and P can be represented as

$$S = \frac{q_{rev}}{T} = \frac{\Delta H_{vap}}{T} \tag{24}$$

and equation 12 becomes the familiar Clausius-Clapeyron equation

$$\frac{dP}{dT} = \frac{\Delta H_{vap}}{T\Delta\bar{V}} \tag{25}$$

If we assume (1) that the partial molar volume of the liquid is negligible compared to that of the vapor, $\bar{V}_l \ll \bar{V}_g$, and (2) that $\bar{V}_g$ can be approximated by the ideal gas law

$$\bar{V}_g = \frac{RT}{P} \tag{26}$$

The Clausius-Clapeyron equation takes on the widely used form

$$\frac{d\ln P}{d(1/T)} = -\frac{\Delta H_{vap}}{R} \tag{27}$$

Therefore vapor pressure measurements of a liquid-vapor system as a function of temperature can be used to calculate the heat of vaporization of the liquid from equation 27.

The validity of assumptions 1 and 2 can be tested by integrating equation 27. During the integration, we make a third assumption, namely, that the ΔH_{vap} is constant over the temperature range in which the integration is carried out. This is probably the most erroneous of the three assumptions since we have seen that in many processes the enthalpy is a function of temperature, for example, as in the Kirchhoff equation:

$$\frac{d(\Delta H)}{dT} = \Delta C_P$$

Integration of equation 27 gives

$$\ln P = A - \frac{\Delta H_{vap}}{RT} \tag{28}$$

where A is the integration constant.

In the literature there are at least 50 empirical equations describing the relationship between vapor pressure and temperature. Three of them in particular resemble equation 28.

Antoine's equation

$$\log P = A - B/(t + C) \tag{29}$$

Rankin's equation

$$\log P = A - B/T + D \log T \tag{30}$$

Nernst's equation

$$\log P = A - B/T + 1.75 \log T + ET \tag{31}$$

where $B = \Delta H_{vap}/2.3R$, t is the temperature in degrees centigrade, T is the absolute temperature, and C, D, and E are constants of the system investigated. A value of C equal to 273.1 would make Antoine's equation identical to the Clausius-Clapeyron equation 28. However, C is usually smaller than 273.1.

Both Rankin's and Nernst's equations would be identical to equation 28 if the higher terms beyond the first two could be neglected.

The fact that equation 28 approaches the empirical equations is somewhat startling because of the approximation involved in its derivation. Neglecting the molar volume of the liquid may introduce an error of 1/1000. More serious is the assumption of ideal behavior, especially with polar liquids and at appreciable pressures. The third assumption of constant ΔH_{vap} over a temperature range is the most serious since the heat of vaporization decreases with increasing temperature and should reach zero at the critical temperature (Fig. 8–1).

Equation 28 is still in fairly good agreement with empirical observation because of the cancellation of two errors. This can be seen if we introduce the compressibility factor, Z, into equation 25.

$$\Delta Z = P\bar{V}_g/RT - P\bar{V}_l/RT \tag{32}$$

$$\frac{d \ln P}{d(1/T)} = -\frac{\Delta H_{vap}}{R \Delta Z} \tag{33}$$

In integrating equation 33 the ratio of $\Delta H_{vap}/\Delta Z$ is usually assumed to be constant over the range of integration (Partington). This is a much more plausible assumption than the three made previously, since ΔH and ΔZ have similar temperature dependence.

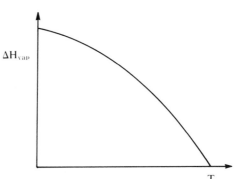

Figure 8-1 *Typical change in heat of vaporization with temperature.*

Under these conditions, therefore, it is not surprising that equation 28 is the limiting equation for many empirical equations, such as equations 29, 30, and 31.

The vapor pressure of a liquid at different temperatures may be determined by a variety of techniques, including static methods (isoteniscope), boiling point methods (such as Ramsay and Young), and dynamic transport techniques (such as gas saturation method). In the present experiment an isoteniscopic technique will be used.

Experimental

The isoteniscope (Fig. 8–2) is composed of a U tube, 8 mm. in diameter and about 15 cm. long, to which two bulbs are attached, A_1 and A_2 (Arm, Daemiker and Schaller). If acetone or other high vapor pressure liquid is used, the length of the isoteniscope should be 30 cm. Through the joint above A_1, the isoteniscope can be attached to a vacuum system (Fig. 8–3). The main features of the vacuum system are: a forepump, C; traps, D; main line with stopcock to provide either vacuum or pressure to the isoteniscope, E; a rotating arm, F, to position the isoteniscope either in the thermostat, T, or outside it; and a Zimmerli gage (Zimmerli, reference 4), Z, to measure the pressure in the system on the vacuum arm of the isoteniscope. This gage is shown in Figure 8–4. The Zimmerli gage incorporated into the vacuum line and the whole vacuum apparatus is built by the instructional staff before the experiment.

To prepare a Zimmerli gage, a 16 mm. diameter tube should be used for A, B, and C in Figure 8–4. The B and C tubes are joined through a capillary, D. Both A and C are connected to the same vacuum line. In a previously cleaned and dried Zimmerli gage clean mercury is added through the side

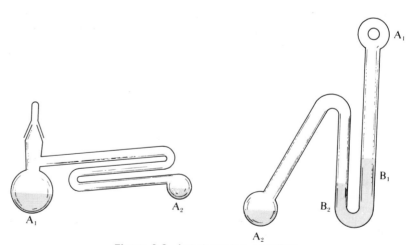

Figure 8–2 *Isoteniscope in two positions.*

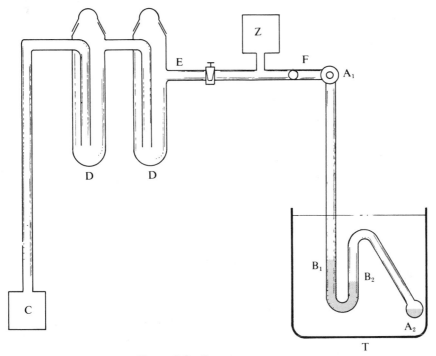

Figure 8-3 *The vacuum apparatus.*

arm that connects it to the system. Tubes A and B are filled about two-thirds full. After the gage is connected to the vacuum system, the trapped air bubbles are removed in the following manner. While the vacuum is on, tilt the gage backward to bring all the visible bubbles to the surface of the mercury in B and A. The gage walls must be trapped sharply to move the bubbles. When no more bubbles are visible, the gage is tilted to the left to permit the mercury to flow from B to D to C, so that air is removed from the reference arm, B. Do not tilt the gage so far that the mercury level in A drops completely, because this may introduce air bubbles into the reference arm, B. Once the mercury level in C is established, return the gage to its vertical position and release the vacuum.

When vacuum is applied to the side arm of the Zimmerli gage the mercury at the top of the bend of the capillary, D, will separate and the mercury level in B will approach that in A. The difference in the mercury level between A and B indicates the absolute pressure of the system. The difference is read with a cathetometer accurate to ± 0.01 torr (mm. of mercury).

The thermostat can be a simple fish tank with a heater, relay, and stirrer. The temperature should be maintained constant, at least within $\pm 0.05^\circ$ C. Sometimes, insulating the three sides of the thermostat helps to maintain the temperature in the desired range. The water in the thermostat should be sufficiently clear so that the pressure reading on the isoteniscope inside the thermostat can be taken with a cathetometer.

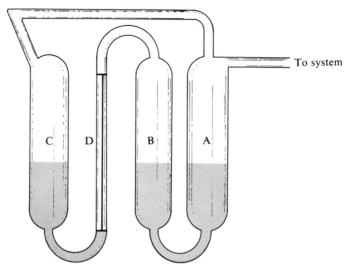

Figure 8–4   *The Zimmerli gage.*

For the experiment, clean the isoteniscope with nitric acid or cleaning solution, rinse it with an ample amount of water, and vacuum dry it. Grease the male joint above bulb A_1. Add about 10 ml. of dimethyl sulfoxide (DMSO) from a freshly opened bottle (99.9 per cent pure) to the A_1 bulb of the isoteniscope while it is in the horizontal position. (As an alternative, dimethylformamide [DMF], dioxane, acetone, and so forth may be used.)

Close the joint on the isoteniscope. With a back and forth movement of the apparatus, transfer the DMSO to bulb A_2. Add enough clean mercury to bulb A_1 to fill the U tube part about 6 to 7 cm. from the bottom. The isoteniscope is still in the horizontal position—there is no mercury in the U tube, only in bulb A_1. Attach the isoteniscope, still in the horizontal position, to the vacuum system through the rotating arm, F, in Figure 8–3. Freeze the DMSO in bulb A_2 by submerging it in a Dry ice–acetone mixture (or in liquid nitrogen when acetone and other substances are used). Open the stopcock and remove any contaminating gases by pumping on the system for ten minutes. At the same time warm the mercury in bulb A_1 to about 50° C. so that residual DMSO may be removed from the mercury. After ten minutes of pumping, close the stopcock leading to the vacuum. Warm the DMSO in bulb A_2 to room temperature. Then slowly open the stopcock and continue pumping on the *liquid* DMSO to remove any remaining traces of foreign condensable vapors.

Apply the Dry ice–acetone bath (or liquid nitrogen) once more to the DMSO in bulb A_2. Close the stopcock. While the DMSO is freezing, rotate the arm of the isoteniscope to bring it into a position in which the mercury will flow slowly from bulb A_1 into the U tube. Open the stopcock and pump on the system, and at the same time rotate the rotating arm, F, to position the isoteniscope in the thermostat, previously set 25° C. Be sure that the U tube arms of the isoteniscope are vertical in the thermostat. Allow sufficient

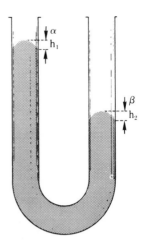

Figure 8-5 *Readings on the mercury columns in the isoteniscope.*

time for thermal equilibrium. This is achieved when the mercury no longer moves in the U tube arm of the isoteniscope. Take pressure and temperature readings at this equilibrium.

To take the pressure readings, measure the positions of the meniscus in the two arms of the isoteniscope with the cathetometer. Also measure the height of the meniscus in both arms in order to make corrections for the capillary depression (Fig. 8–5). Measure the meniscus difference in the Zimmerli gage to obtain the absolute pressure on the vacuum side. To calculate the total vapor pressure of the DMSO at 25° C., add the three pressures: (1) the difference in the meniscus height in the isoteniscope, (2) the difference between the corrections for capillary depression, and (3) the absolute pressure obtained in the Zimmerli gage. For the correction of capillary depression, the diameter of the isoteniscope (8mm.) must be known; then, for each measured height of the meniscus (h_1 and h_2 in Figure 8–5), obtain the correction from the *Handbook of Chemistry and Physics*.

An example of the calculation is given here:

		Correction from Handbook for 8 mm. Tube
Meniscus at α 163.20	$h_1 = 0.80$ mm.	0.29 mm.
Meniscus at β 158.60	$h_2 = 0.60$ mm.	0.20 mm.
4.60 mm.		0.09 mm.

Zimmerli gage reading: 0.02 mm.

Total vapor pressure: 4.60
0.09
0.02
4.71 torrs

Once the vapor pressure at 25° C. is obtained, raise the temperature of the thermostat and obtain five more vapor pressure measurements for DMSO between 25 and 45° C.

Repeat the same experiment for water and obtain as many points on the vapor pressure–temperature curve as time allows.

Plot the P vs. T curves for DMSO and water (or DMF, acetone, dioxane, and so forth).

Plot the log P vs. $1/T$ curves for the same compounds.

Using least squares fitting, obtain the constants of the Antoine, Rankin, and Nernst equations.

What are the average heats of vaporization over the 25 to 45° C. range for DMSO (DMF, acetone, dioxane) and water as calculated from the foregoing equations and from the Clausius-Clapeyron equation (using 25 and 45° C. as the limits)?

How does the ΔH_{vap} calculated from successive pairs of experimental points using the Clausius-Clapeyron equation vary with temperature?

Compare the constants you obtained with the least squares fitting of your data and the calculated heats of vaporation for DMSO (DMF, acetone, dioxane) and water with values for hydrocarbons obtainable from the *Handbook of Chemistry and Physics*. What conclusions can you draw regarding intermolecular forces?

Material and Equipment. Dimethyl sulfoxide (or dimethylformamide, acetone, dioxane, and so forth); 100 cc. mercury; isoteniscope; vacuum apparatus with a Zimmerli gage; thermostat adjustable between 25 and 45° C. and controlled to $\pm 0.05°$ C., with stirrer, heater, and thermoregulator.

REFERENCES

1. G. W. Thomson. In Weissberger, A. (ed.). Techniques of Organic Chemistry. 3rd Edition. Vol. I, Part I, Interscience Publishers, Inc.. New York, 1959, Chapter IX.
2. J. R. Partington. An Advanced Treatise on Physical Chemistry. Vol. III, Section VIII J., Longmans Green & Co., Ltd., London, 1951.
3. H. Arm, H. Daemiker and R. Schaller. Helv. Chim. Acta, **48**, 1772 (1966).
4. A. Zimmerli. Ind. Eng. Chem., Annual Ed., **10**, 283 (1938).

9

PARTIAL PRESSURES AND EXCESS THERMODYNAMIC FUNCTIONS OF LIQUID MIXTURES*

An ideal solution is defined as one that obeys Raoult's law over the whole range of composition. Raoult's law can be given in terms of fugacities or, if one assumes that the vapors of the liquids behave as ideal gases, in terms of partial vapor pressures.

$$f_i = x_i f_i^0 \tag{1}$$

$$P_i = x_i P_i^0 \tag{2}$$

In these equations f_i is the fugacity of component i in a mixture of vapors that are in equilibrium with the liquid mixture, and f_i^0 is the fugacity of component i in equilibrium with a pure liquid at the same temperature and pressure. The symbols P_i and P_i^0 are the partial pressures of component i of the mixture and of the pure liquid, respectively; x_i is the mole fraction of component i in the mixture.

An ideal solution, therefore, would have the total vapor pressure vs. mole fraction diagram represented in Figure 9–1a. Linearity of the partial vapor pressures and hence that of the total vapor pressure is required by equations 1 and 2.

For real solutions, deviations from ideality are found, either positive or negative deviations from Raoult's law (Fig. 9–1b and 9–1c). As can be seen from the diagrams (and it will be proved later in equations 13 and 14), if—in a binary system—one component has positive deviation from Raoult's law, the second component will also have positive deviation. Likewise, if one component has negative deviation, the second component will, too. A purely qualitative explanation would be as follows:

A positive deviation implies that the interaction between components 1 and

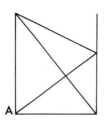 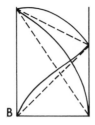 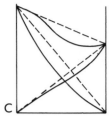

A   B   C

Figure 9–1 *Vapor pressure vs. mole fraction; a, ideal solution; b, positive deviation; c, negative deviation from Raoult's law.*

2 (1–2) is weaker than the interaction between like components (1–1 and 2–2). As a result, more molecules go into the vapor phase than under ideal conditions, in which the interaction 1–2 is the same as 1–1 and 2–2. A negative deviation from Raoult's law implies a stronger interaction in the mixture (1–2) than in the pure components (1–1 and 2–2).

An ideal solution has no heat of mixing and no volume change upon mixing. A nonideal solution has either or both. The degree of nonideality can be expressed quantitatively by the thermodynamic excess functions. This concept was introduced by Scatchard in 1931. A thermodynamic excess function, such as excess Gibbs free energy, is the difference between the Gibbs free energy obtained for a mixture and that obtained for an ideal solution at the same concentration.

$$\Delta G^E = \Delta G_M - [RT x_1 \ln x_1 + RT x_2 \ln x_2] \tag{3}$$

where ΔG^E represents the excess of free energy, and ΔG_M represents the free energy of mixing at a concentration of x_1 and x_2 mole fractions of components 1 and 2, respectively.

The bracketed term in equation 3 represents the free energy of mixing of an ideal solution. The free energy of mixing of a real solution, on the other hand, is given by the following equation:

$$\Delta G_M = RT x_1 \ln \gamma_1 x_1 + RT x_2 \ln \gamma_2 x_2 \tag{4}$$

where γ_1 and γ_2 are the activity coefficients of components 1 and 2 at the specified concentrations.

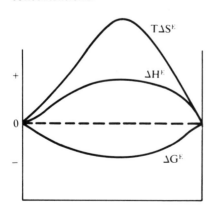

Figure 9–2 *Excess free energy, excess enthalpy, and excess entropy as a function of composition.*

Since in an ideal solution the enthalpy of mixing is zero, the experimentally obtained enthalpy of mixing is automatically the excess enthalpy of mixing. The excess entropy can be calculated if the excess free energy and enthalpy functions are known.

For isothermal mixing

$$\Delta G^E = \Delta H^E - T\,\Delta S^E \tag{5}$$

Such excess functions for a particular system are given in Figure 9–2.

In order to calculate the excess free energy function, partial vapor pressure data as a function of composition must be known. To calculate additional excess thermodynamic functions, the heat of mixing of the solution (obtained in the calorimetric experiment, Chapter 6) must also be known.

For a binary system in which the liquid mixture is in equilibrium with its vapor, the Gibbs-Duhem equation gives the condition for a small change in the composition of the system at constant temperature and pressure.

$$n_1\,d\mu_1 + n_2\,d\mu_2 = 0 \tag{6}$$

where n_1 and n_2 are the number of moles of components 1 and 2, and μ_1 and μ_2 are the respective chemical potentials. By dividing equation 6 by $(n_1 + n_2)$, we can express the Gibbs-Duhem equation in terms of mole fractions—x_1 and x_2.

$$x_1\,d\mu_1 + x_2\,d\mu_2 = 0 \tag{7}$$

Since $d\mu_1$ is a thermodynamic property (exact differential), at constant temperature and pressure we can write

$$d\mu_i = \left(\frac{\partial \mu_i}{\partial x_i}\right)_{T,P} dx_i \tag{8}$$

Substituting this into equation 7 gives

$$x_1\left(\frac{\partial \mu_1}{\partial x_1}\right)_{T,P} dx_1 + x_2\left(\frac{\partial \mu_2}{\partial x_2}\right)_{T,P} dx_2 = 0 \tag{9}$$

Taking into consideration that $x_1 + x_2 = 1$ and $dx_1 + dx_2 = 0$, we obtain another form of the Gibbs-Duhem equation in equation 10:

$$\left(\frac{\partial u_1}{\partial \ln x_1}\right)_{T,P} - \left(\frac{\partial \mu_2}{\partial \ln x_2}\right)_{T,P} = 0 \tag{10}$$

But the chemical potential of component i in a mixture is given by

$$\mu_i = \mu^0 + RT \ln f_i \tag{11}$$

where μ^0 is a constant in the standard state, and f_i is the fugacity of the component i. Hence,

$$d\mu_1 = RT\,d \ln f_i \tag{12}$$

Substituting equation 12 into equation 10 gives

$$\left(\frac{\partial \ln f_1}{\partial \ln x_1}\right)_{T,P} - \left(\frac{\partial \ln f_2}{\partial \ln x_2}\right)_{T,P} = 0 \tag{13}$$

This is the general form of the Gibbs-Duhem-Margules equation, applicable to any binary mixture regardless of whether it is an ideal mixture.

If we assume that the vapors in equilibrium with the liquid mixture behave as ideal gases, the Gibbs-Duhem-Margules equation takes the form

$$\left(\frac{\partial \ln P_1}{\partial \ln x_1}\right)_{T,P} = \left(\frac{\partial \ln P_2}{\partial \ln x_2}\right)_{T,P} \tag{14}$$

where P_1 and P_2 are the partial pressures of components one and two in the vapor phase.

In isoteniscopic experiments, usually the total vapor pressure of the mixture is measured at constant temperature. The total vapor pressure is given by Dalton's law:

$$P = P_1 + P_2 \tag{15}$$

Differentiating P with respect to x_1 gives

$$\frac{dP}{dx_1} = \frac{dP_1}{dx_1} + \frac{dP_2}{dx_1} \tag{16}$$

But, at constant temperature and pressure, the Gibbs-Duhem-Margules equation 14 can be written as

$$\frac{x_1}{P_1}\frac{dP_1}{dx_1} = \frac{dP_2}{dx_2}\frac{x_2}{P_2} \tag{17}$$

and since

$$dx_1 = -dx_2 \tag{18}$$

$$\frac{dP_1}{dx_1} = -\frac{dP_2}{dx_1}\frac{P_1}{P_2}\frac{x_2}{x_1} \tag{19}$$

Substituting equation 19 into equation 16 we obtain

$$\frac{dP}{dx_1} = \frac{dP_2}{dx_1}\left(1 - \frac{P_1 x_2}{P_2 x_1}\right) \tag{20}$$

The Margules equation 21 expresses the variation of vapor pressure of liquid mixtures with composition. If fugacities are used instead of vapor pressures, the partial fugacity of component 1 is given by

$$f_1 = x_1 f_1^0 \exp [\beta_1 x_2^2/2 + \cdots] \tag{21}$$

A similar equation can be written for the fugacity of component 2:

$$f_2 = x_2 f_2^0 \exp [\beta_2 x_1^2/2 + \cdots] \tag{22}$$

The f_1^0 and f_2^0 are the fugacities of the pure components 1 and 2.

These two equations are *simplified empirical* equations in which all but the first terms in the exponents are neglected. Margules proposed this form of the equation because it fits the experimental data well, but no theoretical justification was given.

If the constants β_1 and β_2 were zero, the equations would be reduced to Raoult's law since the exponent would become 1. The sign and the magnitude of β_1 and β_2 describe each system's deviation from Raoult's law.

Taking the logarithms of equations 21 and 22, we obtain

$$\ln f_1 = \ln x_1 + \ln f_1^0 + \beta_1 x_2^2/2 \tag{23}$$

and

$$\ln f_2 = \ln x_2 + \ln f_2^0 + \beta_2 x_1^2/2 \tag{24}$$

Differentiating the first equation with respect to $\ln x_1$ and the second with respect to $\ln x_2$, and keeping T, P constant, gives

$$\left(\frac{\partial \ln f_1}{\partial \ln x_1}\right)_{T,P} = 1 - \beta_1 x_1 x_2 \tag{25}$$

$$\left(\frac{\partial \ln f_2}{\partial \ln x_2}\right)_{T,P} = 1 - \beta_2 x_1 x_2 \tag{26}$$

But, according to equation 13, the left sides of equations 25 and 26 are equal. Hence,

$$\beta_1 = \beta_2 \tag{27}$$

and equations 21 and 22 will have the simplified form:

$$f_1 = x_1 f_1^0 \exp [\beta x_2^2/2] \tag{28}$$

$$f_2 = x_2 f_2^0 \exp [\beta x_1^2/2] \tag{29}$$

If the vapor pressures are not too large, the fugacities can be taken as equal to vapor pressures and

$$P_1 = x_1 P_1^0 \exp [\beta x_2^2/2] \tag{30}$$

$$P_2 = x_2 P_2^0 \exp [\beta x_1^2/2] \tag{31}$$

Substituting the values in equations 30 and 31 into equation 16, we obtain:

$$\frac{dP}{dx_1} = [P_1^0 \exp (\beta x_2^2/2) - P_2^0 \exp (\beta x_1^2/2)][1 - \beta x_1 x_2] \tag{32}$$

The value of β can be determined graphically in two special cases, $x_1 = 0$ and $x_2 = 0$. For $x_1 = 0$,

$$\frac{dP}{dx_1} = P_1^0 \exp [\beta/2] - P_2^0 \tag{33}$$

and for $x_2 = 0$,

$$\frac{dP}{dx_2} = -P_2^0 \exp [\beta/2] + P_1^0 \tag{34}$$

Therefore β can be obtained from the slope of the total vapor pressure vs. mole fraction curve at either end, $x_1 = 0$ or $x_2 = 0$, using equations 33 or 34.

The β value thus obtained can be used in equations 30 and 31 to calculate the partial pressures at any concentration of the liquid mixture. Furthermore, the concentration derivatives of the partial pressures can be calculated from equation 20.

Once the partial vapor pressures at a concentration are known, the activity coefficient necessary to calculate the excess free energy of mixing can be obtained as follows:

The activity coefficient is defined by

$$\gamma_1 = \frac{a_1}{x_1} \tag{35}$$

for component 1, and a similar expression can be written for component 2. But the activity, a_1, or fugacity, f_1, (equation 11) can be written as

$$a_1 = \frac{P_1}{P_1^{\,0}} \tag{36}$$

where P_1 is the vapor pressure of component 1 in the mixture, and $P_1^{\,0}$ is the vapor pressure of pure component 1 at the same temperature. The activity coefficient can be written as

$$\gamma_1 = \frac{1}{x_1} \frac{P_1}{P_1^{\,0}} \tag{37}$$

A combination of equations 3 and 4 gives the excess free energy as

$$\Delta G^E = RTx_1 \ln \gamma_1 + RTx_2 \ln \gamma_2 \tag{38}$$

Therefore, the excess free energy of mixing at constant temperature can be obtained from the partial vapor pressures and the subsequently calculated activity coefficients. This excess free energy of mixing is *per mole of solution* formed.

From calorimetric measurements Δh_M (heat of mixing per mole of solution formed) was obtained in a previous experiment (Experiment 6) as a function of concentration. Since this is the same as the excess enthalpy of mixing per mole of solution, the excess entropy per mole of solution can be calculated from equation 5.

Experimental

Prepare eight mixtures of dimethyl sulfoxide and water. (As an alternative, mixtures of dimethylformamide, dioxane, acetone, or other substances and water may be used.) The vapor pressures of the mixtures and the pure liquids must be measured with the isoteniscope. The temperature of the measurement should be the same as that used in the determination of heat of mixing, 25° C., so that the data obtained earlier can be used here.

The isoteniscope (Fig. 9–3) is composed of a U tube (8 mm. diameter) about 15 cm. long to which two bulbs are attached, A_1 and A_2. (If acetone is used the length of the U tube should be 30 cm. to accommodate the high vapor pressure of pure acetone.) Through the joint above A_1, the isoteniscope can be attached to a vacuum system. The main features of the vacuum system (Fig. 9–4) are: a forepump, C; traps, D; main line with stopcock to provide either vacuum or pressure to the isoteniscope, E; a rotating arm, F, to position the isoteniscope either in the thermostat, T, or outside it; and a Zimmerli gage, Z, to measure the pressure in the system on the vacuum

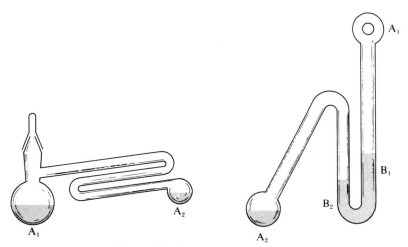

Figure 9–3 *Isoteniscope in two positions.*

arm of the isoteniscope. This gage, shown in Figure 9–5, is built and incorporated into the vacuum line by the instructor before the experiment.

To prepare a Zimmerli gage, a 16 mm. diameter tube should be used for *A*, *B*, and *C* in Figure 9–5. The *B* and *C* tubes are joined through a capillary, *D*. Both *A* and *C* are connected to the same vacuum line. In a previously

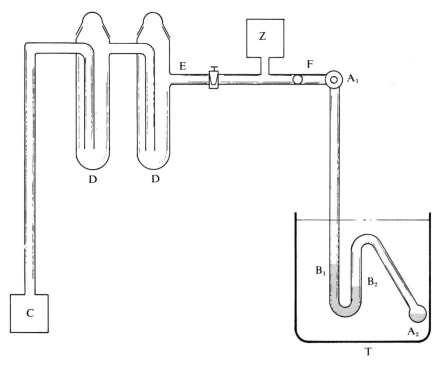

Figure 9–4 *The vacuum apparatus.*

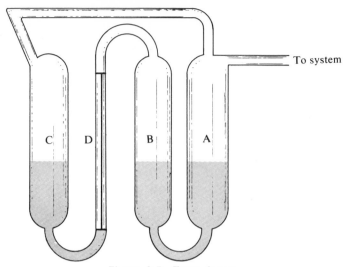

Figure 9–5 *Zimmerli gage.*

cleaned and dried Zimmerli gage clean mercury is added through the side arm that connects it to the system. Tubes A and B are filled about two-thirds full. After the gage is connected to the vacuum system, the trapped air bubbles are removed in the following manner. While the vacuum is on, tilt the gage backward to bring all the visible bubbles to the surface of the mercury in B and A. The gage walls must be tapped sharply to move the bubbles. When no more bubbles are visible, the gage is tilted to the left to permit the mercury to flow from B to D to C, so that air is removed from the reference arm, B. The gage should not be tilted so far that the mercury level in A drops completely, because this may introduce air bubbles into the reference arm, B. Once the mercury level in C is established, the gage should be returned to its vertical position and the vacuum released.

When vacuum is applied to the side arm of the Zimmerli gage the mercury at the top of the bend of the capillary, D, will separate and the mercury level in B will approach that in A. The difference in the mercury level between A and B will give the absolute pressure of the system. The difference is read with a cathetometer accurate to ± 0.01 torr (mm. of mercury).

The thermostat can be a simple fish tank with a heater, relay, and stirrer. The temperature should be maintained constant, at least within $\pm 0.1°$ C. Sometimes, insulating the three sides of the thermostat helps to maintain the temperature in the desired range. The water in the thermostat should be sufficiently clear so that the pressure reading on the isoteniscope inside the thermostat can be taken with a cathetometer.

For the experiment, clean the isoteniscope with nitric acid or cleaning solution, rinse it with an ample amount of water, and vacuum dry it. Grease the male joint above bulb A_1. Add about 10 ml. of dimethyl sulfoxide from a freshly opened bottle (99.9 per cent pure) to the A_1 bulb of the isoteniscope while it is in the horizontal position. (As an alternative dimethylformamide, dioxane, acetone, and so forth may be used.) Close the joint on the isoteni-scope. With a back and forth movement of the apparatus, transfer the DMSO to bulb A_2. Add enough clean mercury to bulb A_1 to fill the U tube part about 6 to 7 cm. from the bottom. The isoteniscope is still in the horizontal position (i.e., there is no mercury in the U tube, only in bulb A_1). Attach the isoteniscope, still kept in the horizontal position, to the vacuum system through the rotating arm, F, in Figure 9–4. Freeze the DMSO in bulb A_2 by submerging it in a Dry ice–acetone mixture (or in liquid nitrogen when acetone and other substances are used). Open the stopcocks and remove any contaminating gases by pumping on it for ten minutes. At the same time, warm the mercury in bulb A_1 to about 50° C. so that residual DMSO may be removed from the mercury. After ten minutes of pumping, close the stop-cock leading to the vacuum. Warm the DMSO to room temperature. Then slowly open the stopcock and continue pumping on the *liquid* DMSO to remove any remaining traces of foreign condensable vapors.

Apply the Dry ice–acetone bath (or liquid nitrogen) once more to the DMSO in bulb A_2. Close the stopcock. While the DMSO is frozen, rotate the arm of the isoteniscope to bring it into a position in which the mercury will flow slowly from bulb A_1 into the U tube. Open the stopcock and pump on the system, at the same time rotating the rotating arm, F, to position the isoteniscope in the thermostat, previously set to 25° C. Be sure that the U tube arms of the isoteniscope are vertical in the thermostat. Allow sufficient time for thermal equilibrium. This is achieved when the mercury no longer moves in the U tube arm of the isoteniscope.

Take pressure and temperature readings at this equilibrium. To take the pressure readings, measure the positions of the menisci in the two arms of the isoteniscope with the cathetometer. Also measure the height of the meniscus in each arm in order to make corrections for the capillary depres-sion (Fig. 9–6). Measure the meniscus difference in the Zimmerli gage to obtain the absolute pressure on the vacuum side. To calculate the total vapor pressure of the DMSO at 25° C., add the three pressures: (1) the differ-ence in meniscus height in the isoteniscope, (2) the difference between the

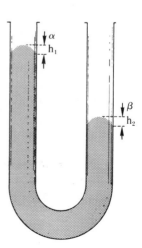

Figure 9–6 *Readings on the mercury columns in the isoteniscope.*

corrections for capillary depression, and (3) the absolute pressure obtained in the Zimmerli gage. For the correction of capillary depression, the diameter of the isoteniscope (8 mm.) must be known; then, for each measured height of the mensicus (h_1 and h_2 in Figure 9–6) obtain the correction from the *Handbook of Chemistry and Physics*.

An example of the calculation is given here:

			Correction from Handbook for 8 mm. *Tube*
Meniscus at α	163.20 mm.	$h_1 = 0.80$ mm.	0.29 mm.
Meniscus at β	158.60 mm.	$h_2 = 0.60$ mm.	0.20 mm.
	4.60 mm.		0.09 mm.

Zimmerli gage reading: 0.02 mm.

Total vapor pressure: 4.60
                              0.09
                              0.02
                           4.71 torr

Repeat the same procedure with water and the eight mixtures of DMSO-water (DMF-water, acetone-water, dioxane-water, and so forth).

Plot the total vapor pressures obtained vs. mole fraction of DMSO (DMF, acetone, dioxane and so forth). From the initial slopes of the plotted curve, obtain the β in equation 33 or 34. Using equations 30 and 31, calculate the partial vapor pressures as a function of composition. Tabulate these data together with the calculated activity coefficient from equation 37 and the excess free energies of mixing from equation 38. Using the excess heat of mixing from the previous experiment, also calculate and tabulate the excess entropies of mixing. Plot all three excess thermodynamic functions vs. mole fraction of DMSO (DMF, acetone, dioxane). Also plot the partial vapor pressures vs. mole fraction.

What kind of interaction is implied by the vapor pressure data?

What is the relative contribution of the excess enthalpy and excess entropy function? What does this imply in terms of molecular interactions?

Material and Equipment. Dimethyl sulfoxide (or dimethylformamide, acetone, dioxane, or other substance); 100 ml. mercury; isoteniscope; vacuum apparatus with a Zimmerli gauge; thermostat adjustable between 25 and 45° C. and controlled to $\pm 0.05°$ C. with stirrer, heater, and thermoregulator; Dry ice–acetone bath or liquid nitrogen.

REFERENCES

1. S. Glasstone. Thermodynamics for Chemists. D. Van Nostrand, Inc., Princeton, N.J. 1949, Chapter 18.
2. I. Prigogine, A. Bellemans and V. Mathot. The Molecular Theory of Solutions, Interscience Publishers, Inc., New York, 1957.
3. G. Scatchard. Chem. Rev., **28,** 321 (1931).
4. G. M. Barrow. Physical Chemistry. 2nd Edition, McGraw-Hill Book Co., New York, 1966, Chapter 19.
5. G. W. Thomson. In Weissberger, A. (ed.). Techniques of Organic Chemistry. 3rd Edition, Vol. I, Part I, Interscience Publishers, Inc., New York, 1959, Chapter IX.
6. H. Arm, H. Daemiker and R. Schaller. Helv. Chim. Acta, **48,** 1772 (1966).
7. A. Zimmerli. Ind. Eng. Chem., Annual Ed., **10,** 283 (1938).
8. T. Boublik et al. J. Phys. Chem., **73,** 2356 (1969).
9. R. L. Benoit and J. Charbonneau. Can. J. Chem., **47,** 4195 (1969).

10

TEMPERATURE DEPENDENCE OF SOLUBILITY OF SOLIDS*

When a solution is saturated with a solute, a thermodynamic equilibrium exists in a heterogeneous system. From the point of view of kinetics this state is reached when the rate of dissolution of the crystals equals the rate of crystallization of solute molecules on existing crystal surfaces. Both rates refer to the same surface area of crystals. From the thermodynamic point of view, such equilibrium means that the chemical potential of the crystal equals the chemical potential of the dissolved molecule.

$$\mu_{\text{cryst}} = \mu_{\text{solute}} \tag{1}$$

For an ideal solution the chemical potential of the solute can be given in terms of its mole fraction, X:

$$\mu_{\text{solute}} = \mu^0 + RT \ln X \tag{2}$$

and

$$X = \frac{n_2}{n_1 + n_2}$$

where n_2 and n_1 are the number of moles of solute and solvent, respectively.

In equation 2, μ^0 is the standard chemical potential and is dependent on temperature and pressure only. It represents the chemical potential of a molecule in a hypothetically pure dissolved state.

Combining equations 1 and 2 and denoting the mole fraction of solute in saturated solution by S, we obtain

$$\ln S = \frac{\mu_{\text{cryst}} - \mu^0}{RT} \tag{3}$$

Differentiating equation 3 with respect to temperature, T, we obtain

$$\left(\frac{\partial \ln S}{\partial T}\right)_P = \frac{H^0 - H_{\text{cryst}}}{RT^2} = \frac{\Delta H^0}{RT^2} \tag{4}$$

114

The ΔH^0 is the *excess* partial molar enthalpy of the solute in its reference state in solution over its molar enthalpy as a crystal at the same temperature. Equation 4 is called the van't Hoff isochore.

If it is assumed that ΔH^0 is constant over a small temperature range, it may be determined from a different form of the van't Hoff isochore

$$\left(\frac{\partial \ln S}{\partial (1/T)}\right)_P = \frac{-\Delta H^0}{R} \tag{5}$$

Hence the slope of a plot of $\ln S$ vs. $1/T$ will equal $-R$ times this excess enthalpy.

If a further assumption is made about the ideality of the solution, the enthalpy term in equation 5 acquires a different meaning.

If Raoult's law is obeyed and the vapor pressure of the solute in saturated solution is equal to the vapor pressure of the crystals, P_{cryst},

$$P_{\text{cryst}} = S \cdot P_L{}^0 \tag{6}$$

where $P_L{}^0$ is the vapor pressure of the supercooled liquid solute and S is its mole fraction in saturated solution. Taking the logarithm of equation 6 and differentiating with respect to T, we obtain

$$\frac{d \ln P_{\text{cryst}}}{dT} - \frac{d \ln P_L{}^0}{dT} = \frac{\ln S}{dT} \tag{7}$$

and if the Clausius-Clapeyron relationship is used equation 7 becomes

$$\frac{\Delta H_{\text{fusion}}}{RT^2} = \frac{\Delta H_{\text{sublimation}} - \Delta H_{\text{vaporization}}}{RT^2} = \left(\frac{\partial \ln S}{\partial T}\right)_P \tag{8}$$

Comparing equations 4 and 8, we see that the assumption that the solution obeys Raoult's law results in the identification

$$\Delta H^0 = \Delta H_{\text{fusion}} \tag{9}$$

Whether the equality expressed in equation 9 is correct may be determined by measuring the temperature dependence of the solubility of a compound in two solvents that have similar intermolecular interactions.

In the present experiment we will investigate the solubility of benzoic acid in water and in dimethyl sulfoxide (DMSO)-water (dimethylformamide [DMF]-water, acetone-water, dioxane-water) the concentration of which is 0.02 mole fraction with respect to the organic solvent.

Experimental

Weigh 4.2 to 4.5 grams of benzoic acid on a platform balance and transfer it to about 500 ml. of distilled water that has been heated to 50° C. in a 1 liter beaker. Use the motor stirrer to dissolve the solid rapidly and maintain

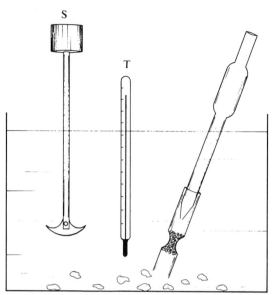

Figure 10–1   *Experimental apparatus, including stirrer, S, thermometer, T, and sample withdrawing tube.*

equilibrium between solid and solution during the entire experiment (Fig. 10–1). To prevent too rapid cooling of the solution, place a small burner flame under the beaker so that a cooling rate of about 1° C. every three to five minutes is obtained.

Place a small plug of cotton in the bottom of the sample withdrawing tube and push it up to the constriction with a glass rod. Place a rubber policeman over the upper end. Put the tube in the beaker containing the initial solution for use when needed.

When crystals of benzoic acid appear and the temperature has dropped to about 45° C., connect a sample withdrawing tube to a 25 ml. pipet. Withdraw a 25 ml. aliquot and transfer it to a flask. Record the mean temperature during this operation. Rinse the pipet free of solution with 10 ml. of ethanol from the wash bottle. Add a few drops of phenolphthalein indicator and titrate with standard 0.1N sodium hydroxide. (This solution should be standardized against benzoic acid.) The end color should last 30 to 60 seconds.

When the temperature drops 3° C., withdraw another sample, record the mean temperature and titrate as before. Secure five or more different samples at 3° C. intervals.

Repeat the same solubility measurements in the DMSO-water (DMF-water, acetone-water, dioxane-water) mixture. Be aware that the solubility of benzoic acid is greater in the organic solvent-water mixture than it was in pure water. Prepare 200 ml. of DMSO-water (DMF-water, acetone-water, dioxane-water) mixture that is 0.02 mole fraction with respect to the organic solvent. Add sufficient benzoic acid to saturate the mixed solvent at 50° C.

Then cool it to 45° C. and obtain the solubilities of benzoic acid at no less than five different temperatures in the 25 to 45° C. range as before.

Calculate the solubility of benzoic acid in terms of mole fractions. One may assume that the density of the solutions is 1.0 gram/cc. For more accurate results the densities of the solutions should be obtained at the corresponding temperatures. Plot the solubility of benzoic acid against temperature for both solvents. Plot the log of the solubility against the reciprocal of the absolute temperature $(1/T)$ for both solvents. Obtain the slope of the curve and calculate the excess partial molar heat of solution, ΔH^0, using equation 5.

Do the two ΔH^0 values in the different solvents agree? Compare ΔH^0 to ΔH_{fusion} of benzoic acid obtainable from the *Handbook of Chemistry and Physics*. What conclusion can you draw regarding the enthalpy measured in these experiments?

Material and Equipment.   25 grams benzoic acid; ~50 ml. dimethyl sulfoxide (or dimethylformamide, acetone, dioxane, or other substance); 0.1N sodium hydroxide solution; phenolphthalein indicator; ethanol: beaker of 1 liter capacity; hot plate; stirrer; thermometer; sample withdrawing tube; 25 ml. pipet.

REFERENCES

1. E. A. Moelwyn-Hughes. Physical Chemistry. 2nd Revised Edition, Pergamon Press, Inc., New York, 1964.
2. R. D. Vold and M. J. Vold. Determination of solubility. In Weissberger, A. (ed.). Techniques of Organic Chemistry. 2nd Edition, Vol. I, Part I, Interscience Publishers, Inc., New York, 1949.

V

ATOMIC AND MOLECULAR STRUCTURE

11

EMISSION SPECTRUM OF LITHIUM (I)

Carbon or metal electrodes connected to a source of direct voltage can produce an arc if they are first brought into contact and then separated. The arc carries a current of several amperes, and with a potential range up to 50 volts between the electrodes, the temperature of the anode is raised enough to volatilize metals. Furthermore, electrons in the outer orbitals of the atom are raised to higher energy levels because of the collisions of the metal atoms with high speed electrons in the arc. The excited electrons quickly return to lower energy levels. The energy difference between levels appears as a quantum of radiation and the frequency of the radiation, v, is related to the separation of the two energy levels, ΔE, by

$$\Delta E = hv \tag{1}$$

where h is Planck's constant. The wavelength, λ, is related to the frequency by

$$\lambda - \frac{c}{v} \tag{2}$$

where c is the velocity of light.

Early observations on the emission spectra of elements, especially of the hydrogen atom, clearly indicated that the emission lines form series in which the lines at longer wavelength (or lower frequency) have greater intensity than those at shorter wavelength. Furthermore, at the shortest wavelength the distinct lines of these series seem to merge into a continuum (Fig. 11–1). Such series were observed very clearly in the emission spectrum of the hydrogen atom and they were named after their discoverers (Lyman series in the far ultraviolet; Balmer series in the visible; Paschen, Brackett, and Pfund series, all in the infrared region, in the order of increasing wavelength). Only the Lyman and the Balmer series can be observed without difficulty. The others overlap and, therefore, the separation of the lines into the individual series requires some ingenuity.

6592.8 Å

4861.3 Å

4340.5 Å

4101.7 Å   **Figure II-I**   *Balmer series of the hydrogen atom.*

The lines within a series obey the following relationship

$$v = \frac{R}{n_2{}^2} - \frac{R}{n_1{}^2} \tag{3}$$

where R is a constant called the Rydberg constant and n_2 and n_1 are integer numbers. The value for n_1 is always greater than that for n_2 since the frequency must be a positive number. Within a series n_2 is a constant ($R/n_2{}^2$ is the fixed term) whereas n_1 takes successive integral values $n_2 + 1, n_2 + 2 + \cdots$, etc., for each spectral line in the series. The $R/n_1{}^2$ is also called the running term. As $n_1 \rightarrow \infty$, the running terms approach zero; hence, the limit of the series is $R/n_2{}^2$. This is the value toward which all series converge.

In the following paragraphs, first the Bohr theory of the atom will be applied to the spectra of hydrogen-like atoms because it conveys a sense of historical development and it provides a good contrast between a simple approach proved to be unworkable and the more complex quantum theory.

The Bohr theory of the atom gained its early acceptance because it successfully related the proposed structure of the hydrogen atom to the energies observed in the emission spectra. Furthermore, it yielded a calculated value for the Rydberg constant that within the experimental uncertainty was in agreement with the observed values.

For the hydrogen Bohr postulated that only certain distinct circular orbits are available for the electron. This condition is specified by the restriction that the angular momentum of the electron be an integral multiple of $h/2\pi$. Bohr also assumed that classical mechanics is valid on the atomic scale:

$$mvr = nh/2\pi \tag{4}$$

Mechanical stability requires that the forces acting on the electron balance each other. Thus, the coulombic attraction must be equal to the centrifugal force, since these are the only two forces acting on the electron.

$$\frac{Ze^2}{r^2} = \frac{mv^2}{r} \tag{5}$$

From equations 4 and 5 we have the possible radii of the orbits as

$$r = \frac{n^2h^2}{4\pi^2me^2Z} \tag{6}$$

where m is the mass of the electron, e is the charge of the electron, and Z is the atomic number (1 for the hydrogen atom). Because both the electron and the nucleus rotate about a common center of gravity, the reduced mass, μ, must be used rather than the mass of the electron

$$\mu = \frac{mM}{m + M} \qquad (7)$$

where m is the mass of the electron and M is the mass of the nucleus. For the hydrogen atom μ is very nearly equal to m.

The total energy is the sum of the kinetic and potential energies; for circular orbits this is

$$E = -\frac{Ze^2}{r} + \tfrac{1}{2}(mv^2) = -\frac{Ze^2}{r} + \frac{Ze^2}{2r} = -\frac{Ze^2}{2r} \qquad (8)$$

Substituting the values for r obtained in equation 6 and using the reduced mass of the electron, we obtain

$$E = -\frac{2\pi^2\mu e^4}{h^2} \cdot \frac{Z^2}{n^2} \qquad (9)$$

An emission line results, according to this theory, when an electron jumps from an excited energy state to a lower energy state

$$(n_1 \rightarrow n_2; n_1 > n_2)$$

Hence,

$$\nu = \frac{1}{hc}(E_1 - E_2) = \frac{2\pi^2\mu e^4}{ch^3} \cdot Z^2\left(\frac{1}{n_2{}^2} - \frac{1}{n_1{}^2}\right) \qquad (10)$$

Therefore, the empirical Rydberg constant for the hydrogen atom is given now in terms of known values of e, μ, c, and h.

$$R = \frac{2\pi^2\mu e^4}{ch^3} Z^2 \qquad (11)$$

The Bohr theory and the subsequent incorporation of relativistic principles and elliptical orbitals by Sommerfeld gave a good account of the behavior of hydrogen and hydrogen-like atoms. The spectra of He^+, Li^{++}, and Be^{+++} are identical in all details to the hydrogen atom spectrum except for a strong displacement to shorter wavelengths. The Rydberg constants of these hydrogen-like atoms differ from that of the hydrogen atom by the changes in the values of μ and Z.

The lines in the Lyman series of the hydrogen atom and in the corresponding series in the hydrogen-like atoms could be calculated by using the proper Rydberg constants and letting $n_2 = 1$ and, for the Balmer series, $n_2 = 2$.

The advance of quantum mechanics and the visualization of the electron orbitals in atoms in terms of radial distribution functions rather than set orbitals with geometric constants (radii of a circle or ellipsoid) changed our understanding of atomic structure. The orbitals have to be considered not only in terms of principal quantum numbers, n, but also in terms of azimuthal quantum numbers. From studies of atomic spectra of the elements certain selection rules emerged. There is no limitation in terms of the principal quantum number for any transition to occur. However, quantum mechanics predicts that the intensity of any transition between identical azimuthal quantum numbers would be extremely small. Therefore, the selection rule

$$\Delta l = \pm 1 \qquad (12)$$

is operative in analysis of emission spectra.

The spectra of the alkali metals are somewhat more complex than those of hydrogen-like atoms. As with the hydrogen-like atoms, there is only one emission electron in the alkali metals, and the spectrum of the neutral atom is designated by Roman numeral I. In contrast, spectra of charged ions obtained with discharge spark are designated by Roman numerals II, III, and so forth, depending on the number of charges on the ion.

The spectra of the alkali metals are similar in many respects to that of the hydrogen atom, but they are displaced to longer wavelength and the overlap between the series is quite large. Still, on the basis of decreasing intensities spaced at regular intervals, the lines belonging to one series can be separated.

The four most intense series are called *principal*, *diffuse*, *sharp*, and *fundamental*. The fundamental series appears in the infrared region and therefore is not observed in the usual spectrograms.

In Figure 11–2 the energy levels of the lithium(I) atom are sketched. The transitions between energy levels that result in the emission lines of the different series are indicated by arrows. The numbers accompanying the arrows represent the wavelengths of the emission lines of lithium(I). These series cannot be represented by the simple formula

$$v = R\left(\frac{1}{n_2^{\,2}} - \frac{1}{n_1^{\,2}}\right)$$

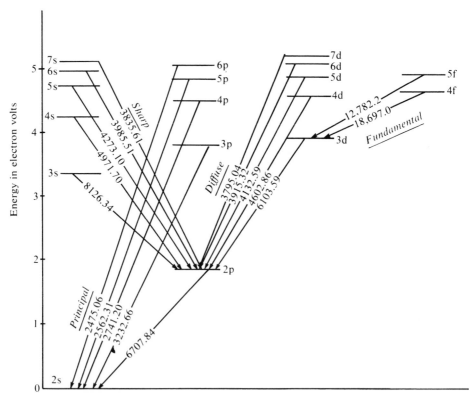

Figure 11–2 *Energy levels and wavelengths of the lithium(I) emission lines (Å) resulting from transitions between the energy levels.*

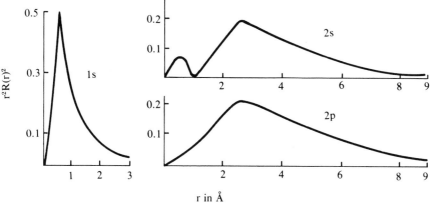

Figure II-3 *Radial distribution functions for hydrogen atoms.*

However, since they also converge to a limit, there must be a constant or fixed term in their mathematical representation that indicates the frequency of the limit of the series. The other term, the running term, is representable by $R(m + p)^2$, where m can take any values from $n_2 + 1$ and p is a constant number ($p < 1$), called the quantum defect or *Rydberg correction* for the principal series. The number n^* is called the effective principal quantum number

$$n^* = m + p \tag{13}$$

Similar correction terms can be applied to the sharp, diffuse, and fundamental running terms.

In essence, the empirical equations for the four series are

$$\text{Principal} = \text{Tps} - \frac{R}{(m + p)^2} \tag{14}$$

$$\text{Sharp} = \text{Tss} - \frac{R}{(m + s)^2} \tag{15}$$

$$\text{Diffuse} = \text{Tss} - \frac{R}{(m + d)^2} \tag{16}$$

$$\text{Fundamental} = \text{Tfs} - \frac{R}{(m + f)^2} \tag{17}$$

In these equations Tps, Tss, and Tfs are the fixed terms toward which limit the series converge and p, s, d, and f are the respective Rydberg corrections.

The necessity for these correction terms can be understood by comparing hydrogen-like atoms to alkali metals. In both cases there is only one emission electron. For hydrogen-like atoms coulombic forces could be postulated between the nucleus and the electron. In the case of alkali metals the nucleus is shielded by completely filled electron shell(s) and the emission electron moves about a *core* rather than a point charge (nucleus). From this point of view the running number m cannot be the true principal quantum number of an excited state.

The Rydberg corrections become greater and greater the more the emission electron approaches the core, i.e. if it moves on the so-called penetrating orbits, which have finite probability to contain electrons near the nucleus (Fig. 11–3). Moreover, the Rydberg corrections appear not only in the running term but also in the fixed term. The limit of each series, i.e., the fixed terms can be expressed as follows:

$$T_{ps} = R/(1 + s)^2; \quad T_{ss} = R/(2 + p)^2; \quad \text{and} \quad T_{fs} = R/(3 + d)^2 \tag{18}$$

If we use the symbols mP for $R/(m + p)^2$, mS for $R/(m + s)^2$, mD for $R/(m + d)^2$, and mF for $R/(m + f)^2$, we can represent the four main series as

$$\begin{array}{lll}
\text{Principal} & = 1S - mP & (m = 2, 3 \cdots) \\
\text{Sharp} & = 2P - mS & (m = 2, 3 \cdots) \\
\text{Diffuse} & = 2P - mD & (m = 3, 4 \cdots) \\
\text{Fundamental} & = 3D - mF & (m = 4, 5 \cdots)
\end{array} \tag{19}$$

Experimental

The apparatus consists of three main parts—electric arc, optical path, and recording equipment.

An electric arc can be produced between two metal or carbon electrodes if sufficient potential is provided and the electrodes are brought into contact. A D.C. current of several amperes is needed, and the electrodes must have good contact with the metal electrode clamps. The electrodes are usually housed in a movable carriage. In this way the electric arc produced can be moved both laterally and vertically so that the beam of light exiting from the housing falls upon the slit. The housing protects the operator from accidentally touching the electrodes and from the intense light. The potential should not be applied until the housing is in place around the electrodes.

The optical path is provided by an adjustable slit or pinhole the size of which can be varied in order to obtain the proper intensity of the emission lines. A prism or a grating or both may serve to disperse the emitted light into its components. The prism usually provides greater intensities, the grating better resolution. Use of a quartz prism permits spectra as low as about 1800 Å. This is also the lower limit of most commercial instruments because air absorbs radiation. However, with high vacuum arrangements and a grating system, work below this limit is possible. Between the slit and the prism there is a shutter that controls the exposure time.

Most emission spectra are recorded on commercial photographic plates of the proper size. (Kodak Spectrum Analysis No. 1 plate is an example.) The gelatin in commercial photographic plates limits the spectra that can be recorded to those above 1900 Å, because below that the radiation is absorbed by the coating. The upper limit of spectra that can be recorded with photographic plates is around 13,000 Å.

Usually the flat cassette camera containing the photographic plate is adjustable in height so that five or six spectra can be recorded on one plate.

1. First, load the flat cassette camera with a photographic plate *in the dark room*. Be sure that the *emulsion side* of the plate faces the sliding front

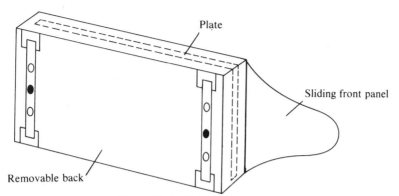

Figure II–4 *Flat cassette camera.*

panel. After loading, be sure that both the sliding front panel and the removable back of the camera are tightly secured in place.

2. Put the flat cassette camera in its place on the spectrograph. The front shutter of the spectrograph should be closed.

3. Record the wavelength scale on the plate: Move out the sliding front panel to the maximum position to the right. (With one hand hold the flat cassette camera in its proper place. Otherwise the camera may move slightly when you slide the front panel; this results in light leaks, which will ruin the plate.)

Turn the wavelength scale into position so that it lies flat against the photographic plate. Since both the photographic plate and the wavelength scale are made of glass, use care to avoid breaking them.

Turn the lamp switch on for five seconds to record the wavelength scale. Turn the light switch off. Turn the wavelength scale down to its resting position and slide the front panel back to cover the plate.

4. Move the flat cassette camera to a new position about 6 mm. from the first exposure. The camera is now ready to take the first arc exposure.

5. Fill the carbon electrode that ends in a cup with lithium chloride, and place it in the lower jaw of the electrode holder. Place the second carbon electrode, the one with the conical end, in the upper jaw of the electrode holder. (Be sure that good contact is established between the electrodes and the jaw; otherwise no current will flow. Occasionally, the jaws of the electrode holders should be cleaned with sandpaper.)

6. Maneuver the upper and lower jaws in such a way that when they make contact, the contact point is at about the level of the slit. Move the housing carriage to position the electrodes laterally in the optical path of the slit. Once this *rough* adjustment is done move the upper electrode 2 to 3 cm. from the lower one. Close the housing of the electrodes.

7. Turn the power on. Slowly lower the upper electrode so that it makes contact with the lower one. When contact is made, electricity will flow and the voltmeter may register between 10 and 100 volts and the ammeter between 4 and 10 amperes. Separate the electrodes slightly to draw an arc between them. Then make the fine adjustment in focusing the

light beam on the slit by moving the housing carriage forward, backward, and laterally and also by moving the jaws with the electrodes up and down.

Once the focusing is done, shut off the power and separate the electrodes by moving the upper one. Wait five to ten minutes, until the electrodes and the jaw cool.

8. Prepare a new lithium chloride sample in new electrodes as in step 5. Be sure that the electrodes are positioned the same way as they were before. The cup filled with lithium chloride may provide a bluish arc for three to four minutes, during which time you should be prepared to record four or five spectra. The differences between these spectra will be in the slit width and exposure time.

Record the first five spectra with a 100 μ slit width setting and exposure times of 3, 6, 12, 18, and 24 seconds. The exposure time is controlled by opening the shutter for the desired period and closing it.

9. Turn on the power and start the arc by lowering the upper electrode to make contact. Draw it away slightly to provide a stable arc. Keep the arc current about 6 amperes by adjusting the step rheostat. Before each spectrum is recorded, remove the sliding front panel of the camera as in step 3. The shutter should be opened for the desired time and closed. Push back the sliding front panel to its original position, and move the flat cassette camera into a new position to record the next spectrum. The slit remains set at 100 μ.

10. After all five spectra and the wavelength scale have been recorded shut off the power, move the electrodes apart, and take the flat cassette camera into the darkroom.

11. Develop the plate with Kodak D/19 developer for five minutes, rinse it in the stop bath, and fix it for ten minutes. After fixing, wash it for 30 minutes and let it dry.

12. Repeat the experiment recording the wavelength scale and five arc spectra of lithium(1) with slit widths of 50, 70, 100, 150, and 200 μ. Allow 6 seconds' exposure time for each spectrum. Develop and fix the plate.

If none of these ten spectra provides optimum conditions, repeat the experiment and select new slit widths and exposure times that will improve the spectra.

Record the spectral lines obtained and compare them with the values given in Figure 11–2. Estimate the standard error of your measurements. Comment on improvements necessary to obtain better results. Convert the wavelengths given in Figure 11–2 into cm^{-1}. By using the differences between successive emission lines, such as $v_1 - v_2 = \Delta v_{12}$; $v_2 - v_3 = \Delta v_{23}$, the fixed terms in equations 14, 15, 16, and 17 can be eliminated. Thus, for example, the combination of equations 13 and 14 gives

$$\Delta v_{12} = R\left[\frac{1}{(n^* + 1)^2} - \frac{1}{(n^*)^2}\right]$$

and

$$\Delta v_{23} = R\left[\frac{1}{(n^* + 2)^2} - \frac{1}{(n^* + 1)^2}\right]$$

Eliminating R between the two proceeding equations one obtains a cubic equation

$$(2\Delta\nu_{23} - 2\Delta\nu_{12})n^{*3} + (5\Delta\nu_{23} - 3\Delta\nu_{21})n^{*2} + (10\Delta\nu_{23})n^{*} + 4\Delta\nu_{23} = 0$$

which can be solved for n^{*}. Similarly, one should obtain the effective quantum numbers for the sharp, diffuse and fundamental series. The Rydberg corrections are then computed from the effective quantum numbers and the fixed terms are obtained for each series. Using the Rydberg corrections and the fixed terms, calculate the Rydberg constant for a number of spectral lines within the series.

What relationship do these Rydberg corrections have to the individual electronic orbitals under consideration? How good a constant is the Rydberg constant when these empirical corrections are used? What is the ionization potential of lithium, in volts, obtainable from the spectra?

Material and Equipment. Lithium chloride; carbon electrodes; photographic plates; three developing trays; developer (Kodak D/19); acid fixer; timer; darkroom; emission spectrograph; electrode compartment.

REFERENCES

1. G. M. Barrow. Physical Chemistry. 2nd Edition, McGraw-Hill Book Co., New York, 1966, pp. 69–80, 264–280.
2. W. J. Moore. Physical Chemistry. 3rd Edition, Prentice-Hall, Inc., Englewood Cliffs, N.J., 1962, pp. 457–512.
3. G. Herzberg. Atomic Spectra and Atomic Structure. Dover Publications, Inc., New York, 1944.

12

INFRARED SPECTRUM*

Transitions between molecular energy levels that are due to vibrational motions occur in the infrared region (400 to 5000 cm^{-1} or 2 to 25 μ). The vibrations of a system consisting of N atoms are a complex phenomenon, and are easier to understand if one begins with a diatomic molecule and proceeds to a more complicated system. A diatomic molecule or any diatomic part of a molecule can be looked upon as two particles connected by a spring (chemical bond). The classical solution of a problem is to assume that the vibration is an harmonic oscillation and the spring behaves according to Hooke's law:

$$f = -kx \tag{1}$$

where f is the force necessary to remove a particle to a distance x from its equilibrium position and k is the force constant or spring constant. In this representation we assume that only one particle vibrates and the spring connects it to a stationary wall.

The work done (which will be stored as potential energy, U) is

$$dU = f \, dx \tag{2}$$

Substituting equation 1 and 2 and integrating, we obtain

$$U = \tfrac{1}{2}kx^2 \tag{3}$$

Newton's law describes the motion of this single particle

$$f = ma \tag{4}$$

which in terms of equation 1 is equal to

$$-kx = m\left(\frac{d^2x}{dt^2}\right) = m\ddot{x} \tag{5}$$

where the acceleration of $a = \ddot{x} = d^2x/dt^2$. The solution of equation 5 is periodic.

$$x = A \cos (2\pi vt + C) \tag{6}$$

where A is the maximum amplitude, ν is the frequency of vibration, and C is a constant called the phase factor, which is determined by the choice of zero time. With the proper choice, C can be taken as zero, and therefore it is omitted from equations that are derived from equation 6.

If this picture is extended to a diatomic molecule in which both atoms can be displaced, equation 3 becomes

$$U = \tfrac{1}{2}k(x_2 - x_1)^2 \tag{7}$$

where x_2 and x_1 are the displacements of atoms 2 and 1 from their equilibrium positions. The displacements can be written according to equation 6.

$$\begin{aligned} x_1 &= A_1 \cos (2\pi\nu t + C) \\ x_2 &= A_2 \cos (2\pi\nu t + C) \end{aligned} \tag{8}$$

This indicates that the amplitudes of atoms 1 and 2 may be different, but if a vibrational solution is to be found for a diatomic molecule, the frequency of the vibration must be the same for both atoms.

The motion equation 5 for a diatomic molecule has two solutions:

$$\nu = 0$$

$$\nu = \frac{1}{2\pi}\left(\frac{k}{\mu}\right)^{1/2} \tag{9}$$

where μ, the reduced mass, is equal to

$$\mu = \frac{m_1 m_2}{m_1 + m_2} \tag{10}$$

For the solution $\nu = 0$, also $A_1 = A_2$, and this turns out to be a translational motion. For the second solution

$$\frac{A_1}{A_2} = -\frac{m_2}{m_1} = \frac{x_1}{x_2} \tag{11}$$

and if $m_1 = m_2$ and $x_1 = -x_2$ it indicates equal displacement from equilibrium in the opposite direction; it indicates a vibration.

Before proceeding to molecules with more than two atoms we must point out that even our simplest model, namely, a particle connected to a stationary wall by a spring, does not necessarily have only simple harmonic oscillations. Let us assume, for instance, that the spring (elastic bar) by which the particle is connected to the stationary wall has a rectangular cross section (Fig. 12–1). If the particle is displaced from the equilibrium position in the x direction, it will perform a simple harmonic oscillation with a frequency according to equation 9.

$$\nu_x = \frac{1}{2\pi}\left(\frac{k_x}{m}\right)^{1/2} \tag{12}$$

If it is displaced in the y direction, the frequency of the simple harmonic oscillation will be

$$\nu_y = \frac{1}{2\pi}\left(\frac{k_y}{m}\right)^{1/2} \tag{13}$$

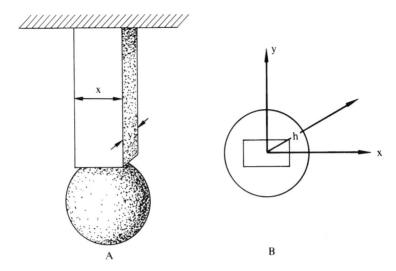

Figure 12-1 *A particle suspended on an elastic bar: A, front view; B, top view.*

However, since the cross section of the elastic bar is not square, $k_x \neq k_y$ and there-fore $\nu_x \neq \nu_y$.

If, on the other hand, the suspended particle is displaced in neither the x nor the y direction but in the direction of h in Figure 12–1B, the motion of the particle will be not a simple harmonic oscillation but a complicated motion, the so-called Lissajous motion shown in Figure 12–2.

Such complicated motions can be resolved into simple motions, however; in the case just discussed two harmonic oscillations are superimposed at right angles to each other. The simple motions into which complicated motions can be resolved are the *normal modes* of vibrations.

For a molecule that has N atoms the number of vibrations will be $3N - 6$. Each atom of the molecule can move in any direction, but its direction can be resolved into three components in the Cartesian system, x, y, and z. Therefore, the total number of motions will be $3N$. Of these $3N$ motions *three* will be *translational* when all the atoms are displaced in the x, y, or z direction. Similarly, of the $3N$ motions three will be rotational if the molecule is not linear (two if it is linear), namely, rotations around the three axes x, y, and z. This leaves $3N - 6$ vibrational motions

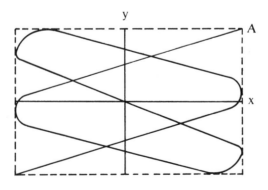

Figure 12–2 *Lissajous   motion   for* $\nu_y/\nu_x < 1$.

Symmetric stretch   Antisymmetric stretch   Bending

Figure 12-3 *Normal modes of vibration of a nonlinear triatomic molecule.*

or degrees of freedom for vibrations of a molecule of N atoms. This also means that such a molecule will have $3N - 6$ (or $3N - 5$ in case of linear molecules) *normal modes of vibrations;* that is, any complicated motions of the molecules can be resolved as superposition of these normal modes of vibration.

Of the $3N - 6$ normal modes, $N - 1$ are stretching modes of vibration and $2N - 5$ ($2N - 4$ if the molecule is linear) are deformation modes, as illustrated with a triatomic molecule (Fig. 12-3).

Some of the normal motions may be degenerate (having the same frequency); an example is the doubly degenerate bending motion of a linear triatomic molecule (Fig. 12-4).

Each normal mode of vibration has its associated frequency; the stretching modes are of higher frequency than the deformation modes, since the force constants associated with stretch are higher than those of the deformation. The normal modes of vibrations that correspond to transitions between two successive vibrational levels ($\Delta v = 1$) are called the fundamental modes of vibrations. In addition to the fundamental bands in the spectra, there are bands of overtones, which are multiples of the fundamental bands and correspond to transition of $\Delta v = 2$, 3, and so forth. The fundamental bands are assigned frequencies ν_1, ν_2, and so forth, and the overtones are designated $2\nu_1$, $3\nu_1$, and so forth. There also are combination bands in the spectra, which may be *sum* ($[\nu_1 + \nu_3]$, and so forth) or *difference* ($\nu_2 - \nu_4$) bands.

It is the task of the spectroscopist to assign all the bands in the infrared spectrum to different vibrations: fundamentals, overtones, and combinations. It is a difficult task even with simple molecules, necessitating use of the symmetry elements of the molecule as determined by group theory calculations. This aspect is beyond the level of the discussion presented here.

Fortunately, evaluation of all the force constants present is not necessary for structural analysis, even for complicated molecules, because certain bands in the spectra are associated with characteristic group vibrations. This means that the same band, at roughly the same frequencies, will appear in many different compounds that contain a characteristic group, let us say $-CH_2$. This may seem contradictory

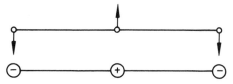

Figure 12-4 *Doubly degenerate bending motion of a linear triatomic molecule.*

to our earlier statement that the whole molecule with all its N atoms vibrates as a unit and the vibration of one part of the molecule cannot be isolated from the vibrations of the rest of the molecule. However, it must be emphasized that in *any normal mode of vibration of a molecule* (and, therefore, in any complex vibration of the whole molecule, which is simply the superposition of a few normal modes of vibration) *the nuclei of all atoms oscillate with the same frequency and, in general, in phase; only the amplitudes of the individual atomic oscillations are different owing to the different force constants.*

Since at least four considerations influence the frequency associated with the vibration at transition of a group, it is surprising that so many group frequencies have been found empirically (see Bellamy). The four factors influencing the spectral position of a group vibration are: (1) mass of the atoms of the molecule, (2) force constants of chemical bonds, (3) symmetry properties of the molecule, and (4) coupling of vibration with other motions, i.e., with rotation (Coriolis interaction). Only when these factors are relatively constant from molecule to molecule can good group frequencies be assigned. For instance, the $\diagdown$C=O group has a very good constant group frequency for the stretching mode. Because it is a terminal group, only the C has chemical bonds to other parts of the molecule, and a minimal number of force constants therefore are involved. Also, its band appears in the part of the spectrum that is free of other group vibrations and usually does not couple with other motions.

Empirically, Beer et al. (reference 5) recommend four criteria for the establishment of group frequencies in the spectra of homologous series: (1) position of the band, (2) intensity of the band, (3) half-width of the band, and (4) behavior of the band upon dilution in nonpolar solvents. If the bands, according to these criteria, behave similarly in the homologous series, they can be considered as originating from the same group mode of vibration.

A large number of group frequencies have been established empirically (see, for example, the table in Bellamy), and they can be used in the tentative assignment of spectral bands to different modes of vibrations.

Even with the approximate assignment of group frequencies a structural analysis is not unambiguous. For one thing, the environment has to be taken into consideration. It is well known that the infrared spectrum of a pure compound varies depending upon whether the gaseous, liquid, or solid state of the compound is used. Shifts of characteristic group frequencies take place because of intermolecular interactions. For that matter, gas phase spectra are best for assignments because under these conditions, in high resolution, the rotational-vibrational spectrum can be observed. In condensed phases, rotational transitions are smeared out and only wide bands of group vibrations appear. Use of dilute solutions in nonpolar solvents may minimize the interaction effects encountered in the condensed phase. Most standard spectra are taken in such solvents as carbon tetrachloride and carbon disulfide.

Environmental effects (solute-solvent interaction) can be seen in the change of spectral band position when one solvent is substituted for another. The change in the frequency of a band, $\Delta\nu$, can be the measure of such interactions. As an extreme example, hydrogen-bonded interaction causes great shifts in spectra. If a compound has a terminal hydroxy group and polar solvents capable of hydrogen-bonding with the hydroxy group are used, a great decrease in the stretching vibrational frequency of the hydroxy group is found, with appropriate widening of the band and increase in the integrated absorption band intensity. Such changes in the different bands can pinpoint the loci and degree of interactions.

In addition to the frequency of the absorption bands, their intensity is also important. The integrated intensity of a band (the area under a peak) is related to concentration and path length by the Beer-Lambert law:

$$\bar{A} = \int \alpha\,(\bar{\nu})\,d\bar{\nu} = \frac{l}{cl} \int \ln \frac{I_0}{I} \tag{14}$$

where $\bar{A}$ is the integrated absorption coefficient, C is the molar concentration, l is the thickness of the cell, $\int \ln I_0/I\,d\bar{\nu}$ is the area under the appropriate absorption band, and $\bar{\nu}$ is the frequency in cm.$^{-1}$

The proportionality constant or absorption coefficient, $\bar{A}$, is related to the change in the dipole moment vector of the group with displacement; this change must be nonzero for absorption in the infrared region.

$$\bar{A} = \frac{\pi N}{3c^2\,(1000)\,\mu_{\mathrm{red}}} \left(\frac{d\mu}{dq}\right) \tag{15}$$

where $\bar{A}$ is the absorption coefficient to be obtained from the spectra (equation 14), N is Avogadro's number, c is the velocity of light, μ_{red} is the reduced mass of the dipole in question, and $(d\mu/dq)$ is the change in dipole with displacement.

Experimental

Dimethyl sulfoxide (DMSO) spectra will be obtained in four different forms. *First*, a small drop of DMSO is placed on a silver chloride plate and spread evenly with second silver chloride plate; the two plates are clamped tightly together so that only a small film of DMSO remains between them; this sample is put in the sample beam. Two similar silver chloride plates are used as a blank in the reference beam. *Second*, a thin film of 95 per cent DMSO and 5 per cent water mixture is placed between silver chloride plates in the same way. *Third*, 0.25M of DMSO in spectral-grade carbon tetrachloride is placed in a 0.5 mm. sodium chloride cavity cell. The carbon tetrachloride in the second *matched* 0.5 mm. sodium chloride cavity cell serves as the blank in the reference beam. *Fourth*, 0.25M solution of DMSO in spectral-grade chloroform is placed in a 0.5 mm. sodium chloride cavity cell, and chloroform in the matched 0.5 mm. sodium chloride cavity cell serves as a blank.

A clean hypodermic needle and syringe are used to fill the cavity cells. The syringe and the needle are rinsed twice with the liquid to be placed in the cavity cell. Then the cavity cell itself is rinsed twice with the solution (or solvent) and finally filled. The cavity cell is capped to avoid evaporation of the solvent during the spectral run. At the end of the run the cavity cell is emptied with the aid of the syringe and a new sample is inserted as described.

L = prism, P = grating

T = detector, F-S = mirrors

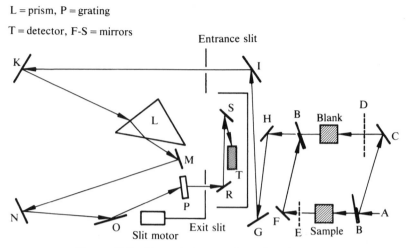

Figure 12-5 *Schematic diagram of an infrared apparatus.*

(The samples are prepared in the same way for investigation of dimethyl-formamide, urea, acetone, dioxane, or some other substance. For urea, since it is solid at room temperature, a potassium bromide disc or pellet is used instead of a thin film between two silver chloride plates. The pellets are prepared by grinding together 5 mg. urea with 200 mg. potassium bromide for about ten minutes in a mortar. The sample is dried at 120° C. for about 30 minutes. The pellet is pressed in a die in a hydraulic press with pressure of approximately 10,000 psi.

In certain instances, for example when urea in carbon tetrachloride is used, the solubility of the substance in the solvent does not reach 0.25M as specified earlier. For these cases spectra can be obtained with much smaller concentrations, such as 0.005M solutions.)

The spectra are taken in a double beam procedure in the appropriate infrared spectrophotometer. Detailed instructions for the operation of the infrared spectrophotometer are given in the manual furnished by the manufacturer. A more detailed account of the electronic components of the apparatus may be found in Part XIV of this book.

The following directions for obtaining a normal infrared spectrum apply to the two spectrophotometers most widely used in laboratories at present: the Model 337 Infracord spectrophotometer of the Perkin-Elmer Corporation, Norwalk, Connecticut, and the IR 4-11 of Beckman Instruments, Fullerton, California.

The Infracord spectrophotometer is relatively simple to operate and is used for standard spectra of organic compounds. Its range is from 400 to 4000 cm.$^{-1}$ It scans this range in two parts that overlap slightly: 400 to 1333 cm.$^{-1}$ and 1200 to 4000 cm.$^{-1}$

Chart papers for each part of the range are supplied, and the chart paper must be changed in order to cover the whole range. To put the chart paper on the drum, set the scan switch to reset. Only when the switch is in the reset position can the cylinder be rotated to the desired wavenumber. The cylinder can be removed easily to facilitate placement of the chart paper. Place chart paper of the selected range (1200 to 4000 cm.$^{-1}$) on the drum so that the line on the cylinder is aligned with the line on the top of the chart paper. The chart paper should be flush with the lip of the cylinder top. Close the lever and check the alignment again. Place the cylinder back into its shaft. Later models, such as Perkin-Elmer 257, require only one chart paper for the whole range.

The instrument must be warmed up for at least five minutes before operation. The infrared source begins to glow a few seconds after the power is turned on. After the warm-up period, check the instrument balance. Setting the transmission to about 50 per cent, block both the sample and reference beams suddenly with cardboard. The pen should not drift when this is done. If upward or downward drift occurs, the balance control at the bottom of the rear control panel must be adjusted to prevent any drift while the spectrum is being taken. Then adjust the instrument to 100 per cent transmission. Set the wavenumber to 4000 cm.$^{-1}$ and with no obstruction in the sample and reference beams center the pen on the 100 per cent transmission line.

With the slit set at N, normal slit width, insert a standard polystyrene film in the sample holder.

Turning the scan switch from reset to stop to slow and finally to fast scan, obtain a portion of the polystyrene spectrum in order to inspect the normal operation of the instrument.

Then obtain the spectrum of DMSO (DMF, acetone, dioxane) between the two silver chloride plates, recording the spectrum on a new chart paper first between 4000 and 1200 cm.$^{-1}$ and then between 1333 and 400 cm.$^{-1}$.

The spectrum should be obtained with N, normal slit width, and fast scan on the plateau regions but the machine must be switched to slow scan every time the pen traces an absorption band.

The spectra of the other samples are obtained similarly. In certain cases when the sodium chloride cavity cells are not matched the base line may appear below the 100 per cent transmission line, tracing a straight line. Under such conditions the base line may be established on the 60 or 50 per cent transmission line by adjusting the 100 per cent adjust control knob or by placing a wire mesh in the reference beam.

In operating any infrared spectrophotometer, such as the Beckman IR 4-11, the following considerations must be observed (letters A to S refer to Fig. 12–5):

1. At the start sufficient time should be allowed for the instrument to warm up. For that reason most instruments are not shut off completely; a residual current (about 0.2 ampere) is left on to keep the Nernst glower (the source of radiation), A, on. If the power is off completely, two to three hours of warm-up time is needed.

2. The glower current is turned on to about 0.6 ampere or to the amount specified by the manufacturer. Usually at the same time the chopper is turned on for double beam operation. The chopper, B, is a half mirror with

a set frequency of rotation; it allows the radiation source to pass through the sample (the empty half of the chopper) and reflects it to a mirror, C, that passes it through the blank.

In front of the blank there is usually a compensating comb, which shuts tight and does not allow light to pass through when the apparatus is set for single beam operation. For double beam operation this comb is automatically operated by a feedback mechanism from the amplifier, and it allows a programmed amount of light to pass through a reference beam.

3. When the glower is on, the appropriate slit must be selected. The slit can be either operated manually or programmed. The programmed slit changes with the scanning wavelength. The standard program is satisfactory for average spectra. The slit switch is turned to select and the slit dial is adjusted to the desired range—$1\frac{1}{2}$ times the standard for fast run or $\frac{3}{4}$ times the standard for slow, high resolution spectra. The programmed slit then operates both the entrance and exit slits, and it changes with the wavelength. In manual operation the operator may select the desired slit width at each wavelength or wavenumber.

4. Before the amplifier gains are adjusted, the period switch should be turned to 2. This permits a relatively fast response when the gain is adjusted. With nothing in the double beam path, select a wavelength of 1000 cm.$^{-1}$. Switch to single beam operation and adjust the coarse and fine gain control in order to obtain 100 per cent transmittance. Then switch back to double beam operation and readjust to 100 per cent transmittance by applying the manual comb, E, in the sample beam.

5. The balance of the instrument should be checked to be sure that no shift to higher or lower transmittance occurs because of imbalance. This is done by partially blocking the two beam paths, first one and then the other, with cardboard or stainless steel and maneuvering the transmittance to about 50 per cent. Then block both beams by suddenly dropping both plates, and watch the needle to see if it drifts. Any drift should be corrected by adjustment of the balance knob.

6. After this adjustment, the proper transmittance scale is selected (0 to 100 per cent). The chart gears are selected for the proper spacings of the graph (63-63 gears for 2 inches/micron), and they are inserted in the chart paper drive mechanism. The scan speed is selected according to requirements: fast scan, 200 cm.$^{-1}$/minute, for preliminary spectra; medium fast, 80 cm.$^{-1}$/minute, for standard; and slow scan, 20 cm.$^{-1}$/minute or slower, for high resolution spectra.

7. The sample and the blank are inserted in the proper cell holders and the spectrum is taken.

In taking the spectra, first obtain the DMSO spectrum with fast scanning. On the second trial, scan fast with standard slit width over the places where no absorption bands appear and change to slow scan for better resolution where the absorption bands do appear. This is especially important in obtaining $\Delta\nu$ values from the intermolecular interactions.

From the spectrum of DMSO (DMF, acetone, dioxane) as a thin film and from the spectrum of the carbon tetrachloride solution, assign the absorption bands obtained to different modes of group vibrations. Use the table given by Bellamy. Check your assignment with values given in the literature (Horrocks and Cotton for DMSO; Yamaguchi et al. for urea; Mahlerbe and Bernstein for dioxane; DeGraaf and Sutherland for methyl-formamide; and Francis for acetone). Calculate the relative force constants for a number of stretching modes of vibrations, such as S=O, C—S, and C—H for DMSO (C—H, N—H for DMF; C—H, C=O for acetone; and so forth) by using equation 9 and assuming that the reduced mass, μ, is that of a diatomic molecule, where m_1 is the mass on one end of the spring (chemical bond) and m_2 is the mass on the other end of the spring. Calculate the absorption coefficients of the different bands from the carbon tetrachloride and chloroform solutions spectra using the Beer-Lambert law (equation 14).

Relate these coefficients roughly to the change in transition dipole moments of the appropriate groups from equation 15.

Calculate the $\Delta\nu$, the shifts due to intermolecular interaction for certain group frequencies in the different spectra. Use the spectrum obtained with carbon tetrachloride as the reference.

What kind of interaction between chloroform and DMSO and between water and DMSO (chloroform and DMF; water and DMF, and so forth) can be inferred from these frequency shifts? What groups are most active in such interactions? How do the assignments of the group frequencies in DMSO agree with the assignment of similar groups in other compounds? What kind of information can you obtain from the calculated change in dipole moment vectors $(d\mu/dq)$ with regard to the different groups?

Material and Equipment. Dimethyl sulfoxide (DMSO) (or dimethyl-formamide (DMF) urea, acetone, dioxane); spectral-grade carbon tetra-chloride (or carbon disulfide); chloroform; infrared spectrophotometer; two 0.5 mm. matched sodium chloride cavity cells; four silver chloride plates about $4 \times 1 \times 0.2$ cm. (in case of solid sample, spectral-grade potassium bromide, mortar, pestle, drying oven, hydraulic press, die assembly).

REFERENCES

1. G. Herzberg. Molecular Spectra II. Infrared and Raman Spectra of Polyatomic Molecules. D. Van Nostrand Co., Inc., Princeton, N.J., 1956.
2. L. J. Bellamy. Infra-red Spectra of Complex Molecules. Matheson Co., London, 1954.
3. H. A. Szymanski. IR. Theory and Practice of IR Spectroscopy. Plenum Press, New York, 1964.

4. W. D. Horrocks, Jr., and F. A. Cotton. Spectrochim. Acta, **17**, 134 (1961).
5. M. Beer, H. B. Kessler and G. B. B. M. Sutherland. J. Chem. Phys., **29**, 1097 (1958).
6. A. Yamaguchi, T. Miyazawa, T. Shimanouchi and S. Mizushima. Spectrochim. Acta, **10**, 170 (1958).
7. F. A. Mahlerbe and H. J. Bernstein. J. Am. Chem. Soc., **74**, 4408 (1952).
8. D. E. DeGraaf and G. B. B. M. Sutherland. J. Chem. Phys., **26**, 716 (1957).
9. S. A. Francis. J. Chem. Phys., **19**, 942 (1951).
10. J. G. David and H. E. Hallam. Trans. Faraday Soc., **65**, 2838 (1969).

13

ELECTRONIC SPECTRA OF CHARGE TRANSFER COMPLEXES; DMSO-I$_2$

Mixtures of certain substances give rise to an absorption band in the ultraviolet region that is not present in the spectra of either of the pure substances. One such classic mixture is benzene and iodine, which has a strong absorption band at 297 mμ. Such a spectral band is interpreted as a result of a strong interaction between the electron donor benzene molecule and the acceptor iodine molecule which, upon the absorption of light, goes into an excited state in which a π electron of the benzene is transferred to the iodine forming the species $C_6H_6{}^+$ and $I_2{}^-$.

The quantum mechanical requirements for charge transfer interactions are that the donor and the acceptor molecular orbitals must have nearly the same energy, and the geometry of the complex must permit overlapping of the donor and acceptor orbitals. Usually the interactions are between high energy donor and low energy acceptor (empty) orbitals. These interactions are associated with high exothermic heats—3 to 11 kcal./mole.

The classification of electronic transitions makes distinctions between three general types. Each type of transition starts with the normal ground state of the molecule, N. If the *electron* that changes its state is both initially and finally (in the excited state) in the *valence* shell, the transition is called a $V \leftarrow N$ transition. (In the notations the higher energy state is written first, and the direction of the arrow indicates whether absorption or emission occurs.) The second type of electronic transition refers to the transition of an electron from an atomic (nonbonding) orbital or from an underlying orbital ($n < n_{\text{valence}}$) to the valence shell; it is designated by the notation $Q \leftarrow N$. The third type of transition, $R \leftarrow N$, refers to a series of transitions similar to those appearing in atomic spectra in which the position of the appearing spectral lines can be described by a Rydberg type of expression consisting of a fixed term and a running term

$$\nu = A - \frac{R}{(n+a)^2} \tag{1}$$

A is the fixed term toward which the series converges, and it is characteristic of the initial state. R is the Rydberg constant, n is the principal quantum number of the excited state, and a is a Rydberg correction term characteristic of the initial state.

The I_2-DMSO (dimethyl sulfoxide) interaction may be looked upon as an analogue to the well characterized n donor-acceptor complexes such as $(CH_3)_3NI_2$. The charge transfer bands of n (nonbonding electron) donors and iodine complexes appear around 250 to 300 mμ, depending on the nature of the donor.

Such donor-acceptor complexes have two distinct spectroscopic features: (1) the appearance of a strong new absorption band in the near-ultraviolet (charge transfer band) and (2) a blue shift of the I_2 peak in the visible region.

In quantum mechanical terms the charge transfer transition is a $V \leftarrow N$ electronic transition for which the simplified wave equation can be written in terms of the molecular orbital approach as

$$\psi_N = a\psi_0(D, A) + b_n\psi_n(D^+ - A^-) + \cdots \tag{2}$$

and

$$\psi_V = a_n^*\psi_n(D^+ - A^-) - b^*\psi_0(D, A) + \cdots \tag{3}$$

where D and A represent the donor and acceptor molecules, respectively; a, b, a^*, and b^* are the weighing coefficients relating the fractional contribution of a specific form to the total bonding of the complex. Usually $a \gg b_n$ and $a_n^* \gg b^*$. Such a transition in terms of energy profiles would look like that in Figure 13-1.

The second observation with I_2 complexes is that in the visible spectrum the absorption band shifts toward shorter wavelengths with an increase in complexing. This has been attributed to the fact that while the I_2 absorption in the free state is due to the excitation of a delocalized electron, this delocalized electron encounters increasing crowding in the excited I_2 state when tight complexes are formed.

One way to prove spectroscopically that such a tight complex is formed is with Job's method of continuous variation. In this method the concentrations of the interacting species (donor and acceptor) are varied in a noninteracting solvent. However, the sum of molar concentrations of the interacting species is kept constant.

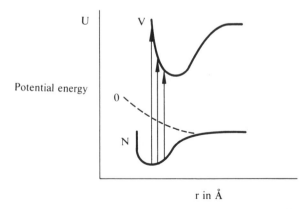

Figure 13-1 *Potential energy diagram of charge transfer band transitions in strong complexes. The arrows indicate the frequencies at which the charge transfer transitions occur.*

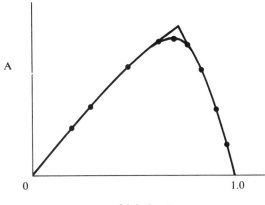

A

0   1.0

Figure 13–2 *Intensities of the charge transfer band of a complexing mixture with Job's method of continuous variation.*

Mole fraction

The spectra of the solutions are taken, and the intensities (absorbance) of the charge transfer absorption band at specified wavelengths are plotted against the mole fraction. A curve similar to that in Figure 13–2 is obtained.

The absorbance, A, is a function of the molar concentration of the absorbing species, c, and is read directly from the chart paper. The Beer-Lambert law expresses this relationship: $\ln I/I_0 = A = \epsilon c l$, where I_0 is the intensity of the incident radiation, I is the intensity of the transmitted radiation, ϵ is the molar absorption coefficient, and l is the length of the pathway in the sample.

The position of the peak of such a curve (Fig. 13–2) indicates the molar ratio present in the complex.

Spectroscopic techniques are uniquely suited to obtaining the equilibrium constant of such complexing, since the measurement itself does not disturb the equilibrium.

The general equation for equilibrium between a donor (DMSO) and acceptor (I_2) molecule can be written as

$$D + I_2 \rightarrow DI_2 \qquad (4)$$

The only assumption made in this derivation is the nature of the complex (1:1). However, if Job's method of continuous variation shows that the nature of the complex is different, the same derivation, with the correct complexing ratio, can be followed.

The equilibrium constant, K_e, is

$$K_e = \frac{(C_C)}{(C_D - C_C)(C_I - C_C)} \qquad (5)$$

where C_C is the concentration of the complex (DI_2) at equilibrium, C_D is the initial concentration of the donor, and C_I is the initial concentration of the I_2, all in moles per liter.

Two regions show absorption—the charge transfer band in the ultraviolet region, which was used for the evaluation of the nature of the complex, and the blue shift region. In the latter region the donor (DMSO) does not absorb and the total absorbance, A, is related to the concentrations of I_2 and the complex in a 1 cm. cuvette.

$$A = \epsilon_C C_C + \epsilon_I (I_2) \qquad (6)$$

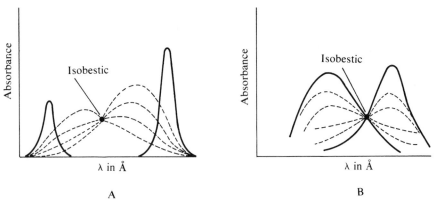

Figure 13–3 *The appearance of the isobestic point in spectra with A, no overlap, and B, overlap of the absorption bands of the two species in equilibrium.*

where ϵ_C and ϵ_I are the molar absorption coefficients for the complex and for the I_2, respectively, and (I_2) is the molar concentration of free I_2 at equilibrium.

When only one complex is formed (in 1:1 ratio), then

$$C_I = C_C + (I_2) \qquad (7)$$

This can be proved by the presence of an isobestic point, a crossing-over point observable in a spectrum when two absorbing species are in equilibrium. If the two species have nonoverlapping absorption bands (case *A* in Figure 13–3), the isobestic point appears in the region of zero absorptivity between them. If the absorption bands of the two species do overlap, the isobestic point appears at a frequency where the product of the molar absorption coefficients (absorptivity) and the molar ratio of each species are equal (in a 1:1 complexing, where the absorptivities are equal).

In essence, the isobestic point is the portion of the spectrum where the absorbance is independent of the ratio of the two concentrations of the equilibrium species. The presence of an isobestic point is not proof of the presence of only two compounds, since a third equilibrium compound may have zero absorbance in the region investigated. However, the absence of an isobestic point is definite proof of the presence of more than two absorbing species. Therefore equation 7 is applicable if an isobestic point is present and is not applicable if it is absent.

If an isobestic point is observed, the combination of equations 4, 5, 6, and 7 may lead to the Benesi-Hildebrand equation in the case of nonoverlapping absorption bands:

$$\frac{bC_I}{A} = \frac{1}{K\epsilon_C}\frac{1}{C_D} + \frac{1}{\epsilon_C} \qquad (8)$$

where b is the thickness of the cell (1 cm.).

For the case of overlapping absorption bands, the modified form is known as the Ketelaar equation:

$$\frac{bC_I}{A - A_0} = \frac{1}{K(\epsilon_C - \epsilon_I)}\frac{1}{C_D} + \frac{1}{\epsilon_C - \epsilon_I} \qquad (9)$$

where A_0 is the absorbance of the initial concentration of the $I_2(C_I)$ in the blue shift region. Therefore, if $bC_I(A - A_0)$ is plotted against $1/C_D$ according to equation 9, the slope is $[1/K(\epsilon_C - \epsilon_I)]$ and the intercept is $[1/(\epsilon_C - \epsilon_I)]$, and the equilibrium constant as well as the absorption coefficient of the complex can be evaluated. Similarly, according to equation 8 a plot of bC_I/A vs. $1/C_D$ yields a straight line in the case of nonoverlapping absorption bands, and again the equilibrium constant can be evaluated from the slope and the intercept.

A general equation that includes both of these specific cases was derived by Rose and Drago.

Experimental

PART I

To test whether the DMSO-I_2 system will form a specified tightly bound complex, the continuous variation method of Job will be applied to the charge transfer absorption band.

Make up ten solutions of DMSO-I_2 mixture in carbon tetrachloride in which the total molarity of the DMSO-I_2 will be constant—6×10^{-3}M. The DMSO-I_2 mole fraction should be varied from 0.1 to 0.9 in these mixtures. (Example I: 0.5×10^{-3}M DMSO and 5.5×10^{-3}M I_2; example II: 1.0×10^{-3}M DMSO and 5.0×10^{-3}M I_2; and so forth.)

About 10 ml. of total solution will be required. Since the DMSO itself absorbs weakly in the near-ultraviolet region, "difference" spectra will be obtained with the reference cell containing the same amount of DMSO in carbon tetrachloride as the sample cell but no I_2.

An example of how a solution should be made up follows: Prepare 100 ml. of a stock solution of 0.1M DMSO in carbon tetrachloride. Similarly prepare 100 ml. of a stock solution of 0.01M I_2 in carbon tetrachloride. These stock solutions will also be used in Part II of the experiment.

To prepare a solution that is 1×10^{-3}M DMSO and 5×10^{-3}M I_2, take 0.1 ml. of the DMSO stock solution and add 5 ml. of the I_2 stock solution and bring it to 10 ml. volume with carbon tetrachloride. A blank containing 0.1 ml. of DMSO stock solution in 10 ml. total volume is also prepared. The DMSO-I_2 solution is run against this blank.

Other DMSO-I_2 solutions and their blanks should be prepared similarly.

The spectrum of each solution run against the blank must be obtained. The operating instructions are given in Part III. From the absorption spectra plot the absorbances vs. the mole fraction. Obtain four such curves for four frequencies between 270 and 300 mμ.

PART II

For the blue shift in the visible portion of the spectrum prepare seven solutions, each containing 1×10^{-3}M I_2 in carbon tetrachloride. The amount of DMSO in the seven solutions should vary from 0 to 0.1M; use the stock solutions prepared in Part I. For example, to prepare a solution containing 1×10^{-3}M I_2 and 1×10^{-2}M DMSO, take 1 ml. of I_2 stock solution and 1 ml. of DMSO stock solution and after mixing bring the volume to 10 ml. Obtain the spectra of the seven solutions run against the carbon tetrachloride blank. Measure the temperature of the solutions. From the absorbancies in the blue shift region, calculate the equilibrium constant by using a plot derived from equation 8 or 9, whichever is suitable for your experiment.

PART III

The spectra can be taken with any spectrophotometer that has a coverage from the ultraviolet to the visible range. Manually operated instruments such as Beckman B or D.U. and automatic recording instruments such as Perkin–Elmer 400 and Cary 14-16 have two regions, visible and ultraviolet; for some instruments light sources in different housings must be used (Beckman B or D.U.), whereas for others one just switches to the region desired (Cary). In general, the operation principle is very simple whether the spectrophotometer is operated manually or is automatic.

The transmittance must be adjusted to 100 per cent (or the absorbance to zero) when the blank is used. This is accomplished by varying the slit width at different wavelengths, manually or automatically. Then the absorbance of the sample relative to the blank is measured over the wavelength range indicated (260 to 300 mμ in the ultraviolet and from 400 to 600 mμ in the visible range).

Detailed rules for operating the instrument are provided by the manual furnished by the manufacturer. A more detailed account of the electronic components of the apparatus may be found in Part XIV of this book.

What kind of interaction exists between DMSO and I_2 according to Job's continuous variation technique? What is the equilibrium constant and the free energy of the complexing? What is the energy involved in the charge transfer process (in electron volts or in calories)? If the absorbance of the I_2 in the visible region could be visualized as a result of a transition of a delocalized π electron, what would be the region of the delocalization (length of the one-dimensional box in the "particle in the box" problem)? Pay attention to the quantum number necessary for this evaluation!

What would be the effect of complexing of DMSO with I_2 upon this region of delocalization?

Material and Equipment. Dimethyl sulfoxide (DMSO); I_2; spectral-grade carbon tetrachloride; spectrophotometer; two matched silica cells, 1 cm. path; two 100 ml. volumetric flasks; 20 10 ml. volumetric flasks; one 2 ml. pipet graduated to 0.05 ml.; two 10 ml. pipet graduated to 0.1 ml.

REFERENCES

1. R. S. Mulliken and W. B. Person. Ann. Rev. Phys. Chem., **13,** 107 (1962).
2. R. S. Mulliken. J. Chem. Phys., **61,** 20 (1964).
3. H. A. Benesi and J. H. Hildebrand. J. Am. Chem. Soc., **71,** 2703 (1949).
4. J. A. Ketelaar. Rec. Trav. Chim., **71,** 1104 (1952).
5. N. J. Rose and R. S. Drago. J. Am. Chem. Soc., **81,** 6138 (1959).
6. R. S. Drago, B. Wayland and R. L. Carlson. J. Am. Chem. Soc., **85,** 3125 (1963).
7. P. Klaebo. Acta Chem. Scand., **18,** 27 (1964).
R. P. Bauman. Absorption Spectroscopy. Wiley & Sons, Inc., New York, 1962.

14

POLARIZABILITY; REFRACTIVE INDEX*

When a molecule is placed in an electrical field between two condenser plates, the applied electrical field is reduced. The reason is that the field polarizes the molecule in such a fashion that the positive end of the molecule aligns itself with the negative plate and vice versa. This total polarization of the molecule can be measured by the dielectric constant, ϵ, which is the ratio of the capacitances of the dielectric compound in question to that of vacuum at the same geometric setting of the condenser.

$$\frac{C}{C_o} = \epsilon \tag{1}$$

The total molar polarization of the compound is given by equation 2.

$$P_T = P_I + P_P = \frac{4\pi N\alpha}{3} + \frac{4\pi N\mu^2}{9kT} = \frac{\epsilon - 1}{\epsilon + 2} \cdot \frac{M}{\rho} \tag{2}$$

where P_T is the total molar polarization; P_I is the induced molar polarization; P_P is the molar polarization due to the permanent dipole moment, μ; M is the molecular weight; and ρ is the density of the compound in question. The α is the polarizability of the molecule (in cubic centimeters), which indicates how much distortion is induced by the field in the nuclear and electronic charge distribution. The N is Avogadro's number, k is the Boltzmann constant, and T is the absolute temperature.

The dielectric dispersion, i.e., the change of the dielectric constant with the frequency of the field, is given schematically in Figure 14–1. There are three plateaus in this dispersion curve. At very low frequencies (radio frequencies) both the induced and the permanent polarization contribute to the total polarization. At higher frequencies the orientation of the permanent dipole cannot follow the field; hence its contribution decreases until the second plateau is reached. At this point only the induced polarization is present. This induced polarization is the contribution of two

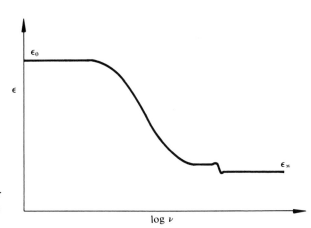

Figure 14-1 *Variation of the dielectric constant with the frequency of the applied field.*

kinds of polarization: P_A, the atomic polarization due to the displacement of the nuclei under the influence of the field, and P_E, the electronic polarization that represents the distortion of the electronic cloud induced by the field. At higher frequencies (beyond the infrared region) the contribution of P_A diminishes to zero, and the dielectric constant measured at such high frequencies is due to electronic polarization alone. When the refractive index, n_R, of a compound is measured in the visible region, the light can be considered a field of high frequency, and the Maxwell relationship holds.

$$n_R{}^2 = \epsilon_\infty \tag{3}$$

where n_R is the refractive index measured and ϵ_∞ is the dielectric constant at high frequencies.

The atomic polarization, which can be measured in the infrared region, is difficult to determine experimentally. In the cases in which it has been measured, it was found to be 3 to 10 percent of the value of the electronic polarization, P_E. Hence as a first approximation, to evaluate the polarizability of a molecule one can neglect the contribution of the atomic polarization and calculate α from the following equation.

$$P_E = \frac{4\pi N\alpha}{3} = \frac{n_R{}^2 - 1}{n_R{}^2 + 2} \cdot \frac{M}{\rho} = R \tag{4}$$

The second part of this equality is called the Lorentz-Lorenz equation and R is the Lorentz-Lorenz molar refraction.

Since the refractive index is both temperature and frequency dependent, it is necessary to specify the experimental conditions. R_D^{20} refers to a molar refractivity with the sodium D line as the light source at 20° C. The molar refractivity is a characteristic of the molecular structure and it was long thought to be an additive property of its parts. However, the increments are dependent not only on the atoms but also on the types of chemical bonds between the atoms; hence the property is both additive and constitutive.

Upon mixing two liquids one would expect the additivity to be operative; hence

$$R_{\text{add}} = X_1 \frac{n_1{}^2 - 1}{n_1{}^2 + 2} \cdot \frac{M_1}{\rho_1} + (1 - X_1) \frac{n_2{}^2 - 1}{n_2{}^2 + 2} \cdot \frac{M_2}{\rho_2} \tag{5}$$

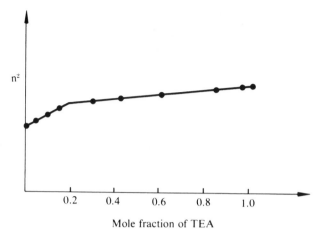

Figure 14–2 *Refractive index of the solution of triethylamine and dimethyl terephthalate in toluene vs. mole fraction.*

The experimental molar refractivity has the following formula:

$$R_{exp} = \frac{n^2 - 1}{n^2 + 2} \cdot \frac{X_1 M_1 + (1 - X_1) M_2}{\rho} \tag{6}$$

In equations 5 and 6 the subscripts 1 and 2 refer to components 1 and 2 and the quantity without subscript is the property of the mixture measured. The X_1 is the mole fraction of compound 1.

The difference between the experimental and the predicted values demonstrates the deviation from additivity.

$$\Delta R = R_{exp} - R_{add} \tag{7}$$

When ΔR exceeds 0.1 cc., one may assume that strong interaction forces between components 1 and 2 cause electronic deformations. Similarly, when the refractive indices of the mixtures vs. mole fraction or volume fraction are plotted, the data often fit straight lines. In cases of strong interaction between two components two straight lines are obtained (Fig. 14–2) and their intersection gives the molar ratio corresponding to the complex.

In using equation 4 in attempting to obtain a polarizability value, α, for either of the pure liquid components or for a mixture, two important facts should be kept in mind. First, the induced polarization, hence the polarizability as obtained from refractivities, is in error because of the neglect of atomic polarization. Second, the single value of α obtainable from equation 4 represents merely an *average* of the polarizabilities in various directions.

Experimental

The Abbé and Pulfrich type refractometers are the ones most commonly used. Both measure the critical angle of refraction relative to a glass prism. When a light passes from one isotropic medium to another $(m \to M)$ both

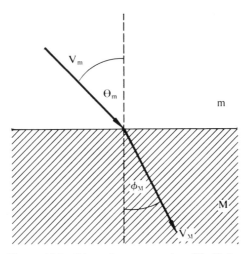

Figure 14-3 *Schematic representation of Snell's law.*

its direction and its velocity will change unless it is normal to the boundary separating m from M.

The relation between the angle of incidence, θ_m, and the angle of refraction, ϕ_M, is given by Snell's law, equation 8.

$$n = \sin \theta_m / \sin \phi_M = V_m / V_M \qquad (8)$$

In this case n is the refractive index of M relative to m. Usually the reference substance, m, is air or, more properly, vacuum, which has a refractive index of 1.00. In measuring critical angles of refractions one must consider that a liquid sample, m, is resting on the surface of a prism, P, which has a greater refractive index than m.

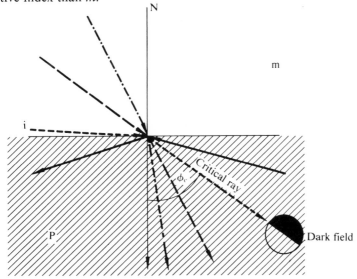

Figure 14-4 *Critical angle of refraction.*

Rays entering medium m at different angles have corresponding angles of refraction according to Snell's law (equation 8). If we follow different angles of incidence, θ_m, from the normal to the glancing incidence, i, we find that the angle of refraction increases (Fig. 14–4). At the glancing incidence we obtain the critical angle of refraction, which is the last angle that lets through light in the prism. Any higher angle of refraction will provide a dark field. This can be seen by following the reverse path of an angle of refraction that is higher than the critical angle. The light moving on the reverse path will be unable to enter the medium, m, but it will be reflected from the boundary back to the prism. In practice, not one point but the whole prism surface is illuminated with the glancing incident light and a telescope collects all the refracted rays with the same ϕ into one line in the focal plane. Thus a sharp boundary is seen in the telescope between the dark field and the bright field illuminated by the monochromatic light source. If white light is used as a source the critical boundary becomes diffuse because of the dispersion. The diffuse critical boundary is sharpened by the use of a compensator (two Amici prisms) and the refractive index obtained is that for the *sodium D light* in spite of the white light source (Fig. 14–5).

Make up ten different mixtures of dimethyl sulfoxide (DMSO) and water (dimethylformamide [DMF] and water, acetone and water, dioxane and water, and so forth). The densities of these solutions are obtained as described in Experiment 19. Be sure that the thermostat is stable at $25 \pm 0.1°$ C. This thermostated water is pumped through the refractometer continuously for at least five minutes before the first measurements are taken.

Clean the surface of the prism with ethanol and wipe it with soft lens paper. Use lintless paper so as not to leave lint on the surface of the prism. Rinse it with water and wipe it with lens paper. Place about 0.5 ml. ware. on the surface. Spread it evenly and clamp the two prisms together. If the prism surface is not completely covered with liquid no clear separation of

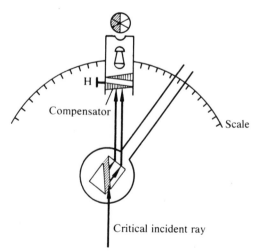

Figure 14–5 *Schematic diagram of an Abbé refractometer.*

dark and light field can be achieved. Allow one to two minutes for thermal equilibrium. Tilt the mirror or move the light source to give maximum illumination. Move the scale to 1.33. Looking through the telescope, adjust screw H to achromatize the boundary by rotating the compensator (Fig. 14–5). The boundary is achromatized when it becomes sharp and divides the field into dark and light halves. Focus the eyepiece if necessary. Adjust the scale so that the boundary passes through the cross hair. Take a few readings by removing the boundary and bringing it back to the cross hair. Average the readings.

Repeat the procedure with the other samples, finally proceeding to pure DMSO (DMF, acetone, dioxane). In the case of pure DMSO close the clamps quickly after applying the sample to the prism because DMSO is hygroscopic.

Calculate the R_D^{25} for water and DMSO (DMF, acetone, dioxane). Compare the values obtained with the values from the *Handbook of Chemistry and Physics*. Also calculate these values by assuming that the additivity rule of atoms and bonds is applicable (reference 3 or 4).

Calculate R_{exp} and R_{add} from equations 5 and 6 for the different mixtures. Plot R_{exp} and ΔR vs. mole fraction and also n^2 vs. mole fraction. Comment.

Calculate the average polarizabilities and plot vs. mole fraction and volume fraction.

To what extent is the additivity rule applicable to pure DMSO and water? Do the different plots indicate a compound formation? If there is a compound formation does it appear at the same composition on the different plots?

Material and Equipment. Dimethyl sulfoxide (DMSO) (or dimethyl formamide [DMF], acetone, dioxane, or other compound); thermostat with heater, thermoregulator, and stirrer; circulating pump; refractometer; thermometer; lens paper, ethanol.

REFERENCES

1. N. Bauer, V. J. Fajans and S. Z. Lewin. Chapter XVII. In Weissberger, A. (ed.). Techniques of Organic Chemistry. 3rd Edition, Vol. I, Part II, Interscience Publishers, Inc., New York, 1960.
2. C. H. Giles, T. J. Rose and D. G. M. Vallance. J. Chem. Soc. (London), **1952**, 3799 (1952).
3. Handbook of Chemistry and Physics. 45th Edition, Chemical Rubber Co., Cleveland, Ohio, 1964, p. E105.
4. Handbook of Chemistry and Physics. 50th Edition, Chemical Rubber Co., Cleveland, Ohio, 1969–70, p. E232.

15

DIPOLE MOMENT;
DIELECTRIC CONSTANT*

A dipole is an entity in which a positive electrical charge is separated by a relatively short distance from an equal negative charge. The dipole moment measures the magnitude of the dipole and is the product of the charge, q, and the separation, r.

$$\mu = qr \tag{1}$$

The dipole moment acts along the line connecting the two separated charges; hence it is a vectorial quantity. In a simple diatomic molecule the dipole moment of the molecule lies along the chemical bond, and by convention it is visualized as directed from the positive end of the molecule toward the negative end. In a more complex molecule the dipole moment can be taken as the vector sum of the bond dipoles. Therefore, the determination of the molecular dipole moment can give insight into the geometric structure of the molecule. The unit of the dipole moment is the debye, which is 10^{-18} e.s.u. cm. Two unit charges separated by 1 Å distance yield a dipole moment of $\mu = (4.80 \times 10^{-10}$ e.s.u.$) (1 \times 10^{-8}$ cm$) = 4.8$ debyes. It appears to be relatively easy to predict the dipole moment of an ionic diatomic molecule once the internuclear separation is known from some independent measurement, such as rotational spectrum. However, the predicted dipole moments are usually much higher than those found experimentally because the calculations are done for an undistorted ionic structure. In reality the cation has a distorting effect upon the electron cloud of the anion and the result is that the dipole is shortened. Another way of presenting the problem is to assume only a certain degree of ionic character in the chemical bond of the diatomic molecule and to take the remaining percentage of the bond as a purely covalent structure with zero dipole moment. The chemical bond is then the linear combination of these two forms. The percentage of ionic character in a bond is given by

$$X = \frac{\mu_{obs}}{\mu_{ionic}} \times 100 \tag{2}$$

In polyatomic molecules the situation is more complicated. Theoretically, one may build models of different geometric structures and calculate the resulting molecular dipole moment from the bond dipole moments by vectorial summation. The structure that gives the dipole moment value closest to that found experimentally may be taken as the true structure. However, this application must be used with care. The bond moments measure the electrical asymmetry of certain sections of a molecule and they are definitely affected by their immediate environment. Hence *group* dipoles, rather than *bond* dipoles, can be used with better results. For example, consider the calculations of the dipole moments of *m*-dichlorobenzene and *m*-chloronitrobenzene. It can be assumed that only the aromatic substitutions provide the group moments. From the tables in Smyth (reference 2) one finds that $-Cl = 1.70$ and $-NO_2 = 4.21$ when attached to the aromatic nucleus. The reference axes of the molecules are selected in such a way that the internal x axis is along the C—Cl bond and the y axis is also in the plane of the molecule. The dipole moment of the molecule is given by

$$\mu = (m_x^2 + m_y^2 + m_z^2)^{1/2} \tag{3}$$

Since for *m*-substitution the angle $\theta = 120$ degrees.

$$m_x = m_1 + m_2 \cos \theta$$
$$m_y = m_2 \sin \theta \tag{4}$$
$$m_z = 0$$

Hence,

$$\mu^2 = (m_1^2 + 2m_1m_2 \cos \theta + m_2^2 \cos^2 \theta + m_2 \sin^2 \theta)$$
$$= (m_1^2 + 2m_1m_2 \cos \theta + m^2) \tag{5}$$

and

μ_{calc}	μ_{exp}	
1.80	1.72	for *m*-dichlorobenzene
3.67	3.69	for *m*-chloronitrobenzene

Deviations between calculated and experimental values are often explained in terms of resonance or inductive effects.

Dipole moments are usually evaluated from dielectric constant measurements at low frequencies (radio frequencies). Under these conditions the permanent dipole moment of the molecule is able to follow the alternating applied field and, therefore, contributes to the total molar polarization.

$$\frac{\epsilon - 1}{\epsilon + 2} \cdot \frac{M}{\rho} = \frac{4\pi N\alpha}{3kT} + \frac{4\pi N\mu^2}{9kT} = P_T = P_I + P_P \tag{6}$$

In the previous experiment (15) we saw that the molar refractivity provides the estimation of the induced polarization, P_I. Therefore, from the combined results of dielectric constant measurements at radio frequencies, ϵ, and from the refractive indices, n, the molar polarization due to the permanent dipole moment, P_P, can be evaluated.

$$P_P = \frac{4\pi N\mu^2}{9kT} = P_T - P_I = \left[\frac{\epsilon - 1}{\epsilon + 2} - \frac{n^2 - 1}{n^2 + 2}\right] \cdot \frac{M}{\rho} \tag{7}$$

Again, as it was in the case of refractivities, if one deals with a mixture of non-interacting liquids one would expect the total molar polarization of the mixture to be the sum of its parts.

$$P_{T\mathrm{calc}} = X_1\left[\frac{\epsilon_1 - 1}{\epsilon_1 + 2}\right] \cdot \frac{M_1}{\rho_1} + (1 - X_1)\left[\frac{\epsilon_2 - 1}{\epsilon_2 + 2}\right] \cdot \frac{M_2}{\rho_2} \tag{8}$$

The experimental total molar polarization, on the other hand, is given by

$$P_{T\mathrm{exp}} = \frac{\epsilon - 1}{\epsilon + 2}\frac{X_1 M_1 + (1 - X_1)M_2}{\rho} \tag{9}$$

where ϵ is the dielectric constant, X is the mole fraction, M is the molecular weight, ρ the density, the subscripts 1 and 2 refer to pure components 1 and 2, and a symbol without a subscript refers to the mixture in question.

$$\Delta P = P_{T\mathrm{exp}} - P_{T\mathrm{calc}} \tag{10}$$

The ΔP may be a measure of nonideality of the mixture; hence it is an indication of the magnitude of the interactions between components 1 and 2.

Experimental

Dielectric constant is usually measured by obtaining the capacitance of a cell filled with the substance under investigation and the air capacitance under the same geometric consideration. Two types of instruments are in general use utilizing: (1) the bridge method, and (2) the resonance method.

The *bridge method* uses a bridge similar to those used for measuring resistance and conductance. A Schering bridge is provided in commercial

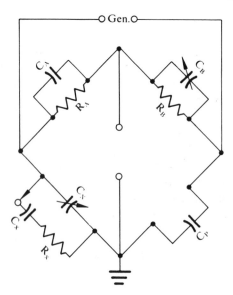

Figure 15–1 *Schering bridge. Substitution method.*

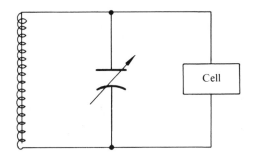

Figure 15–2 *Schematic dia-gram of the oscillometer.*

instruments such as the Type 716 capacitance bridge of General Radio Company, West Concord, Massachusetts. It can measure dielectric constant within a frequency range of 30 cps to 300 kc, and also above this range, up to 1 Mcps, if the operating frequency does not differ from the range selector frequency by more than a factor of 3 (Fig. 15–1).

In the *substitution method* a precision condenser is placed across the "unknown direct" terminal. The capacitance, C', is measured by adjusting the precision condenser, C_P, to balance. The capacitance of the unknown cell, C_x, is then measured by connecting the cell parallel to the precision condenser ("unknown substitution" terminal) and turning the precision condenser down again to balance.

$\Delta C = C' - C$ is the capacitance of the cell.

A typical commercial instrument used for the *resonance method* is the Sargent Model V Oscillometer. The instrument operates on a capacitive retune principle. As shown in Figure 15–2, the sample cell is parallel to calibrated capacitors (represented by a single variable capacitor), which comprise an integral part of the 5 Mc resonant circuit. The addition of a sample produces an increase in capacitance, which decreases the frequency of the circuit in accordance with the formula

$$ v = \frac{1}{2\pi\sqrt{LC}} \tag{11} $$

where v is the frequency, L is the inductance of the oscillator coil, and C is the total series capacitance. In order to return to the resonant frequency it is necessary to remove the exact amount of capacitance that was added by the sample. This is accomplished with calibrated capacitors. A block diagram of the complete circuit is shown in Figure 15–3. The heart of the instrument is a two terminal cathode coupled oscillator, chosen for its extreme stability. In stability tests there is less than 5000 cycles change in any 10 minute period after initial warm-up. This represents less than 0.01

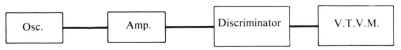

Figure 15–3 *Block diagram of the oscillometer.*

Glass

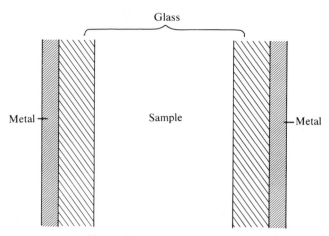

Figure 15–4 *Sample holder tube.*

per cent change in frequency. Almost as important as stability is the oscillator's ability to withstand loading. When the cells supplied with the instrument are used, it is impossible to overload the oscillator, no matter what
material is placed in the cell.

An amplifier is used to isolate the oscillator, as well as to amplify and
limit the input voltage to the discriminator.

The instrument uses a conventional discriminator modified so that it is
insensitive to amplitude changes of the signal voltage. The discriminator
feeds a vacuum tube voltmeter circuit, V.T.V.M., which is used to isolate
the galvanometer from the discriminator circuit, preventing loading of the
discriminator as well as overloading of the galvanometer. A more detailed
account of the electronic components of the apparatus may be found in
Part XIV of this book. Figure 15–4 shows the two plates of a condenser
separated by a dielectric compound comprising the glass walls and the sample.
This, according to the basic definition of a capacitor, is to be treated as three
capacitors in series, one across each glass wall and one across the sample, as
shown in Figure 15–5. The insulation resistances, R_1 and R_2, of the glass
walls are of such magnitude that they may be considered infinite. The R_s,
on the other hand, depends upon the conductivity of the sample.

In dealing with electrolytes R_s appears to be an important factor;
however, most organic liquids have low enough conductivities that the effect
of R_s can be neglected. If we can omit these three shunting resistances in

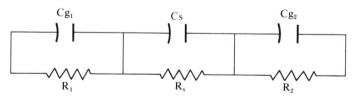

Figure 15–5 *Equivalent circuit of the sample holder.*

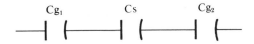

Figure 15–6 *Equivalent circuit.*

Cg_1 Cs Cg_2

the case of organic liquids, the equivalent circuit can be represented simply by Figure 15–6.

Since the dielectric property of the glass wall remains constant, C_{g_1} and C_{g_2} are fixed capacitors and may be represented by a single series equivalent (Fig. 15–7). The capacitance of the cell is given by

$$C = \frac{C_g C_s}{C_g + C_s} \tag{12}$$

and since $C_s = C_o \epsilon$

$$C = \frac{C_g C_o \epsilon}{C_g + C_o \epsilon} \tag{13}$$

where C_g is the capacitance of the glass walls, C_o is the capacitance between the glass walls in vacuum ($\epsilon = 1$), and ϵ is the dielectric constant of the sample placed between the glass walls.

The instrument is zero set when the cell is dry—hence filled with air, the dielectric constant of which is extremely close to unity. Therefore, the capacitance of the cell when the instrument is zero set is

$$C = \frac{C_g C_o}{C_g + C_o} \tag{14}$$

Consequently, the reading of the instrument, (S) is the difference between equations 13 and 14, or

$$\text{Scale reading } (S) = \frac{C_g C_o \epsilon}{C_g + C_o \epsilon} - \frac{C_g C_o}{C_g + C_o} \tag{15}$$

From this expression it can be seen that the capacitance reading of the instrument is not a linear function of the dielectric constant.

The best way to utilize the oscillometric readings for the determination of dielectric constant is to construct a calibration curve with the aid of liquids of known dielectric constants.

Instructions for operating an oscillometer or the circuit with a capacitance bridge are given in the manuals supplied by the manufacturers.

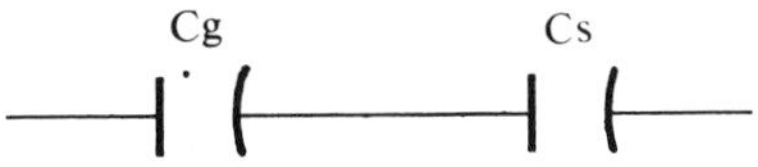

Cg Cs

Figure 15–7 *Equivalent circuit.*

MATERIALS AND MEASUREMENTS

The experiment consists of two parts:

1. Construct a calibration curve, using pure solvents: benzene, *m*-dichlorobenzene, *m*-chloronitrobenzene, acetone, ethanol, methanol, glycol, formamide, hydrazine, and water. Plot instrument readings vs. dielectric constants. The dielectric constant of pure solvents can be obtained from the *Handbook of Chemistry and Physics*.

The purpose of the calibration curve is to cover equally the dielectric constant range from 1 to 80. Therefore, liquid compounds other than those given here can also be used, provided their dielectric constant is known.

2. Prepare ten mixtures of water and dimethyl sulfoxide (DMSO) as in Experiment 14. (Dimethylformamide [DMF]-water, ethanol-water, acetone-water, dioxane-water, or other alternatives may be used.)

Determine the dielectric constant of pure water and DMSO (DMF, acetone, dioxane) and the ten mixtures by taking oscillometric readings in the oscillometer or capacitance reading directly with the substitution method, using the Schering bridge.

Plot the dielectric constant vs. mole fraction and also volume fraction.

Calculate the total molar polarization from equations 8 and 9 and plot both and also the ΔP (equation 10) vs. mole fraction.

Comment on the dielectric properties of the mixtures.

Calculate from equation 7 and 9 the dipole moment of the pure DMSO (DMF, acetone, dioxane) and that of a mixture that shows maximum deviation from ideality.

Consider the DMSO (DMF, acetone, dioxane) molecule as made of two dipole moment vectors (Fig. 15–8). The group dipole moment vector for S=O may be taken as 2.80 debyes and R (a line bisecting the CH_3—S—CH_3 angle) as 1.41 debyes. What angle should the R make with the S=O to account for the observed dipole moment?

In using equations 3, 4, and 5 assume that R vector is along the internal *x* axis and the *y* axis is also in the plane of the molecule. Similar considerations must be given to the dipole moment of the organic molecule if other than DMSO-water mixtures are studied.

(For example, DMF should also be considered as being made up of two

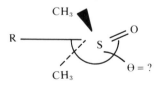

Figure 15–8

dipole moment vectors. The R is a dipole moment vector along the line that bisects the CH_3—N—CH_3 angle and continues along the C—N bond. This should be taken as being along the internal x axis of the molecule. The y axis is also in the plane of the molecule made up of R and the CHO group dipole. The group dipole moment of —CHO is taken as 2.5 debyes and assumed to make a 55 degree angle with R. From the experimentally obtained dipole moment and the data already given, calculate the R dipole using equations 3, 4, and 5. Compare this R value with a calculated value using C—N $= 0.22$ debye, CH_3—N $= 0.62$ debye, and all bond angles $\theta = 110$ degrees within the R group.

For acetone consider that the internal x axis is along the vector R, which bisects the CH_3—C—CH_3 angle. The y axis is also in the plane; hence both R and the C=O vectors are in the x-y plane. The bond moment of C=O can be taken as 2.3 debyes. The angle between C=O and R vector may be taken as 0 degrees. Using equations 3, 4, and 5, calculate the R dipole moment vector from the foregoing deformation and the experimental data. Compare this to a calculated group dipole moment vector using C—C $= 0$, C—H $= 0.4$ bond dipoles and tetrahedral angles.

For 1,4-dioxane consider the chair conformation. If the two oxygen

atoms are in a perfect trans conformation a zero dipole moment would be expected. Any deviation from this may be explained by the fact that the two group dipoles bisecting the two C—O—C angles are not 180 degrees to each other. Consider the internal x axis as lying along one of these group dipoles and the two group dipoles are in the internal x-y plane. Calculate the value of the group dipole by considering that the C—O—C angle is 110 degrees and the C—O bond dipole is 0.74 debye. Then, from the experimental dipole moment of dioxane and the data obtained previously, calculate the angle the two group dipoles form with each other.)

When commenting upon the calculated values (angles, dipole moments, or group dipoles) take into consideration the fact that the primary data used, such as bond dipoles and angles between atomic groups, were taken from tables that list average values of many compounds and may not be correct for the specific compound in question.

Material and Equipment. About 100 ml. of benzene, *m*-dichlorobenzene, *m*-chloronitrobenzene, methanol, acetone, ethanol, glycol, formamide, hydrazine (or dimethyl sulfoxide [DMSO], dimethylformamide [DMF], dioxane, or other compound); oscillometer with sample holder or capacitance bridge such as Type 716 of General Radio Company with sample holder cell.

REFERENCES

1. C. P. Smyth. Dielectric Behavior and Structure. McGraw-Hill Book Co., New York, 1955.
2. C. P. Smyth. Chapter XXXIX. In Weissberger, A. (ed.). Technique of Organic Chemistry. Vol. I, Part III, Interscience Publishers, Inc., New York, 1960.
3. T. B. Hoover. J. Phys. Chem., 73, 57 (1969).
4. A. L. McClellan. Tables of Experimental Dipole Moments. W. H. Freeman & Co., San Francisco, 1963.
5. J. Timmermans. The Physico-chemical Constants of Binary Systems in Concentrated Solutions. Vol. II and IV, Interscience Publishers, Inc., New York, 1959 and 1960.
6. D. C. Dube and R. Parshad. J. Phys. Chem., 73, 3236 (1969).
7. P. Bordewijk, F. Gransch and C. J. F. Böttcher. J. Phys. Chem., 73, 3255 (1969).

16

PROTON MAGNETIC RESONANCE*

The hydrogen nucleon, a bare proton, can be likened to a tiny spinning bar magnet. It has a spin quantum number, $I = \frac{1}{2}$, and the total angular momentum associated with a proton is

$$\bar{M} = [I(I + 1)]^{1/2}\left(\frac{h}{2\pi}\right) \tag{1}$$

For such a particle the magnetic dipole moment, $\bar{\mu}$, is given by the following expression:

$$\bar{\mu} = g_n \frac{e\bar{M}}{2m_p c} ; \qquad |\bar{\mu}| = g_n\mu_n[I(I + 1)]^{1/2} \tag{2}$$

where e is the charge of the electron, m_p is the mass of the proton, c is the velocity of light, μ_n is the nuclear magneton calculated with the mass of the proton, and g_n is a characteristic number for the nuclear species in question (5.58490 for the proton). Rearranging equation 2, one obtains

$$\frac{g_n e}{2m_p c} = \frac{\bar{\mu}}{\bar{M}} = \gamma \tag{3}$$

where γ is called the gyromagnetic ratio.

When the proton is placed in a static uniform magnetic field, H_0, the torque acting on the magnetic dipole moment produces a processional motion of the dipole about the direction of the field.

The angular velocity of this processional motion, ω_0, is

$$\omega_0 = -\gamma H_0 \tag{4}$$

Quantum mechanical considerations show that the proton may have two possible orientations $(2I + 1)$. One can be considered a low energy or parallel orientation (the N pole of the bar magnet nearest to the N pole of the static field) and the other a

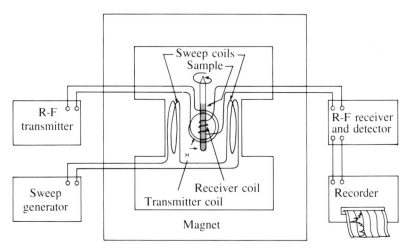

Figure 16–1 *Schematic diagram of an NMR spectrometer.*

high energy or antiparallel orientation. The difference between these two energy states depends on H_0 and $\bar{\mu}$. Thus, a transition between these two states can be produced if electromagnetic radiation of the appropriate frequency is absorbed or emitted.

Because the energy changes only when the frequency of the radiation is exactly the frequency of procession of the nucleon, that is, when the two frequencies are in resonance, the name magnetic resonance (NMR) spectroscopy is used.

$$\omega_0 = 2\pi\nu \tag{5}$$

For a static field of about 10,000 gauss the resonance frequency is in the microwave range, 40 to 60 Mcps. Since it is easier to vary the magnetic field than the microwave frequency, in most of the commerical equipment on the market a set frequency is used (60 Mcps for Varian A60) and the magnetic field is varied by means of a current through a pair of Helmholtz sweep coils. The secondary field produced by the Helmholtz coils is used to increase the field produced by the permanent magnets (see Fig. 16–1).

If a compound is put in a magnetic field that contains hydrogen atoms attached to different groups, the proton magnetic spectrum records the presence of various protons. For example, in the spectrum of pure ethanol at low resolution three peaks appear (Fig. 16–2). The band with the lowest intensity and at the lowest external

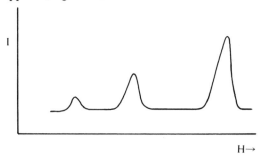

I

H→

Figure 16–2 *The NMR spectrum of ethanol.*

field corresponds to the OH protons, the next band to $-CH_2$ protons, and the largest band and at the highest external field to the $-CH_3$ protons. The intensity of the band measures the number of protons experiencing the same magnetic field. The reason that the OH, CH_2, and CH_3 protons resonate at different frequencies in a uniform field or require different resonance magnetic fields at a set frequency is that each of these protons experiences a somewhat different magnetic field because of electronic shielding. Bare protons all experience the external field and, therefore, would resonate at the same frequency. However, the electronic environment of a nucleus in a magnetic field produces an induced magnetic field that opposes the applied field. Hence the resultant field experienced by the proton is the net effect of the applied and the induced field. Three types of electron circulations can produce an induced field: (1) local diamagnetic, (2) paramagnetic, and (3) interatomic diamagnetic currents.

The resultant effect is that protons in different electronic environments have spectral bands at different fields. A reference compound, such as tetramethylsilane (TMS) is used, and the NMR spectrum is reported in terms of *chemical shifts* from the reference compound.

$$\delta = \frac{H_i - H_r}{H_r} \times 10^6 \tag{6}$$

where H_i is the resonance field of the band of interest, and H_r is the resonance field of the reference substance. The chemical shifts are multiplied by 10^6 and expressed in parts per million (ppm). Since the field strength and the resonance frequency are proportional, chemical shifts can be given in terms of frequency.

$$\delta = \frac{\nu_i - \nu_r}{\nu_r} \times 10^6 \tag{7}$$

Usually the chart papers supplied with commercial instruments have scales printed in both parts per million and cycles per second terms. Since the resonance of TMS is at a higher field than that of most other types of protons, the chemical shifts reported with this reference material give negative values. An alternative reporting of the chemical shifts is given in the literature in terms of τ values

$$\tau = 10.00 + \delta \tag{8}$$

by assigning to TMS a positive shift of 10.00 ppm. Both τ and δ are in parts per million. Although both τ and δ values as expressed in equations 6, 7, and 8 are widely used in the literature, some chart papers are calibrated so that shifts are positive for bands that appear at lower fields than TMS, the latter being taken as zero. Therefore one must be careful to use the same convention in comparing experimental data and data from the literature. Typical chemical shifts of protons are tabulated (see Jackman; Bhacca et al.; Davis).

When the proton magnetic resonance spectrum of ethanol is measured in carbon tetrachloride as a function of concentration, one finds that the OH proton has a chemical shift that depends upon concentration. This is due to the very large paramagnetic shifts (6 to 7 ppm) caused by hydrogen bonding. In pure ethanol, the OH protons are hydrogen bonded, whereas in dilute solution these hydrogen-bonded structures are broken down by the solvent. Since hydrogen bonding causes chemical shifts toward lower fields, breaking up the hydrogen bond by dilution results in

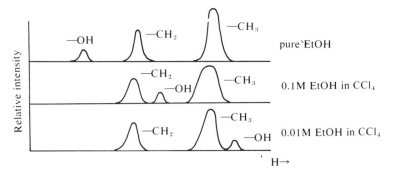

Figure 16–3 *Chemical shifts of the OH protons of ethanol.*

chemical shifts toward higher frequencies. Therefore, the band that corresponds to protons that are hydrogen bonded are both concentration and temperature dependent. For that reason, water is not a good reference material since the chemical shift of the OH proton is temperature dependent.

When one observes proton magnetic resonance in a high resolution apparatus, other features of the spectrum in addition to the chemical shifts become apparent. The bands of the low resolution apparatus, usually broad, become more complex, and they may appear as singlets, doublets, triplets, and so forth. The splitting of the band into multiplets is caused by electron-coupled spin-spin interactions of the protons in neighboring groups. The Ramsey mechanism explains how the bonding electrons can communicate to one proton the spin state of a neighboring proton. Consider a simple covalent bond between two nucleons A and B $(I_A = I_B = \frac{1}{2})$. The spin of the proton can be clockwise or counterclockwise, designated by α and β, respectively. A similar designation can be given to the spin of the bonding electrons. Since there is a high probability that the bonding electrons will be in the vicinity of nucleons A and B, we can visualize the Ramsey mechanism as shown in Figure 16-4. The spins of the bonding electrons have to be opposite according to Pauli's exclusion principle. Also, the nuclear and electron spins tend to pair up when possible; i.e., in the neighborhood of a nucleon with α spin one finds an electron with a β spin for the lowest energy state. Hence the spin orientation of one nucleon (A in step 1) will

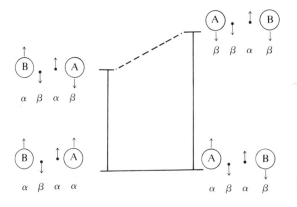

Figure 16–4 *Schematic representation of the electron-coupled spin-spin interaction by the Ramsey mechanism.*

influence the energy required for the *change* in the spin orientation of the other nucleon, *B*.

The electron-coupled spin-spin interaction and the resulting splitting obey a number of simple rules when: (1) the chemical shifts between the interacting groups are larger than the splitting due to the interaction, and (2) each nucleon in one group interacts equally with each nucleon in the other group:

1. Nuclei of an equivalent group (three protons of the methyl groups) do not interact with each other in such a manner as to cause multiplets.

2. The multiplicity caused by a neighboring group is given by $(2nI + 1)$ where I is the spin quantum number, $\frac{1}{2}$ for protons, and n is the number of equivalent protons in the neighboring group. For example, in the proton magnetic resonance spectrum of acetaldehyde, CH_3COH, the aldehyde proton band is a quadruplet because of the neighboring effect of three equivalent methyl protons

$$(2nI + 1) = 2 \times 3 \times \tfrac{1}{2} + 1 = 4$$

3. If more than one neighboring group will interact, the multiplicity of the resulting band will be given by the product $(2n_B I_B + 1) \cdot (2n_C I_C + 1)$. For example, in the proton magnetic resonance spectrum of *n*-butanol, $CH_3 \overset{\beta}{-}CH_2 \overset{\gamma}{-}CH_2-OH$, the β methylene protons should appear in high resolution spectrum as 12 peaks bunched together.

$$(2 \times 3 \times \tfrac{1}{2} + 1) \cdot (2 \times 2 \times \tfrac{1}{2} + 1) = 12$$

The separation of the multiplets is known as the spin-spin coupling constant, J, and it is independent of the magnetic field strength in contrast to the chemical shift.

Multiple band structure (not multiplets) may also arise because of steric hindrance or restricted rotation around a bond. These are especially observable at low temperature, and the multiple band structure becomes a single band when the temperature is raised sufficiently to allow unhindered rotation around the chemical bond in question.

Experimental

Prepare eight solutions of dimethyl sulfoxide (DMSO)-water varying from 0.20 to 0.90 weight fraction of DMSO. Obtain the NMR spectra of pure water and DMSO and the eight solutions prepared. (Dimethylformamide [DMF]-water, acetone-water, dioxane-water, and other alternative systems may be used.)

Most chemistry departments have an NMR spectrometer for research purposes that can be used to perform this experiment. Working with the NMR spectrometer one must follow the instructions given in the manual supplied by the manufacturer. As a first step one uses the TMS standard sample supplied with the instrument.

The homogeneity of the magnetic field is manipulated to obtain a maximum signal reading while the sample is spinning in the field. The instrument is adjusted so that the TMS peak will be placed in the zero mark of the chart paper (or at 10 ppm).

Then the NMR spectrum of pure DMSO (DMF, acetone, dioxane) must be obtained. This permits the measurement of the chemical shift of DMSO (DMF, acetone, dioxane) relative to the TMS sample.

For the measurements of the NMR spectra of DMSO (DMF, acetone, dioxane) and the subsequent mixtures, a single NMR tube should be used for all determinations.

Start with a cool, freshly dried tube (at 120° C.) and add pure DMSO with a syringe to fill the tube to a height of between 5 and 10 cm. In subsequent samples start with the highest DMSO concentration sample and proceed to pure water. In filling the tube with a new sample, rinse the tube at least twice with the new sample before the final filling. Chemical shifts are temperature dependent; therefore care must be taken to reach thermal equilibrium. Since the sample is small this is usually achieved after the sample has been spinning in the magnetic field for three to five minutes.

Once the chemical shift of DMSO is obtained, change the sweep width from 500 and 250 cps and rerecord the spectrum of DMSO as your *reference material* but manipulate the sweep offset in such a way that the DMSO peak is approximately 10 ppm (on the right side of the chart paper). (With DMF and acetone the methyl proton band, and with dioxane the CH_2 proton band, should be shifted toward the 10.00 ppm or zero mark.)

Start to take now the spectra of the other samples on the same chart paper.

Three or four samples can be recorded on the same chart, but care must be taken to change the base line so that successive bands do not interfere with each other. When changing chart paper, be certain that the last sample is recorded on both the old and new chart papers. This assures that the chemical shifts are not obliterated by inadvertent misplacing of the chart paper.

Record the chemical shifts of the OH and CH_3 (OH, NH, CH_2) protons as a function of composition. What is the explanation for the chemical shifts observed for these protons? What other features are apparent in the spectra? To what kind of phenomena can you attribute these features?

Material and Equipment.   Dimethyl sulfoxide (DMSO) (or dimethyl-formamide [DMF], acetone, dioxane, and so forth); NMR spectrometer (such as Varian A 60); three NMR sample tubes.

REFERENCES

1. G. C. Pimentel and A. L. McClellan. The Hydrogen Bond. W. H. Freeman & Co., San Francisco, 1960, Chapter 4.
2. J. N. Shoolery. Chapter 2. In NMR and EPR Spectroscopy. Pergamon Press, Inc., New York, 1960.
3. L. M. Jackman. Application of NMR Spectroscopy in Organic Chemistry. Pergamon Press, Inc., New York, 1959, Chapters 2, 4, and 5.
4. N. S. Bhacca, L. F. Johnson and J. N. Schoolery. NMR Spectra Catalog. Varian Associates, Palo Alto, Calif., 1962.
5. J. C. Davis, Jr. Advanced Physical Chemistry. Ronald Press Co., New York, 1965, Chapter 11.
6. D. M. Porter and W. S. Brey, Jr. J. Phys. Chem., **72,** 650 (1968).
7. D. M. Porter and W. S. Brey, Jr. J. Phys. Chem., **71,** 3779 (1967).

17

RAYLEIGH LIGHT SCATTERING AND DEPOLARIZATION OF BINARY LIQUID MIXTURES*

When a particle such as a gas molecule is subjected to an electric field of strength E, an induced dipole moment, p, is produced in the particle. The magnitude of this dipole moment is proportional to the field.

$$p = \alpha E \qquad (1)$$

The proportionality constant, α, is the polarizability of the particle; that is, the amount of dipole moment induced by a unit field. The electric field of a plane polarized light wave is given by equation 2.

$$E = E_0 \cos 2\pi(\nu t - x/\lambda) \qquad (2)$$

where E_0 is the maximum amplitude, ν the frequency of light, and λ its wavelength; t is the time and x is the location at time t along the line of propagation. Equation 2 shows the periodicity of the electric field; i.e., the same E_0 is obtained at time intervals $1/\nu$.

The combination of equations 1 and 2 gives

$$p = \alpha E_0 \cos 2\pi(\nu t - x/\lambda) \qquad (3)$$

the value of the oscillating induced dipole moment of the particle. This particle in our consideration is much smaller than the wavelength of light. The oscillating dipole itself is a source of radiation, and this new radiation is the *scattered beam*. Its field strength is proportional to the second derivative of the oscillating dipole, d^2p/dt^2. Although the scattered radiation is a spherical wave, its field strength depends on the angle of observation, θ (Fig. 17–1).

170

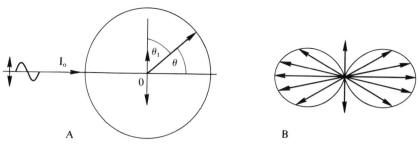

Figure 17-1 *Angular dependence of the intensity of scattered light of plane polarized incident radiation. A, the incident beam, I_0, is polarized in the plane of the drawing; the intensity of the scattered beam depends on the θ angle. The locus of the scattering particle is 0. B, the length of the arrows is proportional to the intensity of the scattered beam, i_s, at different θ values.*

Moreover, the law of conservation of energy requires that the field strength of the scattered radiation should vary with $1/r$ where r is the distance of the observer from the scattering center. Taking the second derivative of equation 3 and introducing $\sin \theta_1/c^2 r$ as the angular and distance dependence, where c is the velocity of light, we obtain

$$E_s = \frac{4\pi^2 v^2 E_0 \sin \theta_1}{c^2 r} \cos 2\pi(vt - x/\lambda) \tag{4}$$

However, experimentally we measure light intensity rather than field strength. The two are related as

$$I \sim E^2 \tag{5}$$

and therefore equation 4 is written as

$$R_\theta = \frac{i_s r^2}{I_0 \sin^2 \theta_1} = \frac{16\pi^4 \alpha^2}{\lambda^4} \tag{6}$$

Equation 6 is the famous Rayleigh equation. It has already been shown in equation 4 that the wavelength of the scattered beam is the same as that of the incident beam. Rayleigh's equation shows that the intensity of the scattered beam, i_s, depends on the inverse fourth power of the wavelength of the incident beam. The λ in equation 6 is the wavelength in a vacuum and was introduced instead of the expression c/v.

The left side of the equation is the ratio of the scattered beam intensity to the incident beam intensity, I_0, multiplied by the square of the distance of observation, r^2. The latter is usually a constant under most experimental conditions. Thus the left side of equation 6 is referred to as the Rayleigh ratio, R_θ. The subscript θ indicates its angular dependence. The molecular parameter on which R_θ depends is the polarizability. The polarizability can be related to the dielectric constant and hence to the refractive index, and equation 6 can be written for the intensity of scattered beam per unit volume as

$$R_\theta = \frac{4\pi^2 (dn/dc)^2 M c}{N \ \lambda^4} \tag{7}$$

where M is the molecular weight of the particle, N is Avogadro's number, c is the concentration in grams/cc. and (dn/dc) is the refractive index gradient. It is obvious,

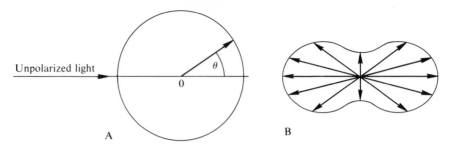

Unpolarized light

A   B

Figure 17–2 *The angular dependence of a scattered beam of unpolarized incident radiation. A, the locus of the scattering particle is 0 and the scattering angle is θ. B, the length of the arrows is proportional to the intensity of the scattered light at different θ values.*

then, from equation 7 that light scattering can be used to determine molecular weights. However, before we progress from ideal gases to solutions one modification has to be introduced. In the derivation of equation 7 we assumed that there is plane polarized light. Most of the light scattering is done with unpolarized light, which can be regarded as the superposition of two plane polarized light beams.

Thus, instead of the $\sin^2 \theta_1$ angular dependence, one has now $(\sin^2 \theta_1 + \sin^2 \theta_2)$ where θ_1 and θ_2 are the angles made by the line of observation with the y and z axes of the Cartesian coordinates.

$$\sin^2 \theta_1 + \sin^2 \theta_2 = 1 + \cos^2 \theta \qquad (8)$$

where θ is now defined as the angle between the incident and the scattered beam. Thus, for unpolarized light, equation 7 becomes

$$\frac{i_{\theta_s}}{I_0} \frac{r^2}{(1 + \cos^2 \theta)} = R' = \frac{2\pi^2 (dn/dc)^2 Mc}{N \; \lambda^4} \qquad (9)$$

where R'_θ is the Rayleigh ratio for unpolarized light.

Comparing Figures 17–1 and 17–2 one sees that although the intensity of the scattered beam of plane polarized light is zero at a 90 degree scattering angle, with unpolarized light at 90 degrees the scattering intensity is a minimum but not zero.

In short, the main feature of the Rayleigh equation is that the scattered light has the same frequency as that of the incident light and its angular intensity depends on the inverse fourth power of the wavelength. Not all scattered light has the same frequency as that of the incident light. For example, the Brillouin scattering appears at $\pm \Delta \nu$ frequencies from that of the frequency of the incident beam. Most of the time, however, the dispersion effects are quite small compared to the scattered light occurring at the frequency of the incident beam, which is called the Rayleigh scattering.

The total Rayleigh scattering from a pure liquid of small molecules is the result of two additive contributions: isotropic and anisotropic scattering. The isotropic scattering, R_{is}, is caused by fluctuations in density. One can visualize that when a tiny probe, a "Maxwell's demon," travels through a liquid, it encounters local fluctuations in density due to the thermal motions. These fluctuations in density cause a fluctuation in the refractive index of the medium, as illustrated in Fig. 17–3. Our microscopic probe then registers the refractive index variations as it travels along the x axis. Since the local refractive index variations are uneven, their magnitude

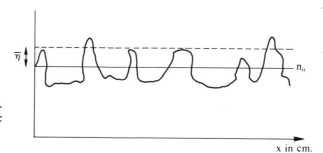

Figure 17–3 *Refractive index fluctuation in an isotropic medium.*

can be described by an *average deviation*, $\bar{\eta}$, from the mean refractive index, n_0. The larger this deviation from the mean, the greater the intensity of scattered light.

In a binary liquid mixture the isotropic scattering is caused by a fluctuation in concentration in addition to the fluctuation in density. This can be visualized easily because the refractive index variation in a medium, as in Figure 17–3, is caused by two effects: the thermal motion of all molecules and the average variation of the components of the mixture our probe encounters as it travels through the medium.

The second contribution to the total Rayleigh scattering comes from the anisotropic scattering, R_{anis}. As its name implies, this contribution exists only if at least some of the molecules of the liquid mixture are *optically* anisotropic. Optical anisotropy means that the polarizability of the molecule is not the same in all directions. It has been shown that the dipole induced in the molecule by the incident light is the source of scattered light and it depends on the polarizability of the molecule (equation 1). Therefore, if a molecule has different polarizabilities in different directions, the induced dipole and, hence, the scattered beam intensity depend on the specific orientation of the molecule. Even in a liquid in which the molecular interactions are small and spherically symmetrical and, therefore, the molecular orientations are more or less random, our probe will encounter changes in polarizability as it travels through the medium. In other words, if the molecules are optically anisotropic the random fluctuation in the molecular orientation will cause fluctuations in the polarizability. Since through dielectric constants the polarizability is related to the refractive index, Figure 17–3 is also applicable to the description of the anisotropic part of the Rayleigh scattering. In the anisotropic case, however, we call the average deviation from the mean orientation $\bar{\delta}$.

Furthermore, if the pure liquid or the liquid mixture is structured, i.e., the molecules have preferential orientations, the fluctuations in orientation may not be random and our microscopic probe may perceive fluctuations as illustrated in Figure 17–4.

Comparison of Figures 17–3 and 17–4 illustrates that the light scattering intensity

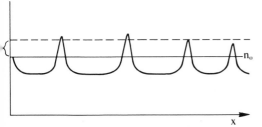

Figure 17–4 *Fluctuation of orientation in a structured liquid.*

is a result not only of the magnitude of the fluctuation but also of the relationships of the individual fluctuations to each other. Two individual fluctuations at infinite distances are not related at all, but at zero distance they double the fluctuation. Any two fluctuations separated from each other between zero and infinite distance have some relationship to each other and this is expressed usually in a correlation function that depends on the distance of separation. The contributions of fluctuations in density and orientation to the refractive index fluctuation are different, as we have seen already ($\bar{\eta}$ compared to $\bar{\delta}$). Moreover, the correlation function relating these fluctuations also has different forms.

Since the two contributions to the total Rayleigh scattering, R_{is} and R_{anis}, indicate different aspects of the liquid structure, it is desirable to separate them. This is accomplished by the Cabannes relation.

$$R_{90is} = R_{90T}(6 - 7\rho_u)/(6 + 6\rho_u) \tag{10}$$

where, at an angle of 90 degrees to the unpolarized incident beam, R_{90T} is the total Rayleigh scattering, R_{90is} is the isotropic contribution to the Rayleigh scattering, and ρ_u is the depolarization. The depolarization can be determined by measuring the scattered intensity of an unpolarized incident beam viewed through a polarizer with its optical axis in the plane of scattering (horizontal) and perpendicular to it (vertical).

$$\rho_u = \frac{i_{s(horiz)}}{i_{s(vertical)}} \tag{11}$$

After the isotropic scattering contribution is calculated from the total Rayleigh scattering and depolarization values by using equations 9 and 10, the anisotropic part, R_{anis}, is obtained from equation 12.

$$R_{90T} = R_{90is} + R_{90anis} \tag{12}$$

For an ideal mixture, a plot of the isotropic scattering due to fluctuation in concentration versus concentration should be parabolic, as illustrated in Figure 17–5.

The fluctuation in density should contribute linearly in ideal solutions. Both the height of the dome of the fluctuation in concentration and the slope of the line due to fluctuation in density depend on the difference in the refractive indices of the

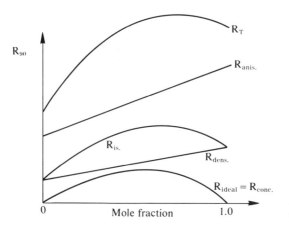

Figure 17–5 The different contributions to the total Rayleigh scattering of an ideal mixture.

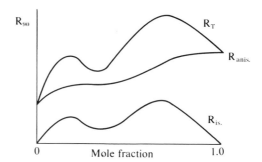

Figure 17-6 *The different contributions to the Rayleigh scattering of a mixture with compound formation.*

components of the mixture. If this difference is large, both the dome of R_{conc} and the slope of R_{dens} will be large.

The plot of the change in the anisotropic contribution to the total Rayleigh scattering with respect to concentration may approach a straight line for perfect random orientation and it may curve for preferential orientation.

If there is very strong compound formation at a certain composition of the mixture, the isotropic and possibly the anisotropic parts of the Rayleigh scattering may show a binodal curve as given in Figure 17–6. If one of the components is optically more anisotropic than the other, the *excess depolarization* may be a useful concept to keep in mind in obtaining information about the optical anisotropy of this component. The excess depolarization of component 2, $\rho_{2/1}$, is given by the equation

$$\rho_{2/1} = \frac{\dfrac{R_{12}}{R_1}\left(\dfrac{\rho_{12}}{1+\rho_{12}}\right) - \left(\dfrac{\rho_1}{1+\rho_1}\right)\phi_1}{\dfrac{R_{12}}{R_1}\left(\dfrac{1}{1+\rho_{12}}\right) - \left(\dfrac{1}{1+\rho_1}\right)\phi_1} \tag{13}$$

where ρ is the depolarization at an angle of 90 degrees, R is the total Rayleigh scattering at 90 degrees, and ϕ is the volume fraction. The subscript 1 refers to component 1 and 12 refers to a mixture of specified composition.

One can plot the excess depolarization of component 2, $\rho_{2/1}$, as calculated from equation 13, against the volume fraction of component 2 (Fig. 17–7). Smooth curves in such plots indicate a uniform liquid structure throughout the concentration range. On the other hand, breaks in such curves may indicate changes in the liquid state structuring at the specified concentration.

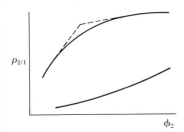

Figure 17–7 *Excess depolarization of component 2, $\rho_{2/1}$, plotted against its volume fraction.*

Experimental

Prepare about 200 ml. of triply distilled tiophene-free benzene for calibration. The light scattering properties of ten liquids will be studied: dimethyl sulfoxide (DMSO), water, and eight mixtures of DMSO-water covering the concentration range of 0.1 to 0.9 mole fraction of DMSO. (Instead of the DMSO-water system, DMF-water, acetone-water, dioxane-water, benzene-nitrobenzene, and other systems may be used.)

A Phoenix-Brice light scattering photometer, either 1000 *A* or 1000 *D*, or any other light scattering apparatus with a rectangular cell, may be employed for the measurements. This apparatus, shown in Figure 17–8, consists of mercury lamp, *A*; a monochromator, *B*, which will select the 546 mμ mercury line; removable neutral filters, *C*, to diminish the intensity of the incident mercury beam, if necessary; a photographic shutter with iris control, *D*; a collimating tube, *F*, which can contain the polarizer; a diaphragm, *G*, which controls the width of the incident beam; a calibrated turntable, *H*, which turns around a stationary center holding up the scattering cell, *K*, or a calibrated reference standard, *L*; a photomultiplier, *M*, which is fixed to the turntable and thereby reads the scattered beam intensity at any angle desired; and a secondary standard, *N*, which is used only to obtain intensity readings at a 0 degree angle. This secondary standard has a fixed position opposite the photomultiplier tube so that it is in the path of the beam only when the photomultiplier tube is at an angle of 0 degrees.

The D.C. power supply output voltage to the photomultiplier tube is controlled by two knobs—a coarse and a fine sensitivity control. The coarse sensitivity control knob has ten fixed positions, which are set by the manufacturer. Usually each coarse setting is related to the next lower one by a factor of 3. For instance, if a certain scattering intensity at a certain θ angle is recorded on the galvanometer as 90 with a coarse setting of 10, the

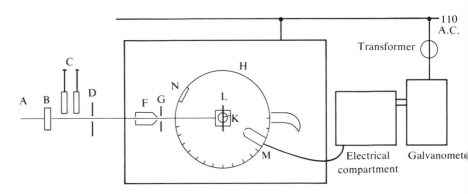

Figure 17–8 *Schematic diagram of a light scattering photometer.*

galvanometer reading will be 30 when the coarse setting is switched to 9. A setting of 10 is the highest sensitivity and one is the lowest. The student should establish this relationship between the settings.

There is a fine sensitivity control knob, which covers the range between the fixed sensitivity settings. This knob should be set with a benzene standard and left in the same position throughout the experiment.

The galvanometer completes the instrument. It operates on 6.3 volts, and the line voltage, therefore, has to be lead through a transformer. The black and red leads coming from the photomultiplier are connected to the galvanometer terminals. On top of the galvanometer there is a mechanical zero adjustment, which controls the angle of the mirror reflecting the light to the galvanometer scale.

Care should be taken not to lean on the galvanometer or on the bench because a slight imbalance causes errors in the readings. *All measurements will be made at a 90 degree scattering angle* and, therefore, the photomultiplier tube position will be set throughout this experiment. A more detailed account of the electronic components of the apparatus may be found in Part XIV of this book.

In order to obtain the depolarization values, the light scattering measurements must be performed with a polarizer in front of the photomultiplier tube. The nose cone of the photomultiplier tube can be removed and a polarizer inserted that has two positions: one with its optical axis horizontal and one with it vertical.

To obtain the R_{90T} values no polarizer is used in front of the photomultiplier tube.

Three measurements will be made on each solution: one without the polarizer, one with the polarizer in the horizontal position, and one with the polarizer in the vertical position. The green filter will be used to obtain the 546 mμ line of the mercury lamp.

Clean the scattering cell thoroughly first with cleaning solution. Then soap and rinse it thoroughly with distilled water and let it dry. Inspect the cell, making sure that its walls are spotless, free from scratches and fingerprints, and generally clean. Filter the triply distilled benzene through an ultrafilter (or through a fine pore-sized sintered glass filter) into the scattering cell. Place the rectangular scattering cell in its position, making sure that for each subsequent measurement the cell will be positioned exactly the same way. Be sure to cover the cell.

The scattering intensities and especially the depolarization measurements will be small. Therefore, to reduce the error, maximum iris opening and maximum slit opening are recommended. Turn the sensitivity switch to high sensitivity (setting of 10) and adjust the other sensitivity switch so that the galvanometer reading with pure benzene at a 90 degree angle and without the polarizer in front of the photomultiplier tube will be about 95 (almost full scale).

Leaving the sensitivity switches in position, obtain three readings on the benzene: i.e., with no polarizer, u; with the polarizer horizontal, h; and with the polarizer vertical, v.

After discarding the sample and drying the cell, filter pure DMSO (DMF, acetone, dioxane, and so forth) through ultrafilter or sintered glass into the cell. Without changing the sensitivity switches again obtain the scattering values with u, h, and v conditions. If the scattering intensity of a solution is greater than that of benzene (i.e., one cannot read the galvanometer), turn the selectivity switch to 9 or 8 and so forth in order to bring the light on the galvanometer on scale. In such cases the galvanometric readings should be multiplied by the appropriate factor.

Obtain the same readings with the eight mixtures, proceeding from the highest concentration to the lowest, and finally with distilled water.

In order to convert the readings taken without the polarizer into Rayleigh ratios the instrument constant has to be calculated from the measurements on benzene. The total Rayleigh scattering of benzene at a 90 degree angle with 546 mμ mercury line is 16.3×10^{-6} cm.$^{-1}$.

If your benzene reading was 95 with a setting of 10 (i.e., $R = 1000$), the instrument constant will be

$$k = \frac{16.3 \times 10^{-6}}{95 \times 10^{-3}} = 1.72 \times 10^{-4}\,\text{cm.}^{-1}$$

All readings obtained on the different solutions are converted to Rayleigh ratios by simply multiplying the reading by the instrument constant. For example, if the reading on DMSO was 85 with a setting of 10 ($R = 1000$). then

$$R_{90\,DMSO} = 1.72 \times 10^{-4} \times 85 \times 10^{-3} = 14.6 \times 10^{-6}\,\text{cm.}^{-1}$$

If the galvanometric reading of a solution was 85 with a setting of 9 ($R = 333$)

$$R_{90soln} = 1.72 \times 10^{-4} \times (85/333) = 43.8 \times 10^{-6}\,\text{cm.}^{-1}$$

Similarly, the depolarization values for benzene are used to calculate the depolarization of the other solutions. The literature value for the depolarization of benzene at a 90 degree angle with 546 mμ mercury line is 0.41 (see reference 2). If, for instance, the depolarization reading obtained on your instrument for benzene was

$$I_h/I_v = \tfrac{30}{90} = 0.333$$

the correction factor $f(\theta)$ to be used with other solutions is

$$\rho_\omega = f(\theta)(I_h/I_v)$$

$$f(\theta) = \frac{0.41}{0.333} = 1.23 \tag{14}$$

This $f(\theta)$ must be used to obtain the depolarization, ρ, for the different solutions from equation 14.

Tabulate your data, giving $R_{90\,total}$, ρ, and the calculated values of R_{is} and R_{anis} obtained from equations 10 and 12 for all your samples.

Plot R_{total}, R_{is}, and R_{anis} as a function of mole fraction. Calculate from equation 12 the excess depolarization of the solutions and plot it as a function of volume fraction.

Do R_{is} and R_{anis} behave the same way regarding concentration? Compare the interpretation of the excess depolarization curve with that of the R_{is} and R_{anis}.

Material and Equipment. 200 ml. triply distilled tiophene-free benzene; dimethyl sulfoxide (DMSO) (or dimethylformamide [DMF], acetone, dioxane, and so forth); Phoenix-Brice type light scattering apparatus; a monochromator green filter for obtaining the 546 mμ mercury line; neutral filters; polarizer; 50 ml. capacity rectangular light scattering cell; sintered glass filter or Millipore ultrafilter.

REFERENCES

1. Y. Sicotte and M. Rinfret. Trans. Faraday Soc., **58**, 1090 (1962).
2. D. J. Coumou. J. Colloid Sci., **15**, 408 (1960).
3. J. Powers and R. S. Stein. J. Chem. Phys., **21**, 1611 (1953).
4. R. L. Schmidt and H. L. Clever. J. Phys. Chem., **72**, 1529 (1968).
5. G. D. Parfitt and M. C. Smith. Trans. Faraday Soc., **65**, 1138 (1969).

18

ELECTRON SPIN RESONANCE SPECTRA

The phenomenon of electron spin resonance (ESR), also called electron para-magnetic resonance (EPR), depends upon the fact that unpaired electrons have angular momentum and magnetic moments that interact with their surroundings. This interaction may reveal the chemical structure of a free radical, transition element ions; it can be used in studying kinetics of electron transfer in oxidation-reduction reactions, in studying defects in crystal structure, and so forth.

An unpaired spinning electron has an angular momentum, $\vec{S}$, which is quantized. Associated with the intrinsic angular momentum of a particle is a magnetic dipole $\vec{\mu}$, which for a spinning electron is given by equation 1.

$$\vec{\mu} = g_e \frac{e}{2m_e c} \vec{S}; \quad |\mu| = g_e \frac{e}{2m_e c} [S(S + 1)]^{1/2} \frac{h}{2\pi} \tag{1}$$

where e is the charge of the electron, m_e is its mass, c is the velocity of light, $\cdot S = \frac{1}{2}$ is the spin quantum number, $h/2\pi$ are the Bohr quantized conditions, and g_e is the spectroscopy splitting factor.

Since the Bohr magneton is defined as

$$\left(\frac{e}{2m_e c}\right)\left(\frac{h}{2\pi}\right) = \mu_B \tag{2}$$

equation 1 reduces to

$$|\mu| = g_e \mu_B [S(S + 1)]^{1/2} \tag{3}$$

For a spinning electron the magnetic moment, $\vec{\mu}$, will point in the opposite direction from the angular momentum, $\vec{M}$, because of the negative charge on the electron (Fig. 18–1).

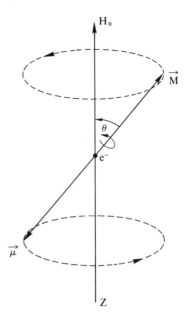

Figure 18-1 *Precession of electronic spin angular momentum, $\vec{M}$, magnetic moment, $\vec{\mu}$, in a magnetic field, H_0.*

If such a spinning electron is put in a magnetic field, H_0, the magnetic moment as well as the spin angular momentum will perform a precessional motion about the direction of the applied field. In a sense the electron behaves as a tiny gyroscope, the precessional motion of which is caused by the torque exerted by the gravitational field of the earth.

The frequency of the precession, ω, depends on the applied field, H_0.

$$\omega = \gamma H_0 = g_e \mu_b \frac{2\pi}{h} H_0 \qquad (4)$$

In equation 4, γ is the gyromagnetic ratio, the ratio of the magnetic moment to the total spin angular momentum.

$$\gamma = \frac{\vec{\mu}}{\vec{S}} = \frac{g_e \mu_B [S(S+1)]^{1/2}}{[S(S+1)]^{1/2} h/2\pi} \qquad (5)$$

If one wanted to obtain information on the nature of the spinning electron, one would look for some property measurable at the precessional frequency. In any macroscopic sample, however, electrons may have a precessional frequency, but they are random with regard to phase differences; hence, no bulk precessing magnetic property would appear. In other words, since the precessional frequency depends only on the magnetic field, the angle between the magnetic dipole and the applied field does not change during the precession and the energy of the system is constant. Such a system that does not emit or absorb energy is spectroscopically uninteresting.

However, if we can cause the magnetic dipole to interact with a second field, we may change the energy of the system so that radiation will be absorbed or emitted, and this can be measured. This second field is a small rotating magnetic field that rotates about the main magnetic field with the same precessional velocity as the

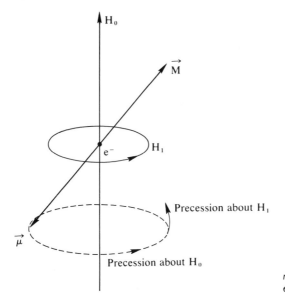

Figure 18-2 *Effect of a rotating magnetic field, H_1, upon the precession of an electron about a static magnetic field, H_0.*

magnetic dipole. Since the magnetic dipole now attempts to precess about both magnetic fields, a complicated motion ensues in which the magnetic dipole precessing about H_0 changes its angle, θ, with the result that it tilts up and down during the precession (Fig. 18-2).

This change in angle results in a change in the energy of the system (Fig. 18-3).

Instead of a rotating magnetic field we produce a linearly oscillating field by applying an oscillating voltage to a coil of wire with its axis perpendicular to that of the static magnetic field. If an electron is placed in a 10,000 gauss static field and the proper quantities are inserted in equation 4, we calculate that the precessional frequency $\nu_e = \omega_e/2\pi = 27.994$ kMcps, which is in the microwave range. Hence, in order to be able to observe an energy change of the system, we have to provide microwaves because the energy change occurs only if the frequency of radiation is exactly the same as the frequency of precession.

Such a situation is described as being "in resonance" and, therefore, the reason for the name electron spin resonance spectroscopy.

The schematic diagram of an apparatus used in ESR spectroscopy is given in Figure 18-4.

The source of radiation is a klystron tube, and it is directed through a wave guide to the sample in a resonant cavity. The resonant cavity enhances the oscillating

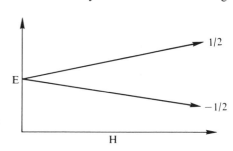

Figure 18-3 *Variation of electronic energy levels in a magnetic field.*

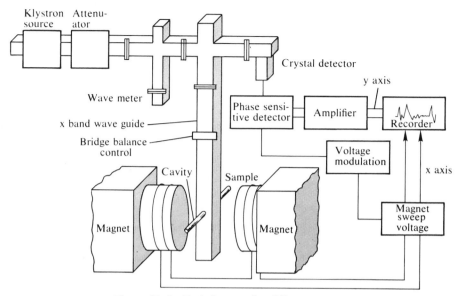

Figure 18-4 *Block diagram of an ESR spectrometer.*

magnetic field component of the microwave along the direction of the sample (Fig. 18–5).

Part of the radiation is reflected from the cavity, and it is detected by a crystal detector. When the magnetic field is such that the resonance condition is achieved, transition occurs between the spin energy levels. Energy is absorbed from the oscillating magnetic field and less energy is detected by the detector. The signals reaching the detector go through a phase detector to select the first harmonic of the absorption signal. After amplification the signal is displayed in an oscilloscope. The signals are usually displayed as the derivative of the absorption curve. Two main effects can be observed in such spectra; i.e., the energy losses and the resonant frequency shifts.

The ESR spectra are of great help in chemical structural investigations because the spinning electron is not isolated from its surroundings but interacts with it. From this interaction we can learn something about the environment of the electron.

In the present experiment two aspects of ESR spectroscopy will be investigated: spin-orbit coupling, and electron spin–nuclear spin interaction. Spin-orbit coupling results from the electron having both *spin* and *orbital* angular momenta. The spectroscopic splitting factor, g_e, for a free electron (not in an orbit around the nucleus) is

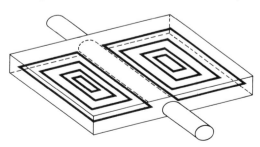

Figure 18-5 *Cavity and sample.*

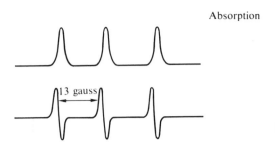

Absorption

Figure 18–6 *The ESR spectra of* $(SO_3)_2NO^-$.

2.0023. For the electron in a free radical or in an ion the spin-orbital coupling produces a change in the spectroscopic splitting factor. For example, the g_e for the stable free radical 1,1-diphenyl-2-picryl hydrazyl (DPPH), which is frequently used as a calibrating material, is 2.0036, very close to that of the free electron. In transition metal ions, however, the g_e may be much greater than 2. In aromatic ions, the g_e is again close to that of a free electron.

Especially in the solid state the spectroscopic splitting factor may give information on the relative orientation of the crystal axes with respect to the applied magnetic field. In every case the spectroscopic splitting factor is determined from the magnetic field, H_0, and the frequency of the oscillating field, ν, when the resonance line is centered in the oscilloscope.

$$h\nu = g_e \mu_B H_0 \qquad (6)$$

where h is Planck's constant, 6.6237×10^{-27}; $\mu_B = 0.9273 \times 10^{-20}$.

The second type of interaction is the electron spin–nuclear spin interaction. If the unpaired electron spends certain time around a certain nucleus, the electron and nuclear magnetic dipoles interact. This produces a hyperfine structure of "fingerprints." For example, the ESR spectrum of the radical ion of $(SO_3)_2NO^-$ shows three bands split by 13 gauss at a frequency of 9500 Mcps and at a magnetic field of 3400 gauss (Fig. 18–6).

The electron spin interacts with the nuclear spin of the N^{14}, which has a unit spin quantum number of $I = 1$. The number of allowed orientations in a magnetic field is $2I + 1$ for each nucleus. Therefore, N^{14} has three possible orientations. This interaction then gives rise to three transitions, as indicated in Figure 18–7. These three transitions appear in the ESR spectra (Fig. 18–6).

Another example is the derivative ESR spectrum of 1,4-naphthosemiquinone, $(C_{10}H_6O_2)$, which shows a triplet and a substructure of quintuplets (Fig. 18–8).

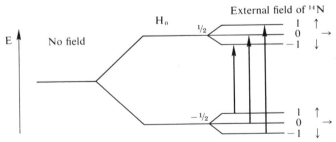

Figure 18–7 *Transition in* $(SO_3)_2NO^-$ *energy levels.*

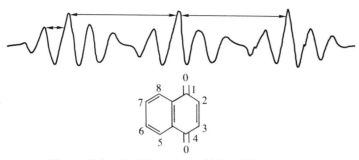

Figure 18–8 *The ESR spectrum of 1,4-naphthosemiquinone.*

The large splitting of the triplet is due to the interaction of the spin of the unpaired electron with the two equivalent protons at positions 2 and 3; the smaller split in the quintuplet indicates the almost equivalent nature of the 5, 6, 7, and 8 protons. This means that in naphthosemiquinone the unpaired electron is completely delocalized and can interact with all the protons of the molecule.

Experimental

The operation of the electron spin resonance spectrometer should follow the instructions given by the manufacturer in the manual. A more detailed account of the electronic components of the apparatus may be found in Part XIV of this book.

MEASUREMENT OF THE g_e VALUE OF A CHAR

Chars are carbons formed in a partial vacuum. The carbon rings formed often have broken bonds, leaving a free radical that behaves very much like nearly free electrons when undergoing ESR absorption.

The purpose of this experiment is to measure the g_e value of a char. In order to determine the g_e value of any material with great precision, both the magnetic field and the resonant frequency must be known precisely. However, we can utilize the calibration with the g_e value of the DPPH free radical. In a sense then we will be basing the g_e value of the char upon the measurements of the g_e value of DPPH. Preparation of a char is best accomplished by dissolving fructose (sugar) in methanol in a quartz test tube and evaporating the methanol at room temperature to leave a clear syrup. The tube should now be plugged with a quartz wool plug (to reduce spattering) and the fructose should be heated to 220° C. in air to drive off the volatile material and prechar the fructose. Now the tube should be connected to a vacuum system and pumped while being heated to 560° C. for about 20 hours. The resulting char exhibits a very narrow line width (0.5 to 1.0 gauss). (It may be noted that cigarette ash also provides a good ESR signal, and may be used instead. The ash should be tightly tamped into a sample test tube.)

Connect the ESR spectrometer and turn on the switches, noting that "grass" must be observed in the oscilloscope when the oscillator-detector-amplifier unit is turned on, and that the magnetic field is on only when current flow is indicated by the power supply meter.

Insert into the probe a sample of DPPH in a test tube to check the apparatus operation. Now take out the DPPH sample and insert a sample 7.5 mm. test tube containing the char.

· Center the resonance signal carefully in the oscilloscope and obtain values of v. Replace the char with the DPPH sample and recenter the resonance signal in the oscilloscope, noting the shift in frequency.

Repeat the centering of the resonance signal with both samples five or six times. Calculate the average resonance frequencies for both samples. From the resonance frequency of the DPPH sample and from the g_e value of DPPH (2.0036), obtain the value of the magnetic field, H, using equation 6.

Using this H value, obtain the spectroscopic splitting factor, g_e, for the char (or ash), again using equation 6 and the experimentally obtained resonance frequency of the char.

What does the magnitude of the g_e of the char indicate with respect to the location of the unpaired electrons in any specific atomic or molecular orbit?

HYPERFINE STRUCTURE OF DPPH

Prepare 0.2M DPPH solution in benzene. Fill the sample tube. Following the operating instructions, obtain the spectrum of DPPH by slowly varying the magnetic field from 70 to 160 gauss. Observe the spectrum in the oscilloscope and record it on the recorder.

Explain the hyperfine structure of DPPH by considering the following interactions:

(1) The electron interacts with only one ^{14}N nucleus.

(2) The electron interacts with two ^{14}N nuclei.

(3) The electron interacts with the phenyl protons only.

(4) The electron interacts with 3-picryl protons and 3-picryl ^{14}N.

(5) The electron interacts with both the hydrazyl ^{14}N and the phenyl protons.

(6) The electron interacts with all the ^{14}N and all the protons in the molecule.

Which of these considerations is reasonable on the basis of the hyperfine structure?

Draw an energy diagram similar to Figure 18–7 for the transitions observed in DPPH.

Material and Equipment. 10 grams 1,1-diphenyl-2-picryl hydrazyl radical (DPPH); benzene; fructose; methanol; electron spin resonance spectrometer; 7.5 mm. test tube; quartz test tube; quartz wool plug; heating mantle; vacuum pump with an acetone–Dry ice cooled trap.

REFERENCES

1. J. C. Davies, Jr. Advanced Physical Chemistry. Ronald Press Co., New York, 1965.
2. NMR and EPR Spectroscopy by NMR–EPR Staff of Varian Associates. Pergamon Press, Inc., New York, 1960.
3. H. M. Assenheim. Introduction to Electron Spin Resonance. Plenum Press, New York, 1967.
4. J. A. McMillan. Electron Paramagnetism. Reinhold Book Corp., New York, 1968

VI

LIQUIDS

19

DENSITY AND PARTIAL MOLAR VOLUMES OF SOLUTIONS*

When n_1 moles of component 1 is mixed with n_2 moles of component 2, the total volume of the solution, V, is

$$V = n_1 \left(\frac{\partial V}{\partial n_1} \right)_{n_2, T, P} + n_2 \left(\frac{\partial V}{\partial n_2} \right)_{n_1, T, P} \tag{1}$$

where $\left(\frac{\partial V}{\partial n_1} \right)_{n_2, T, P} = \bar{V}_1$ and $\left(\frac{\partial V}{\partial n_2} \right)_{n_1, T, P} = \bar{V}_2$ are the partial molar volumes of components 1 and 2, respectively.

The total volume of the solution is, of course, an extensive property of the system. The partial molar volumes or differential molar volumes are usually dependent on concentration. The properties of partial molar quantities have been discussed in Experiments 8 and 10. Under special circumstances the partial molar volumes of a binary mixture are independent of concentration. Under such conditions the partial molar volumes would be equal to the respective molar volumes

$$\bar{V}_1 = V_1^0; \quad \bar{V}_2 = V_2^0 \tag{2}$$

and the volume of mixing, ΔV_M, would be zero. In essence, this is one condition of ideal solutions, the other being that the heat of mixing should be zero. The behavior of real gases is understood in terms of the deviations from ideal behavior; similarly, it is instructive to develop first the behavior of ideal solutions. With the possible exception of mixing isomers, the ΔV_M is seldom zero, although some solutions may approach ideality.

$$V = n_1 V_1^0 + n_2 V_2^0 \quad \text{(ideal)} \tag{3}$$

$$\Delta V_M = (n_1 \bar{V}_1 + n_2 \bar{V}_2) - (n_1 V_1^0 + n_2 V_2^0) \tag{4}$$

Such an approximate additivity rule as in equation 3 may be found in infinitely dilute solutions.

The lack of understanding of liquid structure, in general, hampers the prediction of which liquid pairs would form ideal solutions. Intuitively, one would expect that molecules roughly the same size and shape would form ideal solutions according to equation 3. Experimental evidence indicates that ΔV_M is not zero, and it changes with the concentration, even for similar molecules, such as benzene and toluene.

Since no theory at present can adequately predict the behavior of solutions, it is instructive to develop the case of *ideal solutions*, especially with regard to the parameters obtainable from densities.

The extensive property of volume of solution in equation 3 is eliminated by dividing the equation by V:

$$1 = C_1' V_1^0 + C_2' V_2^0 \tag{5}$$

in which C_1' and C_2' are molar concentrations.

The respective molar volumes V_1^0 and V_2^0 are

$$V_1^0 = \frac{M_1}{\rho_1^0} ; \qquad V_2^0 = \frac{M_2}{\rho_2^0} \tag{6}$$

where M_1 and M_2 are the molecular weights; ρ_1^0 and ρ_2^0 are the densities of pure components 1 and 2. Substituting equation 6 into equation 5, one obtains

$$1 = C_1/\rho_1^0 + C_2/\rho_2^0 \tag{7}$$

where C_1 and C_2 are concentrations in grams/liter.

Using the relationship between concentration and weight fractions,

$$C_2 = X_2 \rho \quad \text{and} \quad C_1 = X_1 \rho \tag{8}$$

where

$$X_1 = \frac{W_1}{W_1 + W_2} \text{ and } X_2 = \frac{W_2}{W_1 + W_2}$$

and combining equations 7 and 8, one obtains

$$\frac{1}{\rho} = \frac{X_1}{\rho_1^0} + \frac{X_2}{\rho_2^0} \tag{9}$$

which is the expression of an ideal solution in terms of densities. In the preceding equations W_1 and W_2 are the masses of components 1 and 2, and ρ is the density of the solution. Equation 9 can be rearranged and the concentration converted to grams/liter by multiplying by ρ_1^0

$$\frac{\rho_1^0}{\rho} = 1 + \left(\frac{\rho_1^0}{\rho_2^0} - 1\right) X_2 = 1 + \left(\frac{\rho_1^0}{\rho_2^0} - 1\right) \frac{C_2}{\rho} \tag{10}$$

and

$$\rho = \rho_1^0 + \left(1 - \frac{\rho_1^0}{\rho_2^0}\right) C_2 \tag{11}$$

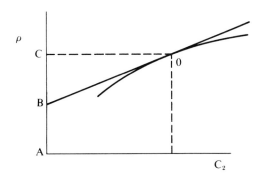

Figure 19-1 *Determination of partial molar quantities by the intercept method.*

The quantity $\left(1 - \dfrac{\rho_1^{\,0}}{\rho_2^{\,0}}\right)$, which is called the *density increment*, is constant for ideal solutions; that is, a plot of ρ vs. C_2 would yield a straight line.*

For nonideal solutions, where $\bar{V}_1 \neq V_1^{\,0}$ and $\bar{V}_2 \neq V_2^{\,0}$, the density increment becomes $(1 - \rho_1/\rho_2)$ and it is not a constant, but it is a function of concentration. However, it can be evaluated from the new form of equation 11 $\rho = \rho_1 + (1 - \rho_1/\rho_2)C_2$ if ρ is plotted vs. C_2 and a tangent is drawn to the curve at the specified concentration. The numerical value of the tangent gives the density increment $(1 - \rho_1/\rho_2)$, and this value multiplied by the concentration is given by the quantity BC in Figure 19-1. Subtracting this value from ρ, which is equal to C in Figure 19-1, one obtains the quantity BA for ρ_1.

Once this is obtained the ρ_2 can also be evaluated from the density increment. The partial molar volumes can be obtained according to equation 12:

$$\bar{V}_1 = \frac{M_1}{\rho_1} \qquad \bar{V}_2 = \frac{M_2}{\rho_2} \tag{12}$$

where M_1 and M_2 are the molecular weights of components 1 and 2, respectively.

Experimental

Clean two pycnometers (Fig. 19-2) of approximately 10 ml. capacity by washing them with cleaning solution and rinsing them 10 to 20 times with distilled water. Dry them in a 120° C. oven and cool them in a desiccator to room temperature.

During the cleaning and drying process prepare eight mixtures of about 30 g. each, containing dimethyl sulfoxide (DMSO) and water in concentrations varying from 0.1 to 0.96 mole fraction of DMSO. Thermostat the solutions at 25° C. (Dimethylformamide [DMF]-water, acetone-water, dioxane-water, or other systems may be used as an alternative.) If the same

---

* It is important to note that $\rho_1^{\,0}$ and $\rho_2^{\,0}$ refer to the densities of the pure components one and two, respectively, and ρ_1 and ρ_2 refer to the densities of components one and two in the mixture. The two are equal ($\rho_1 = \rho_1^{\,0}$) only in ideal solutions.

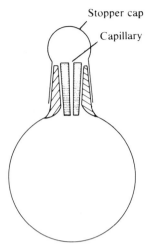

Figure 19-2 *Pycnometer.*

binary system is used in more than one experiment and the experiment on density is the first among these, the student may prepare larger quantities (200 grams) from each solution. These may be stored for possible use in Experiments 8, 9, 10, 12, 14, 15, 16, 17, 20, 21, 22, 25, 35, 36, 37, 38, and 41.

Determine the weight of the dry pycnometers, W_p, and fill them with distilled water making sure the water level in the pycnometer reaches the top of the capillary. Thermostat them at 25° C. for about 10 to 15 minutes.

Obtain the weight of the pycnometer filled with water, W_g, after making certain that the outside of the pycnometers is completely dry. From the density of water at 25° C., $\rho = 0.997044$ gram/cc., calculate the volumes of the pycnometers.

$$V = \frac{W_g - W_p}{\rho}$$

Subsequently weigh the pycnometers filled with different mixtures of DMSO-water and also with pure DMSO. From the volumes of the pycnometers already obtained, calculate the densities of the mixtures and that of DMSO at 25° C.

Plot the densities of the mixtures vs. concentration (the latter in grams/cc. and mole fraction). From the plot and from equation 11, obtain the ρ_1 and ρ_2 as a function of concentration.

Plot ρ_1 and ρ_2 vs. concentration.

Also plot $\bar{V}_1 - V_1^0$ and $\bar{V}_2 - V_2^0$ vs. concentration. ($\bar{V}_1$ and $\bar{V}_2$ are obtained from equation 12.)

What kind of deviation from ideality do you observe in these plots? What are the volumes of mixing at the different concentrations (using equation 4)?

What kind of interactions can explain the behavior of the partial molar volumes vs. concentrations?

Material and Equipment. 250 ml. dimethyl sulfoxide (DMSO) (or dimethylformamide [DMF], acetone, dioxane, and so forth) if solutions are used for this experiment alone or 1750 ml. if solutions are prepared also for future experiments; cleaning solution; two 10 ml. pycnometers; drying oven; desiccator; thermostat with heater, stirrer and thermoregulator.

REFERENCES

1. E. A. Moelwyn-Hughes. Physical Chemistry. 2nd Revised Edition, Pergamon Press, Inc., New York, 1964, Chapter 17.
2. T. R. Partington. Advanced Treatise on Physical Chemistry. Vol. II, Section VIII, Longmans Green & Co., Ltd., London, 1951.
3. J. J. Rowlinson. Liquids and Liquid Mixtures. 2nd Edition, Plenum Press, New York, 1969.
4. J. Timmermans. The Physico-chemical Constants of Binary Systems in Concentrated Solutions. Vol. II and IV, Interscience Publishers, Inc., New York, 1959 and 1960.

20

VISCOSITY OF LIQUIDS*

Viscosity is a molecular property. The coefficient of viscosity (η) is the proportionality constant between the force that causes a laminar flow and the velocity gradient of that flow, (dv/dr), over an area, A, that is parallel to the direction of the flow. This relationship is defined by Newton's law:

$$f = \eta A(dv/dr) \tag{1}$$

Flows that obey this law are called Newtonian flows.

One can visualize that a liquid that flows in a capillary tube is made up of concentric layers, each of which slips by its neighbor with a finite velocity. Near the walls of the capillary the first layer can be considered to be stationary with zero velocity. Each adjacent layer has a greater velocity as we proceed from the wall toward the center of the capillary. In Newtonian flow the greatest velocity is reached at the center; one can visualize a flow profile in a capillary consisting of arrows proportional to their velocity (Fig. 20–1).

In a capillary viscometer (Fig. 20–2) the efflux time of a certain volume of liquid is observed, and by using Poiseuille's equation the viscosity coefficient of the liquid can be evaluated.

$$\frac{V}{t} = \frac{\pi \Delta P R^4}{8\eta l} \tag{2}$$

In Poiseuille's equation certain quantities are determined by the geometry of the instrument used. These are R, the radius of the capillary; l, the length of the capillary; and ΔP, the hydrostatic pressure under which the flow occurs. The latter is fixed because the efflux time is measured from the instant the meniscus reaches reference mark a until it passes reference mark b of the viscometer. Although the average difference of the hydrostatic head $[(a + b/2) - c]$ is set by the geometry of the instrument, the hydrostatic pressure also depends on the density of the liquid in flow.

196

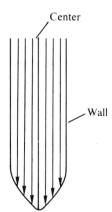

Figure 20–1 *Velocity profile.*

Since it is difficult to measure accurately the radius of the capillary (unless it is supplied by the manufacturer of the viscometer) the best way to determine the coefficient of viscosity of liquids is to compare the efflux time of a liquid having a known viscosity coefficient to that of the liquid in which it is unknown.

$$\frac{\eta_1}{\eta_2} = \frac{\rho_1 t_1}{\rho_2 t_2} \tag{3}$$

In this equation η is the coefficient of viscosity, t the efflux time, and ρ the density of the liquid; the subscript 1 refers to the liquid with a known coefficient of viscosity and 2 to the liquid under examination. Care must be taken so that both efflux times are measured by using the same volume of liquid at the same temperature. (The densities of the liquids must also be measured at the same temperature.)

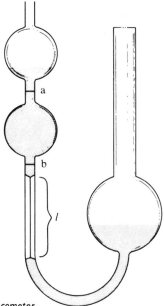

Figure 20–2 *Ostwald capillary viscometer.*

Theories of the viscosity of liquids have been proposed. For our purpose, the best way to examine the phenomenon is from the standpoint of chemical kinetics. This approach was developed by Eyring and his coworkers in the 1930's (reference 2). We investigate neighboring layers in a laminar flow, which are separated by an average distance λ' from each other. The liquid molecules have certain equilibrium positions that are separated from each other by distance λ. The viscous flow is looked upon as a series of transitions between equilibrium states. Each transition involves overcoming an energy barrier, ΔE_v, and the frequency of the transition over the energy barrier is related to the thermal energy:

$$\nu = kT/h \tag{4}$$

The viscosity coefficient is then given by the following expression:

$$\eta = \frac{NkT}{\nu}\left(\frac{\lambda'}{\lambda}\right)^2 \exp\left(\frac{\Delta E_v}{kT}\right) \tag{5}$$

where N is the molecular concentration

$$N = \frac{N_0}{V_m} \tag{6}$$

or in terms of molar concentration

$$\eta = nh\left(\frac{\lambda'}{\lambda}\right)^2 \exp\left(\frac{\Delta E_v}{kT}\right) \tag{7}$$

The empirical expression is the logarithmic form of equation 7

$$\log \eta = A + \frac{B}{T} \tag{8}$$

where A is the log of the preexponential factor in equation 7 and $B = \Delta E_v/R$, in which ΔE_v is the energy of activation per mole. Since the coefficient of viscosity of most liquids decreases with increasing temperature, the slope of the plot of $\log \eta$ vs. $1/T$ yields a positive energy of activation for viscosity. These energies of activation can be compared to energies of evaporation of the liquids. Energy of vaporization is required for the complete transfer of one mole of substance from the condensed (liquid) environment into the gas phase. On the other hand, the energy of activation for viscosity is required to transfer one mole of a substance from one pseudoequilibrium state in the liquid to another. In the latter process only a fraction of the intermolecular interactions are broken; the ratio $\Delta E_v/\Delta E_{vap}$ represents that fraction. Such comparisons, therefore, may indicate how far the molecules must be removed from their environment to reach a new equilibrium position.

The intercept of the $\log \eta$ vs. $1/T$ plot yields the log of the preexponential factor, A. According to equation 7 this is related to the structure of the liquid. If, as a rough approximation, one assumes that the separation of the liquid layers is constant in the different samples, the experimental values of A are inversely proportional to the square of the separation of the equilibrium positions in the liquid.

Experimental

Prepare eight 50 ml. mixtures of dimethyl sulfoxide (DMSO)-water (or dimethylformamide [DMF]-water, acetone-water, dioxane-water, and so forth), the densities of which have been measured at 25° C. previously in Experiment 19. Use a Cannon-Ubbelohde capillary viscometer that was calibrated by the manufacturer and its calibration constant supplied. (A viscometer that can deliver a liquid of 1 centipoise viscosity in about 100 seconds is preferable.) Alternatively, a simple Ostwald capillary viscometer can also be used. Determine its calibration constant at 25° C. by using a liquid of known viscosity. (For carbon tetrachloride $\eta = 0.921$ centipoise or 9.21×10^{-3} grams/cm. sec.). Be certain that the capillary of the viscometer is clean. (A viscometer can be readied by using cleaning solutions and rinsing it with water a few times.) Viscometers that can accommodate 5 or 10 ml. of liquid per sample are recommended.

Using a stopwatch, measure the efflux time of the pure water, the pure DMSO and the eight mixtures in the viscometer. Position the viscometer in the thermostat (25° C.) so that the capillary portion of the viscometer is vertical. Draw the liquid into the capillary arm above the upper mark of the viscometer. Let the liquid flow under atmospheric pressure, and start the stopwatch when the meniscus of the liquid reaches the upper mark. Stop the watch when the meniscus reaches the lower mark.

Repeat the experiment on the same liquid at least twice; if consecutive readings disagree more than 1 second, repeat the experiment three or four times. Using the average efflux time, and using the densities obtained in Experiment 19, calculate the absolute viscosity in terms of centipoises (equation 3). When introducing a new sample into the viscometer, rinse the viscometer (including capillary) at least once with the new sample; this liquid is discarded and the experiment is started with a new aliquot.

Report the viscosities obtained in centipoises vs. mole fractions of the solutions.

Raise the temperature of the thermostat to 45° C. Repeat the viscosity measurements on pure water, DMSO, and at least one of the solutions that had maximum viscosity in the 25° C. experiment (and if time allows, using three additional solutions). Determine the density of the water, DMSO, and the solutions that were measured at 45° C. Using these values, convert the flow times into centipoises and plot the viscosity vs. composition curve also at 45° C.

From the comparative data of the same solution at different temperatures calculate ΔE_v and A from equation 8 and plot these values against composition.

Compare the ratios of $\Delta E_v / \Delta E_{vap}$ for DMSO and water. The heats of vaporization were obtained in Experiment 8. How does the value of λ'/λ change with composition?

Material and Equipment. 250 ml. dimethyl sulfoxide (DMSO) (or dimethylformamide [DMF], acetone, dioxane, and so forth); carbon tetrachloride; Cannon-Ubbelohde or Ostwald capillary viscometer of 5 or 10 ml. capacity with an efflux time of about 100 seconds for water; thermostat with heater, stirrer, and thermoregulator; stopwatch.

REFERENCES

1. P. Andrade. Viscosity and Plasticity. Heffer & Sons, Ltd., Cambridge, 1947.
2. H. Eyring. J. Chem. Phys., **4**, 283 (1936).
3. J. M. G. Cowie and P. M. Toporowski. Can. J. Chem., **39**, 2240 (1961).
4. T. A. Bak and K. Anderson. Acta Chem. Scand., **12**, 1367 (1958).

21

SURFACE TENSION

The most important property of liquid-gas surfaces is their surface tension. Liquids, in most instances, behave as though they were covered by a thin elastic layer. In classic terms the surface tension acts along the surface and tends to make its area as small as possible. A typical model used to demonstrate surface tension is a box that has a sliding cover (Fig. 21–1). The interfacial tension between this cover and the liquid is assumed to be zero. By uncovering a new surface, dA, the cover has to slide from position a to b.

The mechanical work needed to accomplish this is proportional to the uncovered surface and may be equated to the increase in the Gibbs free energy of the system.

$$dG = \gamma \, dA \qquad (1)$$

The proportionality constant, γ, is the surface tension and, therefore, it is defined thermodynamically for a one component system as

$$\left(\frac{\partial G}{\partial A}\right)_{T,P,n} = \gamma \qquad (2)$$

The system can be looked upon as having a total free energy composed of two terms—the bulk and the surface.

$$G_T = G^0 n_0 + G^S A \qquad (3)$$

where G_T, G^0, and G^S are the *total*, the *molal*, and the *surface* free energies, respectively; n_0 is the number of moles of liquid, and A is the surface area.

Since the sliding of the lid is a mechanically reversible process, the heat associated with it is proportional to the surface entropy

$$dq_{\text{rev}} = T \, dS^S = TS^S \, dA \qquad (4)$$

where S^S is the surface entropy per unit area.

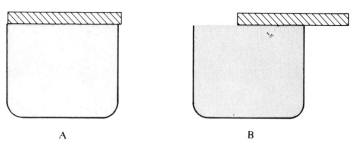

A   B

Figure 21–1 *Liquid surface covered by a theoretical lid.*

The surface tension is a surface free energy term according to equations 2 and 3. Therefore, the temperature dependence of the surface tension can give the surface entropy.

$$(\partial\gamma/\partial T)_P = -S^S \tag{5}$$

The surface tension of most liquids decreases with temperature, and the Eötvös equation describes their behavior:

$$\gamma(M/\rho)^{2/3} = k(T_C - T) \tag{6}$$

where M is the molecular weight of the liquid, ρ is its density, T_c is the critical temperature of the liquid, and T is the working temperature. This equation predicts a linear decrease in the surface tension per mole with increasing temperature; more important, it predicts that the surface tension vanishes at the critical temperature. $[(M/\rho)^{2/3}$ has the dimension of area per mole.]

An important molecular parameter that can be derived from the surface tension was suggested by Garner and Sugden. They named it *parachor*, P, and it was defined as

$$P = M\gamma^{1/4}/\rho \tag{7}$$

The parachor can be considered as a molar volume corrected for the compressive effects due to intermolecular forces. For many compounds it is an additive property of the atoms and functional groups and, therefore, it is similar to molar polarizability. A review of this molecular parameter is given by Quayle.

In a binary system the surface tension varies with the composition. The thermodynamic treatment of this variation was first provided by Willard Gibbs on the assumption that the surface under consideration is infinitely thin (approaches zero). More modern theories take into account a finite thickness of the surface layer but, in essence, they come to the same conclusion (Eriksson). In the following the Gibbs arguments are reproduced.

First, we define the nomenclature: In a binary mixture we have n_1 moles of component 1 and n_2 moles of component 2 in the bulk phase. The corresponding number of moles of components 1 and 2 in the surface phase are n_1' and n_2'. Similarly, the chemical potentials designated by μ_1 and μ_2 refer to the bulk phase, and μ_1' and μ_2' refer to the surface phase.

At equilibrium the Gibbs-Duhem equation is operative both in the bulk and in the surface phase. Therefore,

$$n_1 \, d\mu_1 + n_2 \, d\mu_2 = 0 \tag{8}$$

and

$$n_1' \, d\mu_1' + n_2' \, d\mu_2' + A \, d\gamma = 0 \tag{9}$$

The third term in equation 9 is the contribution due to the surface tension since this is also a free energy term (see equations 2 and 3).

For the equilibrium condition between bulk and surface, the chemical potentials of each component must be the same in both phases. Hence,

$$\mu_1 = \mu_1'; \quad d\mu_1 = d\mu_1'$$
$$\mu_2 = \mu_2'; \quad d\mu_2 = d\mu_2' \tag{10}$$

Therefore, equation 9 can be rewritten as

$$n_1' \, d\mu_1 + n_2' \, d\mu_2 + A \, d\gamma = 0 \tag{11}$$

Rearranging equation 8 to solve for $d\mu_1$ and substituting into equation 11, we obtain

$$-n_1'\left(\frac{n_2}{n_1}\right) d\mu_2 + n_2' \, d\mu_2 + A \, d\gamma = 0 \tag{12}$$

and

$$-\left(\frac{A \, d\gamma}{d\mu_2}\right) = n_2' - n_1'\left(\frac{n_2}{n_1}\right) \tag{13}$$

The term $n_1'(n_2/n_1)$ is the number of moles of component 2 associated with n_1' moles of component 1 in the bulk. Therefore, the right side of equation 13 represents the *number of moles of component 2 in excess in the surface phase as compared to the bulk*. This excess amount can be called the adsorbed amount in the surface phase, Γ_2.

Since the chemical potential can be written in terms of activities,

$$\mu_2 = \mu^0_e + RT \ln a_2 \tag{14}$$

at constant temperature

$$d\mu_2 = RT \, d \ln a_2 \tag{15}$$

and equation 13 becomes

$$-\frac{A \, d\gamma}{RT \, d \ln a_2} = \Gamma_2 \tag{16}$$

If we now define Γ_2' as the excess amount per unit surface area of component 2 in the surface phase compared to the bulk phase

$$\Gamma_2' = \Gamma_2/A \tag{17}$$

and $d \ln a_2$ is written as da_2/a_2, equation 16 becomes

$$\Gamma_2' = -\frac{a_2 \, d\gamma}{RT \, da_2} \tag{18}$$

and in dilute solutions (approximating ideal conditions), $C_2 \rightarrow a_2$, the actual concentrations approach the activities, and

$$\Gamma_2' = -\frac{C_2}{RT}\frac{d\gamma}{dC_2} = -\frac{1}{RT}\frac{d\gamma}{d\ln C_2} \tag{19}$$

Therefore, if one plots the surface tension measured vs. the logarithm of activity (or concentration) of a component, the excess amount of the component in the surface phase can be calculated for each concentration. When $d\gamma/dC$ is positive, the excess quantity is negative. This means that the concentration of the component in question in the surface phase is less than in the bulk. We call these compounds surface inactive or capillary inactive. When $d\gamma/dC$ is negative, one finds excess amounts of the component in question in the surface phase and the component is called surface active or capillary active.

Experimental

The simplest way to measure surface tension is with the capillary rise method. The experimental setup is illustrated in Figure 21–2.

When a capillary tube is dipped into a liquid, the level of the liquid rises above the outer plane surface if it wets the tube walls. Let the upward force per centimeter of contact be γ_1 (Fig. 21–3); the total upward force in the round tube is $2\pi r\gamma_1$ where r is the radius of the tube. This is balanced by a downward force, which equals the volume of liquid in the tube above the outside surface multiplied by its density (ρ) and the gravitational constant (g):

$$2\pi r\gamma_1 = \pi r^2 hg\rho \tag{20}$$

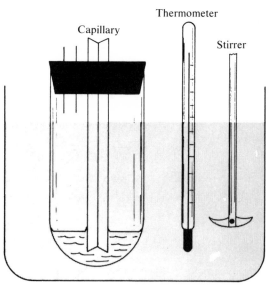

Figure 21–2 *Apparatus for measuring surface tension with the capillary rise technique.*

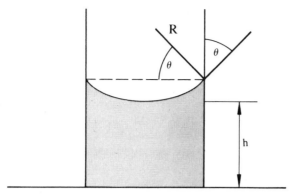

Figure 21-3 *Diagram of capillary rise.*

or

$$2\gamma_1 = rh\rho g$$

Surface tension, γ, is defined as the force per centimeter required to extend the surface in its own plane. From Figure 21–3 the relationship between γ, the true surface tension, and γ_1, the upward force, is

$$\gamma_1 = \gamma \cos \theta \qquad (21)$$

where θ is the contact angle.

Substituting and rearranging,

$$2\gamma = \frac{h\rho g r}{\cos \theta} \qquad (22)$$

If the wetting or contact angle, θ, is zero or very nearly so, $\cos \theta = 1$ and this equation may be written

$$\gamma = \frac{h\rho g r}{2} \qquad (23)$$

The radius, r, can be measured with a microscope micrometer, or *by weighing the mercury necessary to fill a known length of the tube*. Often the tube is calibrated with a liquid of known surface tension, whereupon it may be used to measure the surface tension of another liquid. In this case,

$$\frac{\gamma_a}{\gamma_b} = \frac{h_a \rho_a}{h_b \rho_b} \qquad (24)$$

where h_a and h_b are the capillary rises observed with liquids a and b having densities ρ_a and ρ_b, respectively.

Select a capillary tube about 20 cm. long and 0.30 mm. internal diameter. Soak the capillary in hot nitric acid for several minutes and then thoroughly wash it with water and finally rinse it with distilled water. Determine the radius of the capillary by: (a) filling part of it with mercury, measuring the length filled, and weighing the mercury, or (b) calibrating with benzene, methanol, and toluene according to equation 24. When using the pure liquids, measure the capillary rise with a cathetometer. Assume that there is a zero contact angle.

Using a cathetometer, measure the capillary rise of pure dimethyl sulfoxide (DMSO) and pure water and eight mixtures of DMSO-water between 0.1 and 0.9 mole fraction of DMSO. (Dimethylformamide [DMF]-water, acetone-water, dioxane-water, or other systems may be used as an alternative.)

Perhaps the greatest difficulty in measuring surface tension is caused by the adsorption of surface active substances. These impurities are usually firmly adsorbed and are therefore difficult to remove. They are especially noticeable when the surface tension of an aqueous solution is measured, because they decrease it considerably even when present in infinitesimal quantities, since they collect at the glass-water-air or vapor interface.

In the present method, an error of 1 mm. in a measured height of rise of 10 cm. leads to an error of about 1 per cent. The surface tension is also a function of the temperature; a change of 1° may alter its value by about 0.5 per cent.

The radius of a capillary may vary along its entire length. An error of 3 per cent probably is a reasonable minimum to expect with an uncalibrated tube. The other possible errors are that the diameter of the container is too small and that the capillary is not mounted in a truly vertical position. For more accurate results, it is customary to correct the height of rise for the amount of liquid above the bottom of the meniscus. Because this correction is small, it has been neglected in the following calculations.

The maximum error in the calculated value of the surface tension using the capillary rise method may be obtained by applying the usual procedures of error analysis to equation 24. The method of normal and maximum values will be used although one factor, the change of surface tension with temperature, is not taken into account. Hence, it should be realized that the calculated error will be inexact because of the failure to make this correction. If the temperature coefficient of surface tension and the variation of the thermostat were known, a more accurate evaluation of the maximum error could be obtained.

Assume that the following quantities (and their possible errors) were obtained in order to calculate γ_a, the surface tension of a given sample:

$$\gamma_b = 72.75 \pm 0.05 \text{ dynes cm.}^{-1} \qquad \rho_a = 0.9000 \pm 0.0002$$
$$h_a = 7.0 \pm 0.1 \text{ cm.} \qquad\qquad \rho_b = 0.9970 \pm 0.0002$$
$$h_b = 14.0 \pm 0.1 \text{ cm.}$$

Since the relative error in the value of the density is so much smaller than the others, it may be neglected; then

$$\pm \Delta\gamma_a = \frac{\gamma_b h_a \rho_a}{h_b \rho_b} - \frac{\gamma_b h_a \rho_a}{h_b \rho_b}$$

$$= \frac{72.80 \times 7.1 \times 0.9000}{13.9 \times 0.9970} - \frac{72.75 \times 7.0 \times 0.9000}{14.0 \times 0.9970}$$

$$= 33.57 - 32.84$$

It can therefore be concluded that the surface tension is 32.84 ± 0.73 dynes cm.$^{-1}$

Calculate the maximum error for one of the samples used in this experiment. Assume that the surface tension of benzene, to be used as the standard, is known within ± 0.03 unit.

Calculate the radius of the capillary obtained by the different techniques. Average the values. Using the average value of the capillary radius, calculate the surface tension of the pure liquids and the mixtures. Calculate the parachor for water and DMSO. Plot the surface tension vs. mole fraction. Calculate the excess amounts, Γ'_2, from the plot by using equation 19. Plot Γ'_2 vs. mole fraction.

Is DMSO a surface active agent in water? How good is the approximation of substituting mole fraction for activity in equation 18?

Material and Equipment. Dimethyl sulfoxide (DMSO) (or dimethyl formamide [DMF], acetone, dioxane, and so forth); benzene; methanol; toluene; nitric acid (or cleaning solution); mercury; capillary tubes about 20 cm. long and 0.3 mm. i.d.; test tube with stopper to hold capillary tube; thermostat with heater, stirrer, and thermoregulator; thermometer.

REFERENCES

1. F. B. Garner and S. Sugden. J. Chem. Soc., **1929**, 1298 (1929).
2. O. R. Quayle. Chem. Rev., **53**, 439 (1953).
3. A. W. Adamson. Physical Chemistry of Surfaces, Interscience Publishers, Inc., New York, 1960.
4. J. C. Eriksson. Arkiv. f. Kemi, **25**, 331 (1966).
5. J. L. Lebowitz, E. Helfand and P. Praestgaard. J. Chem. Phys. **43**, 774 (1965).
6. F. B. Sprow and J. M. Prausnitz. Can. J. Chem. Eng., **45**, 25 (1967).

VII

SOLIDS

22

X-RAY DIFFRACTION PATTERNS OF DIMETHYL SULFOXIDE (DMSO) AND WATER*

Atoms in a crystal are arranged in patterns that are repeated periodically in three dimensions. Figure 22–1 represents such a periodical variation in two dimensions.

A lattice is a three-dimensional array of geometric points that represent the positions of the atoms. A unit cell is the smallest unit of structure of the lattice that has the symmetry properties of the entire crystal structure. The unit cells are characterized by their symmetry and dimensions. Among the most important are:

Cubic	$a = b = c$	$\alpha = \beta = \gamma = 90$ degrees
Tetragonal	$a = b \neq c$	$\alpha = \beta = \gamma = 90$ degrees
Orthorhombic	$a \neq b \neq c$	$\alpha = \beta = \gamma = 90$ degrees
Hexagonal	$a = b \neq c$	$\alpha = \beta = 90$ degrees $\quad \gamma = 120$ degrees
Monoclinic	$a \neq b \neq c$	$\alpha = \gamma = 90$ degrees $\quad \beta \neq 90$ degrees
Triclinic	$a \neq b \neq c$	$\alpha \neq \beta \neq \gamma$

where a, b, and c are the unit cell dimensions and α, β, and γ are the angles between b and c, a and c, and a and b, respectively.

When a single crystal is exposed to x-rays, the three dimensional array of atoms in the crystal scatters the electromagnetic radiation in such a way that the scattered waves reinforce each other only in certain directions. These directions or angles of reflection are characteristic of the interatomic separations and their orientation in space. The reflections are governed by Bragg's law:

$$\lambda = 2d_n \sin \theta \tag{1}$$

where λ is the wavelength, d_n is the interatomic spacing, and θ is the angle of reflection

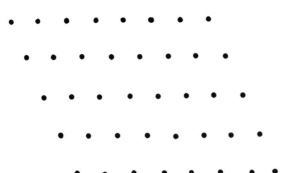

Figure 22–1 *Space lattice in two dimensions.*

at which reinforcement occurs. This can be understood by observing the path of the scattered light in Figure 22–2 in which the path difference (*ABC*) between beam 1 and beam 2 has to be an integer times the wavelength of the radiation (*nλ*) in order to cause constructive interference.

Although the x-rays are scattered by the real atoms (more correctly by the electrons of the atoms) we speak of reflections coming from planes within the crystal. One can regard a lattice array of atoms as an infinite stack of parallel and equally spaced planes containing the atoms. The scattering by atoms in a plane is, therefore, the same as a reflection by the plane.

In Figure 22–3 it is demonstrated that within a two-dimensional lattice one may draw a number of *sets* of parallel equidistant planes.

These planes are identified by sets of index numbers, and the most commonly used index numbers are referred to as Miller indices.

These indices represent the reciprocal of the intercepts of a specified plane with the unit axes. An example for a hexagonal unit cell is provided in Figure 22–4. (One has to remember that the unit cell is the simplest repeating structure and, therefore, the base is taken as one-third of a hexagon.)

The indices *h*, *k*, and *l* by convention represent the reciprocal of the intercepts at the *a*, *b*, and *c* axes, respectively. The plane 001 is parallel to the *a* and *b* axes ($a = b$) and intercepts the *c* axis at unit distance $\left(\dfrac{1}{\infty} = 0; \dfrac{1}{1} = 1\right)$. The 111 plane

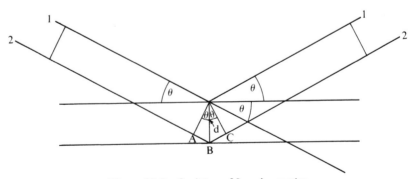

Figure 22-2 *Conditions of Bragg's equation.*

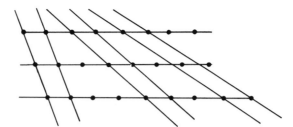

Figure 22–3 *Sets of planes in a two-dimensional lattice.*

intercepts all three axes at unit length. The 112 plane intercepts the $a = b$ axes

at unit length and the c axis at one-half the unit length $\left(\dfrac{1}{\frac{1}{2}} = 2\right)$.

Each set of planes containing atoms has a specific reflection in a single crystal pattern. The location of the reflection depends on the type of unit cell and its dimensions. The relative intensities of the reflections depend on the arrangement of the atoms in the unit cell. The unit cell and the location of the atoms within the unit cell determines the crystal structure.

The purpose of the present experiment is to relate the x-ray diffraction pattern to a unit cell. Our intention is only a partial structural analysis (analyzing the unit cell and indexing the reflections), which one can achieve quickly with the powder technique.

A powder diagram is produced when the x-ray beam interacts with the thousands of tiny crystals in the sample, each of which has different orientation; in the sample, therefore, all possible crystal orientations are present. Many of the crystals are so oriented that a particular set of planes (h, k, l) are at the appropriate Bragg angle (θ) (see equation 1) with respect to the x-ray beam. Such tiny crystals are in position to reflect x-rays. As can be seen from Figure 22–5, the reflection occurs at a 2θ angle, which (referring to Figure 22–2) is the angle between the incident and the scattered beam. Since, however, for each set of planes (h, k, l) all possible orientations are present in the sample, the scattered and reinforced beam from such a set forms a cone. At the intersection of the cone with a flat photographic plate a ring is produced

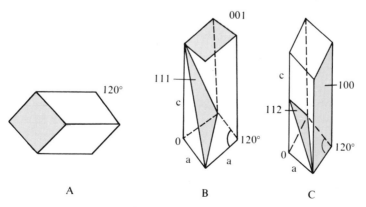

Figure 22–4 *A, base of a hexagonal unit cell; B, the 001 and 111 planes; C, the 100 and 112 planes.*

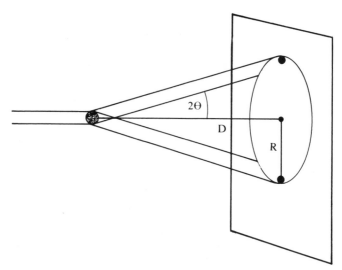

Figure 22–5 *Experimental setup for producing a powder diagram with a flat cassette camera.*

on the plate. Therefore, each set of plane h, k, l produces a ring, and on the photograph concentric rings appear.

If one knows the film to sample distance, D, and the radius of a concentric ring, R, the Bragg angle can be evaluated since

$$\tan 2\theta = \frac{R}{D} \tag{2}$$

Once the Bragg angle is evaluated, the corresponding interplanar distance can be calculated from equation 1 if one knows the wavelength of the monochromatic x-rays.

However, with a flat cassette camera an inaccuracy enters into the calculations since the various parts of the photographic plates are at different distances from the scattering point. This is rectified by wrapping the photographic film around the sample in the form of a coaxial cylinder, the axis of the cylinder being perpendicular to the x-ray beam.

A set of interplanar distances are thus obtained, each corresponding to a certain reflection (ring). The internal relationship of these interplanar distances in a crystal is related to the symmetry of the unit cell. For instance, in a cubic system the interplanar spacing obeys the following equation:

$$d_{hkl} = \frac{a}{(h^2 + k^2 + l^2)^{1/2}} \tag{3}$$

or

$$(h^2 + k^2 + l^2)d_{hkl}^2 = a^2 \tag{4}$$

where h, k, and l are the Miller indices of the planes producing the reflection, d_{hkl} is the spacing between those planes, and a is the edge of the cubic unit cell. Equation

4, therefore, predicts the following relationship between the different interplanar spacings in a cubic cell.

$$1 \times d_{100}^2 = a^2 \qquad \text{or} \qquad d_{100} = a$$

$$2 \times d_{110}^2 = a^2 \qquad\qquad\qquad d_{110} = a/\sqrt{2}$$

$$3 \times d_{111}^2 = a^2 \qquad\qquad\qquad d_{111} = a/\sqrt{3}$$

$$4 \times d_{200}^2 = a^2 \qquad\qquad\qquad d_{200} = a/\sqrt{4}$$

$$5 \times d_{210}^2 = a^2 \quad \text{and so forth} \quad d_{210} = a/\sqrt{5}$$

Therefore, if all the observed interplanar spacings calculated from a powder diagram can be accounted for by such an internal relationship as given here, one can assume that the unit cell is cubic, and the edge of the unit cell can be calculated.

For each type of unit cell there is a special relationship between the interplanar spacings.

For a tetragonal cell

$$\frac{1}{d_{hkl}^2} = \frac{h^2 + k^2}{a^2} + \frac{l^2}{c^2} \tag{5}$$

For a hexagonal cell

$$\frac{1}{d_{hkl}^2} = \frac{4(h^2 + hk + k^2)}{3a^2} + \frac{l^2}{c_2} \tag{6}$$

For an orthorhombic cell

$$\frac{1}{d_{hkl}^2} = \frac{h^2}{a^2} + \frac{k^2}{b^2} + \frac{l^2}{c^2} \tag{7}$$

For a monoclinic cell

$$\frac{1}{d_{hkl}^2} = \frac{\dfrac{h^2}{a^2} + \dfrac{l^2}{c^2} + \dfrac{2hl}{ac}\cos\beta}{\sin^2\beta} + \frac{k^2}{b^2} \tag{8}$$

The question of which symmetry the set of interplanar distances are related to can be answered either analytically or graphically. Analytically, one may first try the simplest system (cubic) to see whether it can account for all the interplanar distances. If the cubic system fails, one proceeds to tetragonal, orthorhombic, and other systems. Charts are also available for the different symmetries. One such chart for a cubic system is reproduced in Figure 22–6.

The experimentally obtained d_{hkl} values are plotted on a strip of paper, using the dimensions of the abscissa on the chart. With the zero point of the paper sliding along the y axis and the edge of the paper parallel to the x axis, the paper is moved up and down until all reflection dimensions plotted on the paper are accounted for by intersecting lines on the chart paper. When this is achieved, the dimension of the unit cell is read directly from the ordinates and the indexing of the experimental distances is read from the intersecting line as shown in Figure 22–6.

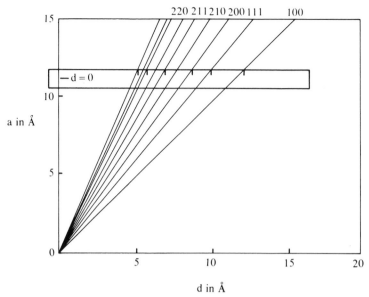

Figure 22–6 *Indexing chart for a cubic system—the edge of cubic cell a (ordinates) vs. the inter-planar distance (abscissa).*

Graphical and straight analytical solutions are useful for unit cells with high symmetry (cubic, tetragonal, hexagonal), but special methods must be used for low symmetry unit cells (see Azaroff and Buerger).

In the present experiment the symmetries of the unit cells will be given (hexagonal for ice and monoclinic for DMSO) and the student has to index the calculated reflections.

Since setting up the x-ray diffraction experiment with refrigeration, aligning samples, and exposing the film for $2\frac{1}{2}$ to 3 hours to x-ray diffraction takes more time than is usually available in one laboratory period, it is recommended that the students be provided with the powder patterns of DMSO and water previously taken by the instructor. The laboratory period will then consist of familiarizing the student with the x-ray equipment, focusing x-ray beams, preparing the sample and aligning it in a powder camera, taking measurements of the exposed films, and calculating inter-planar distances.

The indexing of the powder patterns is to be done as part of the laboratory report.

Experimental

Since both DMSO (melting point 18° C.) and water require refrigeration, a cooling setup can be used with the sample. Dry nitrogen is passed through a coil that is in liquid nitrogen or in an acetone–Dry ice solution. The Rudman design is reproduced in Figure 22–7.

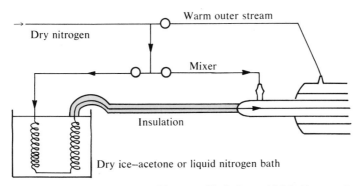

Figure 22–7 *Cooling device. (Courtesy of R. Rudman, Adelphi University.)*

The sample (DMSO, H_2O, and a DMSO:$2H_2O$ mixture) is placed in the proper capillary tube, and it is properly aligned according to procedure B. The sample is crystallized by the cooling device and maintained below the melting point. (The DMSO:$2H_2O$ mixture must be cooled to $-80°$ C. before the viscous liquid solidifies.) Samples are exposed for three hours at 15-milliampere and 35-kilovolt settings. The films are developed in x-ray developer for ten minutes and fixed. *The DMSO and water provide clear powder patterns. The DMSO:$2H_2O$ mixture is amorphous and gives only a faint diffuse halo.* The powder patterns can be reproduced and one provided to each group of students for measurements.

[If systems other than DMSO-water are used, similar procedures may be adopted. However, it must be observed that not all the well known solvents were investigated at low temperatures by x-ray crystallography. For instance, the crystal structures of dimethyl formamide (DMF), acetone, dioxane, and ethanol have not been determined, although acetone hydrate has been investigated (see Quist and Frank). Therefore, as an alternative system one may use methanol-water, formic acid–water, formamide-water, acetic acid–water, or propionic acid–water. In each of these systems, liquid nitrogen must be used as a coolant.]

STUDENT EXPERIMENTS

Procedure A

In any x-ray diffraction system it is necessary to limit the monochromatic x-ray radiation to a thin pencil of a beam. The x-rays emitted as a result of the bombardment of a target material with electrons is characteristic of the target. The radiation emitted contains a variety of wavelengths (white radiation) beyond a few specific wavelengths characteristic of the element of the target. These characteristic wavelengths are called $K\alpha_2$, $K\alpha_1$, and $K\beta$ radiation. In order to use monochromatic wavelengths for diffraction, a filter is needed that will eliminate everything except the desired radiation.

Figure 22–8 *Collimator.*

When copper is the target, a nickel filter will absorb the $K\beta$ radiation and the wavelength of the filtered beam is $CuK\alpha_2 = 1.54434$ Å and $K\alpha_1 = 1.54050$ Å. If a cobalt, iron, or manganese filter is used with a copper target, $K\beta$ radiation is obtained with a wavelength of 1.39217 Å. For our diffraction measurements $CuK\alpha$ is used and an average wavelength of *1.5418 Å* can be assumed.

To provide a thin pencil of a beam, a collimator with pinhole or slit apertures is used (Fig. 22–8).

It is critical that the collimator be aligned properly by moving the carriages up and down and sideways in order to obtain the maximum intensity of radiation.

In the demonstration a fluorescent screen is used in the dark to observe the x-rays. However, lead shields must be placed around the collimator to minimize radiation effects due to stray scattering from the side of the collimator.

Procedure B

Preparation of sample. Samples to be cooled can be introduced into the capillary tube simply by inserting the capillary tube into the liquid.

[If the DMSO-water sample is placed in the capillary and, after sealing and aligning, it is frozen *in situ*, one may get a few large crystals in the path of the beam instead of thousands of tiny crystals. Therefore, not all possible orientations will be available, and instead of concentric rings, there may be relatively short arcs and even the rings may be quite grainy. Since procedure C requires measurements along the equator, these arcs may not lie along the equator because they are not complete rings. To remedy the situation the frozen capillary should either be turned slowly and continuously by a motor driving the pulley or be turned manually every ten minutes. In addition, one may enhance the random orientation by shutting off the nitrogen coolant every 30 minutes and allowing the liquid to melt. It then is recrystal- lized quickly by again turning on the nitrogen coolant.

Another technique to use with DMSO and water may be to freeze the sample in a cooled mortar (or in a cold room). Grind the sample to a fine powder and fill the capillary as described in the next paragraph, being certain to maintain it below melting point temperature throughout the operation.]

The capillary tube is usually between 0.3 and 1.0 mm. in diameter, 8 mm. long, and is made of very thin plastic, or Pyrex or Lindemann glass.

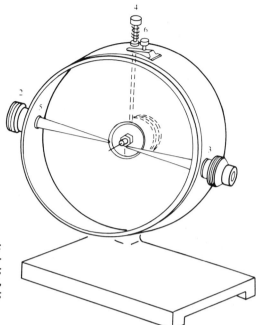

Figure 22–9 *Powder camera: 1, specimen holder; 2, beam-collimating tube; 3, insert mount; 4, specimen alignment screw; 5, inner circumference of camera; 6, external adjusting screw.*

Seal one end with a small alcohol flame. Then gently push or tap the powdered specimen into the open end and make it settle by gently running a file over the tube. Continue this procedure until the tube is filled sufficiently, and then seal it with a tiny piece of soft wax.

Mounting and alignment.

1. Place the prepared specimen into the small insert plug, which has been tipped with a small piece of wax or modeling clay. Use a 114.59- or a 57.3-mm. powder camera. Insert the mounted specimen into the specimen holder (No. 1, Fig. 22–9). Position the prepared specimen so that its axis is parallel to the axis of the specimen holder. It need not be coaxial since true centering is accomplished in step 4.

2. Apply the magnifying lens to the beam-collimating tube (No. 2, Fig. 22–9), and remove the beam-receiving tube. Using a light stand, direct the light through the insert mount (No. 3, Fig. 22–9) and onto the mounted specimen.

3. Rotate the specimen rotating pulley in the back of the camera by hand, and sight through the magnifying lens toward the light. The specimen will move up and down past the round opening at the tip of the beam-collimating tube.

4. Adjust the specimen alignment screw (No. 4, Fig. 22–9) gently until the specimen is in the center of the round opening, and then turn the specimen alignment screw back several turns. Repeat this procedure until the specimen does not deviate up or down from the center of the round opening as the pulley is rotated. When this condition exists, the specimen is aligned. Oscillation should not occur at the ends to insure sharpness of pattern.

Film loading. (This procedure must be done in the darkroom.) Gently lower the film into the camera along the inner circumference (No. 5, Fig. 22–9) so that the holes in the film coincide with the mounts into which the beam-collimating and beam-receiving tubes are inserted. Place one end of the film so that it presses against the fixed pin of the film expanding device; place the other end of the film beneath the movable pin of the film expanding device. Then gently screw down the external adjusting screw (No. 6, Fig. 22–9) of the film expanding device. This insures that the film will be held in close contact with the interior of the camera body. Next, carefully insert the beam-collimating and beam-receiving tubes into the camera body, and replace the camera cover.

The camera can now be taken from the darkroom and placed onto the camera track. Tighten the knurled screw on the base of the camera and proceed with the x-ray exposure.

Exposure. The average exposure time with the 114.59 mm. powder camera is one to three hours, and the 57.3 mm. powder camera gives well defined lines with an exposure time of five to sixty minutes.

Procedure C

Measurement and Calculation. Photographic film shrinks a variable amount after processing. The Straumanis technique corrects for film shrinkage, making the film self-calibrating.

Lay the film flat on the viewing glass of the measuring frame assembly and hold it in place with the spring clamp. Establish a zero degree reference point in the forward reflection region by bisecting the distance between two corresponding diffraction lines, one on each side of the beam-receiving tube hole in the film. To determine the 180 degree reference point, follow the same procedure for the back reflection region, without changing the position of the film. Bisect the distance between the corresponding diffraction lines (one on each side of the beam-collimating tube hole in the film). The distance between the zero degree reference point and the 180 degree reference point should be 180 mm. If this does not measure 180 mm., the correction factor is the ratio of 180 mm. to the actual measurement. For example, if the film length is 178 mm., multiply all measured distances by 180/178 to get their true value. Determine the distance of the various diffraction lines from the zero degree reference point by reading from the measuring scale. Then apply the correction factor. The vernier scale in the measuring slide assembly can be read to 0.05 mm.

The resulting series of numbers divided by 2 are equal to θ, the Bragg angles.

SAMPLE CALCULATION

1. Locate the positions of lines by using a millimeter scale and reading along the center line of the film. The zero mark on the scale need not coincide with any line. Enter the values so obtained as S_1 values in the second column.

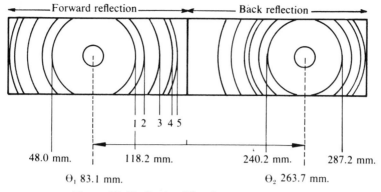

Figure 22–10 *Powder diffraction pattern.*

2. Determine pattern centers θ_1 and θ_2 (Fig. 22–10).

$$\theta_1 = \left(\frac{48.0 + 118.2}{2}\right) = 83.1 \text{ mm.}$$

$$\theta_2 = \left(\frac{240.2 + 287.2}{2}\right) = 263.7 \text{ mm.}$$

$$\theta_2 - \theta_1 = 263.7 - 83.1 = 180.6 \text{ mm.}$$

Therefore, since the correct length should be 180 mm., each measurement must be multiplied by

$$\frac{180}{180.6} = 0.996$$

3. Subtract 83.1 mm. (θ_1) from each S_1 value. Enter the answer in the third column as S (uncorrected).

4. Multiply each value in the third column by 0.996 to get S (corrected). Enter the answer in the fourth column as S (corrected).

5. Divide values of S by 2 and enter the answer in the fifth column. This value indicates the Bragg angle in degrees since 2 mm. equals 1 degree Bragg. (With the 57.3 mm. camera the figure in the fourth column would be employed since 1 mm. on the film equals 1 degree Bragg.)

6. In the sixth column enter corresponding values of sin θ.

7. In the seventh column enter 2 × sin θ.

8. In the eighth column enter values of d from the formula $d = \dfrac{\lambda}{2 \times \sin \theta}$.

9. In the ninth column enter the values for the relative intensity of lines.

Line No.	S_1 mm.	S Uncorrected $(S_1 - \theta_1)$	S Corrected	$\frac{S}{2}(\theta°)$	Sin θ	2 Sin θ	d	$\frac{I}{I_1}$
1	118.2	35.1	34.95	17.47	0.3000	0.600	2.57	1.0
2	123.7	40.6	40.43	20.21	0.3455	0.691	2.22	0.8
3	142.0	58.9	58.71	29.35	0.4901	0.980	1.57	0.8
4	153.5	70.4	70.12	35.06	0.5743	1.148	1.34	1.0
5	157.0	73.9	73.6	36.8	0.5990	1.198	1.29	0.4

Calculate the d values for DMSO and tabulate them in decreasing order. The literature (see Thomas, Shoemaker, and Eriks) reports that DMSO crystallizes in a monoclinic unit cell with the following dimensions: $a = 5.30$ Å, $b = 6.829$ Å, $c = 11.69$ Å, and $\beta = 94°30'$.

With the aid of equation 8 calculate the d_{hkl} values expected from such a unit cell for various khl combinations. Compare these calculated values with the experimental d values of the DMSO pattern. Try to index your interplanar spacings.

A shorter calculation in indexing that will induce only a small error is to assume that the DMSO unit cell is orthorhombic (i.e., $\beta = 90°$) instead of $94°30'$. Therefore, under this condition, equation 7 should be used instead of equation 8.

For ice, the unit cell reported in the literature (see Petersen and Levy) is hexagonal, with $a = 4.51$ Å and $c = 7.35$ Å. Index your x-ray reflections of ice similarly to the foregoing procedure by comparing the values with data calculable from such a unit cell. Use equation 6.

A computer program to calculate the d values and index them is given in Appendix I.

The inability of the DMSO:$2H_2O$ mixture to give a crystalline pattern should be interpreted in the light of previous experience with the system.

[If a system other than DMSO-water is used, the following unit cell dimensions taken from the literature can be used: methanol has an orthorhombic unit cell—$a = 6.43$ Å, $b = 7.24$ Å, and $c = 4.67$ Å (see Taner and Lipscomb); formic acid crystals are orthorhombic—$a = 10.23$ Å, $b = 3.64$ Å, and $c = 5.34$ Å (see Holtzberger et al.); formamide has a monoclinic unit cell—$a = 3.69$ Å, $b = 9.18$ Å, $c = 6.87$ Å, and $\beta = 98°$ (see Ladell and Post); acetic acid has an orthorhombic unit cell—$a = 13.32$, $b = 4.08$, and $c = 5.77$ (see Jones and Templeton); propionic acid possesses monoclinic symmetry—$a = 4.04$, $b = 9.06$, $c = 11.00$, and $\beta = 91°15'$ (see Strieter et al.).]

Material and Equipment   Dimethyl sulfoxide (DMSO) (or methanol, formic acid, formamide, and so forth); liquid nitrogen (or Dry ice–acetone) cooling mixture; tank of compressed nitrogen gas; thin plastic, Pyrex, or Lindemann glass capillary tube, 0.3 to 1.0 mm. i.d. and 8 mm. long; x-ray developer; fixer; x-ray diffraction apparatus with copper target; cooling device; powder camera; Kodak No-Screen Medical x-ray Film $1\frac{3}{8}$ in. $\times$ 25 ft.; millimeter scale.

REFERENCES

1. L. V. Azaroff and M. J. Buerger. The Powder Method in Crystallography. McGraw-Hill Book Co., New York, 1958.
2. R. Rudman. J. Chem. Ed., **44**, 331 (1967).

3. R. Thomas, C. B. Shoemaker and K. Eriks. Acta Cryst., **21,** 12 (1966).
4. S. W. Petersen and H. A. Levy. Acta Cryst., **10,** 70 (1957).
5. A. S. Quist and H. S. Frank. J. Phys. Chem., **65,** 560 (1961).
6. K. J. Taner and W. N. Lipscomb. Acta Cryst., **5,** 606 (1952).
7. F. Holtzberger, B. Post and I. Fankuchen. Acta Cryst., **6,** 127 (1953).
8. J. Ladell and B. Post. Acta Cryst., **7,** 559 (1954).
9. R. E. Jones and D. H. Templeton. Acta Cryst., **11,** 484 (1958).
10. F. J. Strieter, D. H. Templeton, R. F. Shewerman and R. L. Sass. Acta Cryst., **15,** 1233 (1962).

23

DIFFERENTIAL THERMAL ANALYSIS; SECOND AND FIRST ORDER TRANSITIONS IN POLYSTYRENE; KINETICS OF PYROLYSIS

Differential thermal analysis (DTA) is a technique in which heat effects associated with physical changes (phase transitions) and chemical changes (heats of reaction) are monitored as a function of temperature.

Differential is the key word in the designation because, in essence, the difference between the temperature of a standard material and that of the sample is recorded as a function of time. Since the whole process is performed at a constant heating rate, the time scale is directly related to the furnace temperature. The temperature difference

$$\Delta T = T_{\text{sample}} - T_{\text{reference}} \tag{I}$$

can be either positive or negative. In an endothermic process, such as most phase transitions, dehydrations, reduction reactions, and some decomposition reactions, heat is absorbed in the process and, therefore, the temperature of the sample is lower than that of the reference material. Hence ΔT is negative.

In an exothermic process, such as crystallization, some cross-linking processes, oxidation reactions, and some decomposition reactions, the opposite is true and ΔT is positive.

During the heating of a sample, for example, from room temperature to its decomposition temperature, peaks with positive and negative ΔT may be recorded; each peak corresponds to a heat effect associated with a specific process (Fig. 23–1).

For example, the DTA curves for nylon in air show a small endothermic peak

224

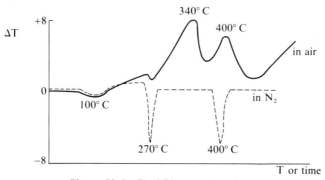

Figure 23-1   *The DTA curves for nylon.*

at 100° C. because of dehydration. The exothermic peaks at 340 and 400° C. are due to oxidation decomposition reactions. The same sample heated in nitrogen atmosphere shows the same endothermic peak because of water loss, but the melting process at 270° C. and the decomposition at 400° C. are endothermic.

The latter indicates the obvious—that thermal degradation in air and in nitrogen are different processes. The phase transition, due to melting in air, shows up only as a minimum in an exothermic process. This is because the exothermic oxidation reaction has started before the phase transition occurs; because the heat due to the exothermic oxidation reaction is greater than that absorbed in the endothermic melting process, the overall effect still is exothermic, but the endothermic peak will cause a minimum.

The question arises as to what kind of information is obtainable from a DTA curve. Obviously, the first and most direct information is the temperature at which a certain process occurs. For example, the melting point of a polymer, such as nylon, is easily obtained. The temperature at which a certain reaction (decomposition) may start is another important parameter; for example, it is 340° C. for nylon in a nitrogen atmosphere. The peak temperatures may be associated with the temperature at which maximum reaction rate occurs, although there is considerable disagreement on this point as we shall see later.

A special case in which the temperature of transformation is of great importance in polymers is the glass transition temperature, T_g. This is the temperature at which amorphous (noncrystalline) polymers are converted from a brittle, glasslike form to a rubbery, flexible form. This is not a true phase transition but one that involves a change in the local degrees of freedom. It can be visualized that above the glass transition temperature certain segmental motions of the polymer are comparatively unhindered by the interaction with neighboring chains. Below the glass transition temperature, such motions are hindered greatly, and the relaxation times associated with such hindered motions are usually long compared to the duration of the experiment.

The operative definition of glass transition temperature is that at this temperature, or within a few degrees, the specific heat, the coefficient of thermal expansion, the free volume, and the dielectric constant (in the case of a polar polymer) all change rapidly.

Since the mechanical behavior of polymers changes markedly at the glass transition temperature, it is an important characteristic of every polymer.

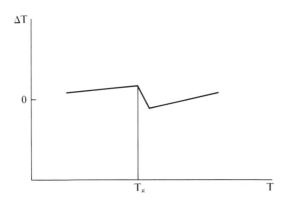

Figure 23–2 *Glass transition in DTA diagrams.*

In the DTA experiment, T_g is manifested by a drastic change in the base line, indicating a change in the specific heat of the polymer (Fig. 23–2). No enthalpy is associated with such transition (for which reason it is also called a second order transition); therefore, the effect in a DTA curve is slight and is observable only if the instrument is sensitive enough.

The second direct information obtainable from DTA curves should be the heat associated with certain processes. That means that the area under a peak in the DTA curve is *proportional* to the enthalpy of the specific process. The difficulty, however, is in evaluating the proportionality constant, K.

$$m \cdot \Delta H = K \int_{t_1}^{t_2} \Delta T \, dt \qquad (2)$$

where ΔH is the enthalpy change of the transformation, m is the mass of the sample, and $\int \Delta T \, dt$ is the area under the DTA curve.

DTA curves are not very reproducible even when using the same instrument. The peak temperatures and the shape and size of the bands are influenced by the instrumental and sample characteristics.

Among the instrumental factors affecting DTA curves are the shape and size of the furnace, the constitution and geometry of the sample holder, the heating rate, and the size and the location of the thermocouple.

Among the sample characteristics influencing DTA curves, the most important are the heat capacity, thermal conductivity, particle size and packing of sample, and the amount of sample.

Because of the multiple sample characteristics, direct calibration of a DTA instrument with materials of known heats of transformation yields only semiquantitative results. Somewhat better calibration can be achieved by producing DTA peaks supplying different amounts of electrical energy. Under such a calibration procedure, Ozawa et al. found that the proportionality constant, K, is

$$K = 2\pi l \lambda_h / \log \left(R_0 / R_i \right) \qquad (3)$$

where R_i and R_0 are the inner and outer radii of the cell holder, l is the length of the cell, and λ_h is the thermal conductivity of the cell holder. Since λ_h is a function of temperature, the proportionality constant is also a function of temperature, usually given in the form of a virial expression. Thus, if DTA curves are used to evaluate enthalpies, the calibration constant as a function of temperature has to be obtained, and even under such conditions the data will not be quantitative.

A third and more important item of information obtainable from DTA curves is related to the kinetics of pyrolytic reactions.

There are a number of theories of differential thermal analysis that can be used to analyze kinetics of pyrolytic reactions. In the following discussion a semiempirical, semitheoretical development is given that was advanced by Kissinger. This can be used because, in spite of the approximations made in the solution of the nonlinear partial differential equation, the final conclusions have been proved nearly correct by the more elaborate analysis of Akita and Kase.

The overall pyrolytic reaction can be described as

$$\text{solid} \rightarrow \text{solid} + \text{gas} \tag{4}$$

and the rate of reaction in general is given by

$$(dx/dt) = A(1 - x)^n e^{-\Delta E/RT} \tag{5}$$

In equation 5, x is the fraction of solid reacting; n is the order of the reaction, A is the preexponential factor of the Arrhenius equation, and ΔE is the energy of activation.

In most pyrolytic reactions that have been studied, the empirical order of reaction, n, is one or a fraction, $\frac{1}{2}$, $\frac{2}{3}$, and so forth, being the result of a combination of mechanistic steps. The most interesting feature of n is that it remains constant for the largest part of the reaction. Kissinger assured and later Akita and Kase proved by analysis that in an unstirred system the maximum reaction rate occurs at or nearly the temperature at which the DTA curve has a peak.

With this assumption, the kinetics of pyrolytic reactions is written as

$$\frac{\Delta E \phi}{RT_m{}^2} = A_n (1 - x)_m^{n-1} e^{-\Delta E/RT} \tag{6}$$

In equation 6, ϕ is the heating rate and $(1 - x)_m$ is the amount of material left unreacted when the peak in the DTA curve is reached. The latter is not readily available from differential thermal analysis, but it can be calculated if the integration of equation 5 is carried out by successive integration of parts, and all terms beyond the second term are neglected. Evaluating $(1 - x)_m$ and substituting the value into equation 6 and differentiating, the following relationship is obtained:

$$\frac{d\left[\ln\left(\dfrac{\phi}{T_m{}^2}\right)\right]}{d(1/T)} = -\Delta E/R \tag{7}$$

According to equation 7, DTA curves must be obtained at different heating rates, ϕ, and the temperature of the sample, T_m, must be recorded when the DTA curves reach their peaks. The $\ln (\phi/T_m{}^2)$ terms obtained with the different heating rates are plotted against the reciprocal of the furnace (reference material) temperature $(1/T)$. In many DTA diagrams, the T and the temperature of the furnace are monitored simultaneously (Fig. 23–3).

Such a plot (Fig. 23–4) yields the activation energy of the transformation in question.

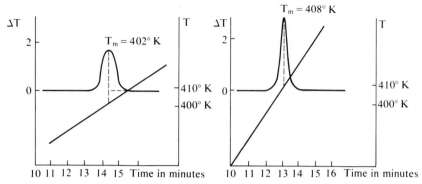

Figure 23–3 *The DTA curves and furnace temperature as a function of time at two heating rates.*

Moreover, acccording to Kissinger, the order of the reaction can be obtained from the shape and asymmetry of the DTA band. The shape index, S, is defined as

$$S = a/b \qquad (8)$$

in which the quantities a and b refer to Figure 23–5.

According to the empirical observation of Kissinger, the order of reaction, n, is related to the shape index, S, by

$$n = 1.26S^{1/2} \qquad (9)$$

Thus both ΔE and n may be evaluated from DTA curves.

Furthermore, once the order of reaction is known and found to be a fraction rather than 0 or 1, the following equation can be used to determine the amount of material left unreacted, $(1 - x)_m$, when the peak temperature, T_m, has been reached.

$$n(1 - x)_m^{n-1} = 1 + (n - 1)\frac{2RT_m}{\Delta E} \qquad (10)$$

Knowing $(1 - x)_m$ and substituting it into equation 6, one can determine the preexponential factor, A.

For the special case in which the order of the reaction is unity $(n = 1)$, the relationship between the amount of material unreacted $(1 - x)$, ΔE, and n at any T is given by

$$\ln\left(\frac{1}{1 - x}\right) = \frac{ART^2}{\Delta E\phi}e^{-\Delta E/RT}\left(1 - \frac{2RT}{\Delta E}\right) \qquad (11)$$

Thus again, A, the preexponential factor, can be obtained by simultaneously solving two equations of the type of equation 11 with two different heating rates, ϕ.

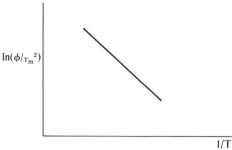

Figure 23–4 *The Kissinger plot of DTA data.*

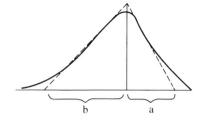

Figure 23–5 *Method of measuring the asymmetry of a DTA band.*

b   a

Experimental

The DTA curves of polystyrene will be obtained at different heating rates. Finely powdered polystyrene samples will be run against a reference material, alumina, Al_2O_3, usually supplied with commercial DTA apparatus.

The preparation of the sample should be as follows: Grind about 1 gram of polystyrene sample into a 100 mesh powder and store it in a desiccator. Similarly, grind 1 to 2 grams of alumina (reference) to a 100 mesh powder, heat it to 1000° C. in a muffle furnace for several hours, and store it in a desiccator for use in the instrument. To save time the alumina powder should be prepared in advance by the instructional staff.

The DTA apparatus usually consists of four separate units: (a) A *furnace* is used to heat a sample block to high temperatures of 1200° C. or more. (b) The furnace is controlled by a *programmer* that provides a specific power input to the furnace so that different heating rates may be employed

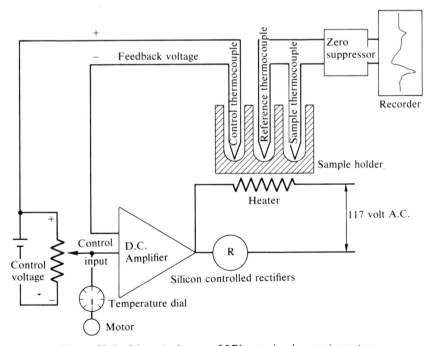

Figure 23–6 *Schematic diagram of DTA control and measuring system.*

(i.e., from 0.5 to 25° C./minute). (c) The *sample holder* is a metal block with insulated handle that fits into the top of the furnace. The cylindrical metal block has six to eight wells to hold quartz crucibles for sample reference material and for control thermocouple. (d) The thermocouples in the sample and reference material are connected through a *zero suppressor* to a 1 mv. recorder. Only the differential temperature is recorded. The zero suppressor enables the operator to shift the recording pen to the center of the chart paper when zero output exists (i.e., when there is no temperature difference between sample and reference material).

Thus, ΔT is recorded as a function of time (chart speed). However, for DTA curves the ΔT is needed as a function of T. Hence, calibration curves have to be established for each *heating rate* in which the T is recorded as a function of time. These calibration curves are to be supplied to the student by the instructor because the laboratory times does not suffice to run both calibrations and experiments.

[The calibration curves should be run at a chart speed of 12 inches/hour. All the wells should be filled with crucibles containing alumina. The *marked junction* of the calibration thermocouple is inserted in the third well (where the actual sample will be later), and the *free junction* of the calibration thermocouple is inserted in a Dewar flask containing ice and water at 0° C. The calibration curves are run at 10, 20, and 25° C./minute rates up to 600° C. each.]

In obtaining the DTA curves of the polystyrene samples the instructions in the instrument manual must be followed strictly. A more detailed account of the electronic components of the apparatus may be found in Part XIV of this book.

About 100 mg. of polystyrene sample is used in one crucible and about the same amount of alumina is used in each of the seven reference crucibles.

Determine the DTA curve of the polystyrene sample in nitrogen at a 20° C./minute heating rate and finally at a 25° C./minute heating rate. Also be sure the length, l, and the outside radius of the metal block, R_0, are measured. The radius of the quartz crucible R_i, should also be measured.

From the DTA curves obtain the glass transition temperature (see Fig. 23–2), the melting point, and the temperatures at maximum reaction rates, T_m, of each peak.

Estimate the enthalpy association with each process indicated by the area under the different peaks using equations 2 and 3. The dimensions of the block, l and R_0, and R_i, the radius of the crucible, were obtained. The thermal conductivity of the block should be obtained from the *Handbook of Thermo-physical Properties of Solid Materials* (reference 6). In the case of a Fisher 260 DTA apparatus the metal block is Inconel (77 percent nickel, 15 percent chromium, and 7 percent iron alloy), and its thermal conductivity on the average over the range of 200 to 600° C. is 0.035 cal./sec. cm. °C.

From equation 7 (or from a plot similar to Fig. 23–4) obtain the energy of activation associated with each process corresponding to a distinct DTA peak.

From the shape of each DTA peak, obtain the order of reaction by using equations 8 and 9.

Calculate the preexponential factor, A, for each process from equations 6 and 10 or 6 and 11.

Knowing T_m, ΔE, n, and A corresponding to each DTA peak, speculate on the different steps in the pyrolytic decomposition of polystyrene.

Material and Equipment. 1 gram of polystyrene; 1 to 2 grams of alumina; tank of compressed nitrogen gas; differential thermoanalyser unit; grinding mill; muffle furnace; desiccator; Dewar flask; ice.

REFERENCES

1. W. W. Wendlandt. Thermal Methods of Analysis. Interscience Publishers, Inc., New York, 1964.
2. P. E. Slade, Jr., and L. T. Jenkins (eds.). Techniques and methods of polymer evaluation. In Thermal Analysis, Vol. I, Marcell Dekker, Inc., New York, 1966.
3. T. Ozawa, M. Momota and I. Isozaki. Bull. Chem. Soc. Japan, **40**, 1583 (1967).
4. H. E. Kissinger. Analyt. Chem., **29**, 1702 (1957).
5. K. Akita and M. Kase. J. Phys. Chem., **72**, 906 (1968).
6. A. Goldsmith, T. E. Waterman and H. J. Hirschhorn. Handbook of Thermo-physical Properties of Solid Materials. The Macmillan Co., New York, 1961.

VIII

PHASE EQUILIBRIA

24

BOILING POINT-
COMPOSITION DIAGRAMS OF
BINARY SYSTEMS*

According to the phase rule, two degrees of freedom exist in a binary system in which two phases (liquid-vapor) are in equilibrium.

$$\phi = C - P + 2 = 2 - 2 + 2 = 2 \tag{1}$$

If one external variable (the pressure) is kept constant, the binary system can be described in terms of boiling point temperature and composition. The composition is usually given in terms of: (a) mole fraction (N_A)

$$N_A = \frac{n_A}{n_A + n_B} \tag{2}$$

where n_A and n_B are the number of moles of components A and B; (b) weight fraction (X_A)

$$X_A = \frac{W_A}{W_A + W_B} \tag{3}$$

where W_A and W_B are the masses of A and B respectively; or (c) volume fraction (ϕ_A)

$$\phi_A = V_A/V_T \tag{4}$$

where V_A and V_B refer to the volumes of A and B and V_T the total volume of the mixture,

$$V_T = V_A + V_B \tag{5}$$

One can apply equation 5 only to ideal solutions. In all other cases the final volume of the mixture has to be determined.

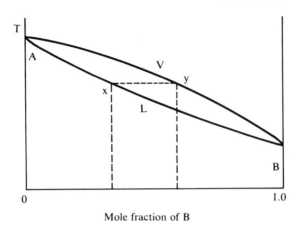

Figure 24-1 *Boiling point–composition diagram.*

In the distillation of binary liquid mixtures, the behavior of the system can be illustrated graphically in terms of boiling point–composition curves. They express the relation between the boiling temperature and the composition of the liquid and vapor in equilibrium at a specified pressure (usually 1 atm.).

The curves for a binary liquid pair in which the boiling points of all the mixtures are intermediate between those of the two components are shown in Figure 24–1. The curve *ALB* is the *liquidus* curve, relating the boiling point to the composition of the liquid mixture. The composition of the vapor for this system is given by *AVB*. To find the composition of the vapor in equilibrium with a definite liquid mixture, a tie line is drawn from the boiling point curve parallel to the abscissa until it intersects the vapor composition curve. Thus, as shown in Figure 24–1, a liquid of composition *x* develops a vapor of composition *y*. A knowledge of boiling point–vapor composition curves is of great value in separating liquids by fractional distillation.

In addition to the system illustrated in Figure 24–1, mixtures exist that have a higher or lower boiling point than either of the components (Figures 24–2 and 24–3). Such mixtures have a maximum or a minimum point in their boiling point curves

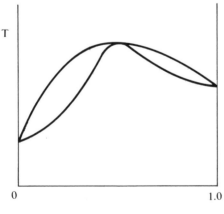

Figure 24-2 *A boiling point–composition diagram for a mixture having a maximum boiling point azeotrope.*

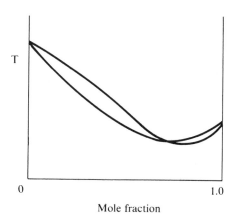

Figure 24-3 *A boiling point–composition diagram for a mixture having a minimum boiling point azeotrope.*

at which the liquidus and vapor curves contact tangentially. The liquid and vapor then have the same composition, and at this point the distillation proceeds with no further change in composition or temperature. The mixture at this composition has an *azeotropic* composition, and is known as the azeotropic mixture (or solution) of that system.

Examples of systems having azeotropes and containing water are the liquid pairs, nitric acid–water, hydrochloric acid–water, hydrogen bromide–water, and formic acid–water. Some nonaqueous systems possessing azeotropes are chloroform–acetone and pyridine–formic acid. In the present experiment the dimethylsulfoxide (DMSO)–water system (or a system of water and dimethylformamide [DMF], acetone, dioxane, and so forth) is investigated in order to determine the boiling point–vapor composition curve.

The method, as employed by various authors, consists of boiling a sample of liquid in a suitable distilling apparatus and removing the vapor by total condensation in a receiver from which the overflow returns continuously to the still.

Experimental

The distillation apparatus consists of a 200 ml. three neck flask. In the middle neck place a thermometer (0 to 200° C.) so that the mercury bulb extends below the neck. Connect a condenser to the distillation flask through one of the side openings (Fig. 24–4). The lower end of the condenser should have a collection tube of about 10 ml. capacity from which the distilled sample can be removed through a stopcock. The third neck should contain a sampler through which the residue of the distillation can be withdrawn and new material can be added to the flask. Pumice or glass beads should be added to the flask to prevent "bumping" during distillation. All connections in the apparatus should be glass connections and no lubricant grease should be used. A heating mantle, which can be controlled with a Variac, is attached to the lower part of the distillation flask. Glass wool

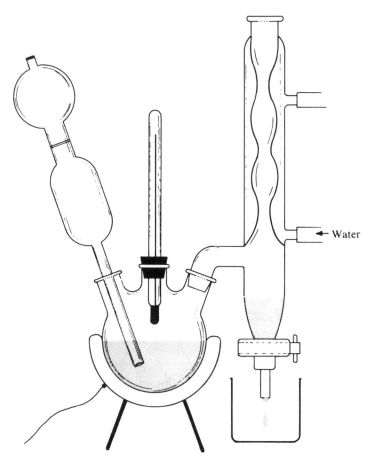

Figure 24-4   *Distillation apparatus.*

insulation should be provided for the upper part of the distillation flask. Place ten test tubes marked D_1 to D_{10} into a rack for collection of the distillates. Similarly prepare ten test tubes marked R_1 to R_{10} for the collected residue samples.

Place 75 ml. of water and 25 ml. of DMSO in the empty flask. Start the distillation and continue it until equilibrium is reached; i.e., the boiling point temperature has become constant. Reject the first 4 or 5 ml. of distillate. Now record the temperature. Turn off the heater and leave it off during the sampling operations.

Withdraw a 5 ml. sample of distillate from the collecting bulb and place it in the test tube marked D_1. Securely stopper the collected sample with a rubber stopper and place it in a 20° C. bath.

Withdraw through the sampler a 5 ml. portion of the residue from the distilling flask and place it in the test tube marked R_1. Securely stopper the tube and place it in a 20° C. bath.

Repeat the whole process five times, adding 25 ml. of DMSO to the flask each time after removal of the distillate-residue (D and R) sample pairs.

Repeat the same process starting with 95 ml. of DMSO and 5 ml. of water and collect the liquidus and vapor samples. Next add 10 ml. of water to the mixture and establish the boiling point and collect the samples. Repeat the last process three more times, each time adding 10 ml. of water to the mixture.

You have now collected ten liquidus and ten condensed vapor samples (five from samples with high DMSO concentration and five from samples with high water compositions) and have recorded the corresponding boiling points.

[When a system other than DMSO-water is used, such as dimethyl-formamide (DMF)-water acetone-water, or dioxane-water, the same approach can be adopted; five samples each of liquid and vapor are collected, starting with a high concentration of water in the distillation mixture and subsequently increasing the concentration of the organic component. Subsequently, five liquid and five vapor samples are collected, starting with a distillation mixture having about 0.98 mole fraction of organic component; every time, after taking samples, dilute the distillation mixture with water to reach a midpoint.]

Prepare ten DMSO-water mixtures (0.1 to 0.96 mole fraction of DMSO). Measure the refractive indices of these ten solutions as well as those of pure DMSO and pure water at 25° C. Establish the standardization curve by plotting the refractive index vs. composition. Measure the refractive indices of the ten liquidus and the ten condensed vapor samples obtained in the distillation process. The refractometer should be thermostated at 25° C.

(If instead of the DMSO-water system another binary mixture is used, a standardization curve of refractive index vs. composition must be established for the system under investigation.) The data obtained in Experiment 14 may be used for this purpose.

Using your standardization curve, plot the boiling points against the corresponding compositions for the liquidus and vapor samples.

Does the system investigated form an azeotrope?

What are the characteristic features of this normal boiling point–composition diagram?

Material and Equipment. 500 ml. dimethyl sulfoxide (DMSO) (or dimethylformamide [DMF], acetone, dioxane, and so forth); glass beads or boiling chips; glass wool; 200 ml. three neck distillation flask with glass joints; condenser; 10 ml. collection tube; thermometer (0 to 200° C.) with glass joints; sample withdrawing flask; 20 test tubes of 10 ml. capacity; test tube rack; 20 rubber stoppers; heating mantle; Variac; thermostat with heater, stirrer, and thermoregulator; circulating pump; refractometer.

REFERENCES

1. E. Hutchinson. Physical Chemistry. W. B. Saunders Co., Philadelphia, 1962, p. 123–130.
2. G. M. Barrow. Physical Chemistry. 2nd Edition, McGraw-Hill Book Co., Inc., New York, 1966, p. 602–607.
3. W. J. Moore. Physical Chemistry. 3rd Edition, Prentice-Hall, Inc., Englewood Cliffs, N.J., 1962, p. 128–130 and 138–141.
4. L. H. Horsley. Azeotropic Data I and II. Adv. Chem. Ser., **6,** (1952) and **35,** (1962).
5. J. Timmermans. Physico-chemical Constants of Binary Systems in Concentrated Solutions. Vol. II and IV., Interscience Publishers, Inc., New York, 1960.
6. J. S. Rowlinson. Liquids and Liquid Mixtures. 2nd Edition, Plenum Press, New York, 1969.

25

COOLING CURVES; PHASE DIAGRAM OF CONDENSED BINARY SYSTEMS*

Phase diagrams that are obtained in open air without regard to the composition of the vapor phase are called diagrams of condensed systems. Under these conditions, the vapor phase is not considered as one of the phases although, obviously, it is in equilibrium with the rest of the system.

Phase diagrams of condensed phases are plotted as temperature vs. composition. Such a diagram for a *simple eutectic system* is given in Figure 25–1A. The two pure components, *A* and *B*, have respective melting points at *M* and *N*. These melting points or freezing points can be determined visually.

However, in most cases one obtains them from cooling curves. When a pure substance is cooled from the liquid to the solid state at a regulated speed, the temperature remains constant during most of the freezing. When the temperature of the system is monitored as a function of time, the freezing appears as a horizontal portion ("halt") of the cooling curve, as in curve I, Figure 25–1B. The small dip that usually precedes the horizontal part represents minor supercooling. If the temperature is read visually from a thermometer at specific intervals, this dip may be missed. If the temperature change is monitored by a recorder, the supercooling effect is evident. The temperature corresponding to the flat portion of the cooling curve is the melting or freezing point of the pure substance.

The cooling curve of a mixture of, let us say, 80 percent component *A* and 20 percent component *B* may look similar to curve II in Figure 25–1B. Instead of long horizontal portions one obtains either short horizontal portions or none at all; then the phase transition may be inferred from the change in the slope ("break") of the cooling curve. In cooling curve II in Figure 25–1B two such phase transitions

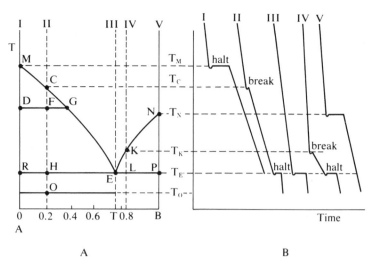

Figure 25–I A, solid-liquid phase diagram for a simple eutectic system and, B, the corresponding cooling curves.

are observable—one at T_C and the other at T_H. The cooling process in the phase diagram, Figure 25–1A, corresponds to a downward movement along II at a specific composition (20 percent B). Similarly, the cooling curve for a composition of 80 percent B and 20 percent A (cooling curve IV) shows two phase transitions—one at T_K and the other at T_L.

One can obtain a series of cooling curves for a simple eutectic system similar to II and IV at different compositions, each of which shows two phase transitions. The temperature at one of those phase transitions varies with composition, but the other is the same at all compositions. This latter temperature is the freezing point of the eutectic mixture (T_H, T_E, T_L). The composition of the simple eutectic mixture is indicated by point E at which the two liquid composition curves ($MCGE$ and NKE) intersect. The cooling curve of a mixture with eutectic composition shows only one phase transition, as is illustrated by curve III in Figure 25–1B.

The phase diagram of a simple eutectic system is thus obtained by connecting the points indicating the temperatures of the phase transitions at different compositions. In Figure 25–1A we can distinguish five different domains. Above the line $MCGEKN$ only a liquid phase exists. In the domain defined by $MCGEHR$, pure substance A is in equilibrium with saturated solutions of A of different concentrations. In the domain $NKELP$ pure solid B exists in equilibrium with saturated solutions of B of different concentrations. In the domain $RHETA$ pure solid A is in equilibrium with solid eutectic (composition T), and in the rectangle of $ELPBT$ pure solid B coexists with solid eutectic mixture.

In any domain in which two phases are in equilibrium, tie lines such as DFG may be drawn. For example, if we begin with a liquid mixture of 80 percent A and 20 percent B and we cool it, we shall proceed along line II. At a temperature corresponding to T_C the first crystals of A appear. Cooling the mixture further, we reach temperature T_F. We now have pure solid A in equilibrium with a liquid of composition G. The proportion of solid A/solution $G = FG/DF$ at this temperature.

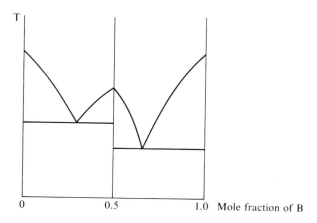

Figure 25-2 *Freezing point diagram showing formation of a 1:1 compound, AB.*

Upon further cooling we reach temperature T_H. Now both phases are solid: pure crystal A and solid eutectic mixture T. The proportion of A to T equals HE/RH. We can cool the two solid phases further, for example, to point T_O, but neither the composition nor the proportion of the two phases will change any more.

If compounds A and B are capable of forming a compound, AB, one obtains a freezing point diagram as given in Figure 25-2. Such a diagram is simply the combination of two simple eutectic diagrams and can be handled as such. More complicated freezing point diagrams result from the combination of three simple eutectic diagrams, such as in the case of sulfuric acid and water, which forms three compounds: mono-, di-, and tetra-hydrate.

Other freezing point diagrams, such as the one given in Figure 25-3, may show domains with solid solutions. The domain $ACHK$ contains solid A with dissolved B in it (solid solution), and the area $DFLB$ contains solid B with dissolved A in it (solid solution). Area CHE represents solid A in equilibrium with liquid; conversely, area EDF represents solid B in equilibrium with liquid. The domain $KHEFL$ contains solid A saturated with B and solid B saturated with A.

A cooling curve for the composition 90 percent A and 10 percent B has two breaks but there is no halt (Fig. 25-4).

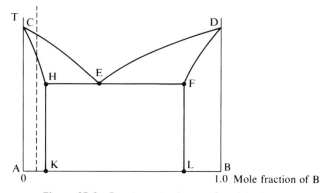

Figure 25-3 *Freezing point diagram for solid solutions.*

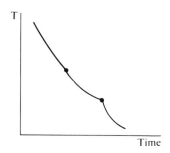

Figure 25–4 *Cooling curve for a mixture forming a solid solution.*

Experimental

The cooling curves for distilled water, pure dimethyl sulfoxide (DMSO) (or dimethylformamide [DMF], acetone, dioxane, and so forth) and seven mixtures of DMSO-water (or DMF-water, acetone-water, dioxane-water, and so forth) will be determined. The mixture should cover the concentration range of 0.1 to 0.9 mole fraction. The pure liquids and the liquid mixtures will be cooled to their freezing points by using liquid nitrogen in the apparatus shown in Figure 25–5.

In its simplest form the time-temperature freezing point apparatus consists of a Dewar flask, A, which is three-fourths filled with liquid nitrogen, B. Into the liquid nitrogen dips a double-walled test tube, D, about 2.5 cm. in diameter and about 30 cm. long. This test tube is filled to a depth of 10 cm. with the material, M. The double wall serves an insulating purpose, and in its simplest form this may be made of two test tubes, one inside the other

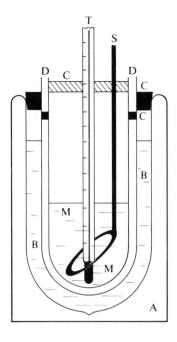

Figure 25–5 *Time-temperature freezing point apparatus.*

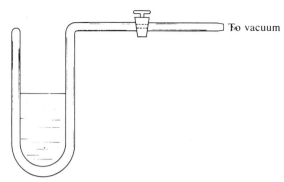

Figure 25–6 *Evacuated double-walled flask.*

with 2 to 3 mm. of air space between them. If a slower cooling rate is desired, the double-walled test tube should be sealed and connected to a vacuum pump, as shown in Figure 25–6.

With the latter setup a slow cooling rate is obtained when the double-walled flask is evacuated; when the test tube is to be warmed to melt a previously used sample and to clean the test tube, atmospheric pressure can be admitted between the walls.

A number of devices may be used to monitor the temperature. The simplest is a calibrated ethanol thermometer, T in Figure 25–5, that registers as low as $-110°$ C. A stirrer, S, is in order to maintain thermal equilibrium throughout the sample. The thermometer and the stirrer in the test tube, as well as the test tube in the Dewar flask, are positioned by cork stoppers, C. In its simplest form the stirrer can be operated manually by moving a glass loop up and down around the thermometer. Alternatively, a motor driven stirrer may be employed, as in Figure 25–7.

However, care has to be taken to ensure that the stirrer has a 1 to 2 mm. clearance from the walls to the test tube, because crystals that start to accumulate at the walls may make stirring difficult. If a motor driven stirrer is

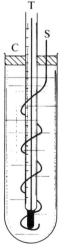

Figure 25–7 *Spiral motor driven stirrer.*

used, it should be hooked to a Variac, since, with cooling, the viscosity of the liquids increases and more power is needed to maintain the same stirring rate.

With the simplest equipment, as in Figure 25–1, the cooling curves should be obtained as follows:

Place the sample into the clean test tube, and position the thermometer and the stirrer in the test tube with a cork. If the double wall is sealed, connect the flask to a vacuum system and evacuate it. Close the stopcock. Position the test tube in liquid nitrogen in the Dewar flask with a cork. Take temperature readings every five seconds while maintaining a constant stirring rate (manually or by motor). Plot the cooling curve as temperature vs. time. After the entire sample is frozen, remove the test tube from the Dewar flask; allow atmospheric pressure to enter into the double wall; melt the sample by allowing it to reach room temperature in a water bath. Remove the thermometer and stirrer. Clean the test tube and the thermometer-stirrer assembly; thoroughly rinse them with distilled water and, finally, with absolute ethanol, and dry them. Repeat the process with a new sample.

If a calibrated ethanol thermometer is not available or if it is inconvenient to read it, a thermocouple may be employed. The most convenient thermocouple for low temperature measurements is made of constantan and copper.

Constantan is a copper-nickel alloy (60 and 40 percent) and when connected to copper wire it generates a thermoelectric potential of approximately 0.03 mv./degree between $-200°$ C. and room temperature. This potential is generated as a result of an electron transfer that takes place when two dissimilar metals are placed in contact with each other. This transfer creates an electric double layer at the junction. The potential of this double layer is measured with reference to a standard potential. The standard potential is the thermoelectric potential generated by the same two wires at a standard temperature, usually $0°$ C. The potential thus is measured between the *ends* of the copper wires—one coming from the sample, the other from the reference ice-water temperature bath. The constantan wires complete the circuit (see Fig. 25–8).

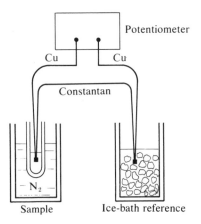

Sample   Ice-bath reference   **Figure 25–8**   *The thermocouple circuit.*

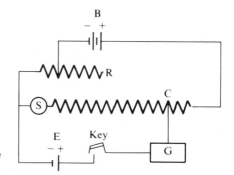

Figure 25–9 *Diagram of a simple potentiometer.*

The advantage of connecting the copper wires to the potentiometer is that the binding posts or junctions of the potentiometer are usually also copper; hence, no thermoelectric potential will be generated at the sites of the connection. The constantan-copper thermocouple is useful between −200 and +300° C., and the *Handbook of Chemistry and Physics, International Critical Table*, and similar references provide tables for converting the measured potentials into temperatures. However, it is advisable for the student to check the calibration with at least two well known temperatures (for example, the normal boiling point of liquid nitrogen, −195.8° C., and of water, 100° C.).

For medium precision measurements of the thermoelectric potential, a potentiometer such as the Leeds and Northrup Student Potentiometer or 8686 Millivolt Potentiometer may be used. These potentiometers, if set on an 0.016 volt scale, can be read with ±0.01 mv. accuracy; with a constantan-copper thermocouple the error is about ±0.5° C. For very precise work a more expensive potentiometer has to be used, one that can be read to within 0.1 μv. Such potentiometers (Brooks-Harris) have a limit of error of 0.01° C., but temperature differences of 0.002° can be measured.

The principle of a potentiometer is illustrated in Figure 25–9.

A storage cell battery, *B*, is connected to the variable resistance, *R*. The sample is connected to the variable resistance with a uniform wire. A constant drop in the potential is maintained over this wire, and the difference in potential between any two points on the wire is proportional to the distance between the two points, *SC*. The emf across the slide wire, *SC*, is compared to a standard potential. For example, a standard cadmium cell, *E*, is used that has an emf of 1.0184 absolute volts at 25° C. and a temperature coefficient of 4×10^{-5} volts degree^{-1}. The negative pole of the standard cell is connected to the negative end of the slide wire. Its positive pole is connected to the positive end of the slide wire at *C* through a tapping key and a galvanometer. The slider, *C*, is adjusted to correspond to the known potential of the standard cell and *R* is adjusted until no galvanometer deflection is observed when the tapping key is depressed.

The description of the more complicated 8686 Leeds and Northrup Millivolt Potentiometer is taken from the Leeds and Northrup manual:

[The 8686 Millivolt Potentiometer is used for precision checking of thermocouple pyrometers. It includes a built-in lamp-and-scale galvanometer, standard cell], and batteries and has provision for recalibrating and internal standard cell voltage. The basic instrument consists of two selector switches and a slide wire, all arranged so that the value of voltage appears at a central readout window as a single row of digits, including scale reading. There is also a switch for selecting the function to be performed. A more detailed account of the electronic components of potentiometers may be found in Part XIV of this book.

This instrument is basically a double branch potentiometer. The main branch consists of nine steps of 10 mv. each, plus the 1010 mv. step for measuring standard cells, all regulated by a step switch. The reference junction compensator also is connected to this branch.

The secondary branch has nine steps of 1 mv. each (controlled by another switch), plus the 1.1 mv. slide wire, coupled in a Kelvin-Varley arrangement.

The −10.1 to 0 mv. range is obtained by turning the switch in the main branch circuit to the negative position and thus reversing the secondary branch circuit.

The instrument is designed for a current of 4.5 milliamperes in the potentiometer circuit, with 4.0 milliamperes passing through the main branch and 0.5 milliampere through the secondary branch.

The reference junction compensator is applied to the emf that is equivalent to the temperature of the reference junction (determined from conversion tables), in opposition to the thermocouple emf.

The voltage drop between the contact on the voltage divider and a fixed point is applied in opposition to the standard cell voltage for standardization of the current in the potentiometer circuit. If the value of this current is correct, the voltage drop balances the standard cell voltage and no adjustment is necessary. However, if the current is not correct, the unbalance that occurs is corrected by adjusting the coarse and fine battery rheostats.

Therefore, in operation the potentiometer circuit current is standardized by adjusting the battery rheostats so that the galvanometer registers zero. The reference junction temperature (0° C.) is maintained in a crushed ice bath in a Dewar flask.

The temperature readings are obtained when the galvanometer is balanced at zero by adjusting the selector switches stepwise and finally the slide wire. The value at the index of the readout window is the value of millivolts being measured. Furthermore, when the switch in the main branch is set on the negative position, the reading on the instrument is negative.]

Thus, the temperature at specific intervals may be measured by using the thermocouple setup, as in Figure 25–8, in connection with a potentiometer, as in Figure 25–9, and reading the emf every five to ten seconds.

Finally, since the reading of the emf at short intervals is cumbersome, the output of the potentiometer may be connected to a recorder. However, temperature drops in DMSO-water and similar systems occur in the range of 100° C. This range has to be recorded continuously on a time scale without referring to different temperature standards during the experiment. Because of the limitation of the recorder, the whole 100° C. range has to be fitted onto the paper with subsequent loss in the accuracy of reading the temperature.

After obtaining the cooling curves of all nine samples, construct a freezing point phase diagram for the chemical system investigated.

Do DMSO (or DMF, acetone, dioxane, and so forth) and water form a compound? Do they form a solid solution? Do they have a eutectic mixture?

Material and Equipment. Dimethyl sulfoxide (DMSO) (or dimethylformamide [DMF], acetone, dioxane, and so forth); tank of liquid nitrogen; 500 ml. Dewar flask; double-walled test tube 30 cm. long and 2.5 cm. in diameter; vacuum pump; hand or motor driven stirrer; Variac; calibrated ethanol thermometer (or thermocouple with potentiometer and possibly a recorder and reference ice bath); cork stoppers; absolute ethanol.

REFERENCES

1. J. E. Ricci. The Phase Rule. D. Van Nostrand Co., Inc., Princeton, N.J., 1951.
2. J. Reilly and W. N. Rae. Physico-chemical Methods. 5th Edition, Vol. I, Mathiesen & Co., London, 1954.
3. D. H. Rasmussen and A. P. MacKenzie. Nature, **220**, 1315 (1968).

26

PHASE DIAGRAM OF A THREE-COMPONENT PARTIALLY IMMISCIBLE LIQUID SYSTEM*

The simplest representation of a three-component system is one in which a liquid system separates into two phases. Such a system has two domains: (a) a domain of perfect miscibility and the presence of only one phase separated from (b) a domain in which two immiscible liquid phases are in equilibrium.

The Gibbs phase rule governs the equilibrium conditions.

$$\phi = C - P + 2 \tag{I}$$

where ϕ is the degrees of freedom, P is the number of phases, and C is the number of components. Therefore, if the two external variables (temperature and pressure) are fixed, one degree of freedom exists when two phases are present, whereas two degrees of freedom exist when there is perfect miscibility. These degrees of freedom are the variables that describe the composition of the system.

The behavior of a ternary system at constant temperature and pressure can be presented in a diagram with three co-ordinates (Fig. 26–1). In order to plot the composition of the system let us assume that there is 20 percent component A, 50 percent component B, and 30 percent component C. The three corners of the triple co-ordinate graph represent points at which we have either 100 percent component A, 100 percent component B, or 100 percent component C. Moving on the line from C to A in Figure 26–1, we trace the positions of a binary system (components A and C) from pure C to increasing concentrations of A, and finally to pure A. Similarly, base line A-B represents the variable composition of binary mixtures of A and B. The line B-C represents the variable composition of the binary system of components B and C.

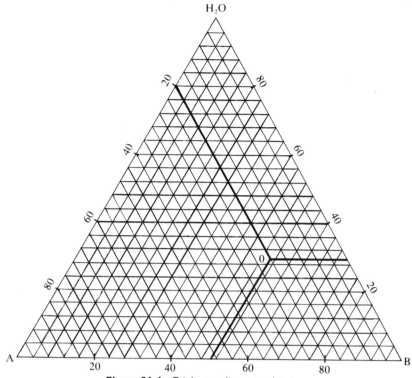

Figure 26–1 *Triple co-ordinate graph paper.*

Now, let us return to our ternary mixture. The starting point is the 20 percent mark on the *C-A* line. This point would indicate a composition of 20 percent *A* and 80 percent *C* in a binary system. However, this is not the case in a ternary system. The point corresponding to 20 percent *A* composition in any ternary mixture will be on a line that originates at this point and runs parallel to the *C-B* line (that is, to the side of the triangle opposite the *A* corner).

Similarly, we locate the 50 percent mark on the *A-B* base line. Again, this corresponds to a composition of 50 percent *B* and 50 percent *A* in a binary mixture. Other mixtures with 50 percent *B* but variable amounts of *A* and *C* are indicated by points on a line originating at this point and running parallel to the *A-C* axis (i.e., parallel to the side of the triangle opposite the *B* corner).

Now the line representing 20 percent *A* and that representing 50 percent *B* intersect at point *O*, which indicates the total composition since *A* and *B* are given in percentages and the composition of *C* is automatically a fixed value. However, to further prove that the *O* is also on the line representing 30 percent *C*, let us find the point corresponding to 30 percent *C* on the *B-C* axis. Using this as a starting point, we draw a line parallel to the *A-B* axis (the side of the triangle opposite the *C* corner). We see that this line also intersects the other two lines at point *O*.

Thus, the composition of a ternary system may be described by one point in a triple co-ordinate diagram.

The phase diagram of a ternary liquid system separating into two phases is

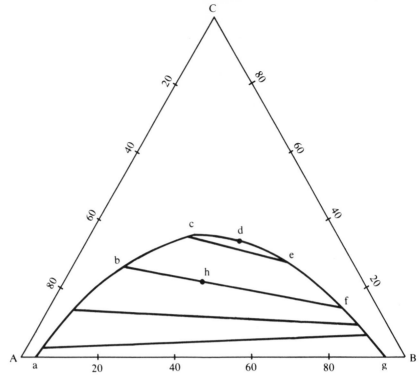

Figure 26-2 *Phase diagram of a ternary system with two immiscible liquids, A and B.*

given in Figure 26-2. The points on the dome (curve *abcdefg*) represent the compositions at which the two phases separate. Above this dome only a single phase, hence complete miscibility, exists; at a composition described by a point under this dome, the system will separate into two phases. The *a* and *g* positions in Figure 26-2 indicate that there is slight miscibility between components *A* and *B*. If no miscibility existed between components *A* and *B*, the *a* position would coincide with corner *A* and the *g* position with corner *B*. The diagram also indicates that it is the third component, *C*, that is really miscible with either *A* or *B* in all proportions. In such a phase diagram the tie lines have a very important aspect: They connect the concentrations of the two phases experimentally found to be in equilibrium with each other. For instance, when a mixture with composition *h* (Fig. 26-2) is prepared it separates into two phases. Phase one, rich in *A* and *C* and poor in *B*, has the composition designated on the diagram by *b*. Phase two, rich in *B* and *C* but poor in *A*, has a composition designated by point *f*. The quantitative ratio of the two phases is given by

$$\frac{\text{Phase one}}{\text{Phase two}} = \frac{fh}{bh} \qquad (2)$$

and, therefore, once a phase diagram is available it can be used to determine the compositions and proportions of the phases that would result when a mixture of specified overall composition is prepared.

We may notice that the dome in Figure 26-2 is not symmetrical and the tie lines are not parallel to each other. This is simply because the solubility of *C* in the two

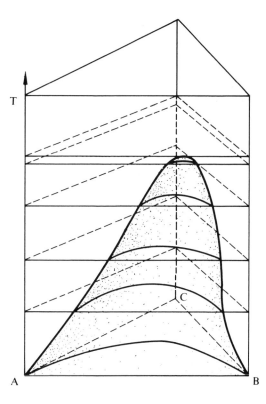

Figure 26–3 *Temperature dependent phase diagram of a three-component system.*

phases (*A* and *B*) is not the same. In whatever direction the tie lines are slanted they connect points of equilibrium compositions. These equilibrium compositions, *b* vs. *f* and *c* vs. *e*, become increasingly similar with each subsequent tie line, starting from the base of the dome and proceeding upward. Similarly, the tie lines become shorter and finally converge to a point, *d*, at which the two phases, which separate, have identical composition. This is called the isothermal critical point or the plait point.

Obviously, the external variables temperature and pressure also influence the phase diagram. If one stacks up isothermal and isobaric phase diagrams (planes) representing different temperatures (let us say, along a temperature axis), a three-dimensional representation of the phase diagrams will be achieved, as shown in Figure 26–3.

Experimental

The phase diagram of a dimethyl sulfoxide–water–benzene system will be obtained. Other ternary systems may be used in which dimethylformamide, formamide, dioxane, acetone, and so forth, replace dimethyl sulfoxide (DMSO). Therefore, in the following instructions the proper organic compound may be substituted for DMSO.

In order to obtain the points on the miscibility curve dome one has to

TABLE 26–1.

BOTTLE NO.	COMPOSITION AT THE END OF TITRATION			REFRACTIVE INDEX OF MIXTURE
	DMSO (ml.)	C_6H_6 (ml.)	H_2O (ml.)	
1	15	0.2	15.5	1.4122
2	..	...	....	......
.				
.				
.				
etc.				

titrate different mixtures of DMSO-benzene with water to the point at which cloudiness appears. This can be accomplished in the following manner: Stopper and number 10 to 15 Erlenmeyer flasks of 100 ml. capacity. From two burets (one containing DMSO and one benzene) prepare mixtures of DMSO and benzene in different proportions. For example, in the first bottle add 15 ml. of DMSO and 0.2 ml. of benzene, in the second bottle add 14 ml. of DMSO and 1 ml. of benzene, and so forth; in the last bottle, combine 1 ml. of DMSO and 14 ml. of benzene.

Titrate each bottle with water from a third buret to the point of slight turbidity. Be very careful because any overtitration will result in separation of the phases.

Measure the refractive index of each titrated mixture. Prepare a table of your results such as Table 26–1.

Since the phase diagram to be drawn will represent an isothermal equilibrium, the following precautions must be observed. After the original mixtures of DMSO and benzene are prepared, they should be thermostated at 25° C. The titration must be done slowly. From previous experiments on the heat of mixing the student should be aware that the DMSO-water reaction is exothermic. Therefore, the water should be added very slowly, with constant swirling of the Erlenmeyer flask in the thermostat, so that thermal equilibrium is achieved before the end point is reached.

Determine the density of each of the components at 25° C.

After the end point of the titration has been reached and *the refractive index of each mixture has been measured*, add a slight excess of water to each bottle. Allow sufficient time in the thermostat for the two phases to separate.

After the phases have separated, take a small aliquot from each phase and measure the refractive indices. Tabulate your data as in Table 26–2.

TABLE 26–2.

BOTTLE NO.	REFRACTIVE INDEX OF C_6H_6 RICH PHASE	H_2O RICH PHASE
1	1.4950	1.3720
2	......	......
.		
.		
.		
etc.		

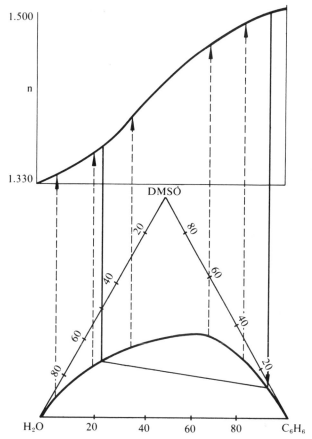

Figure 26–4 *Three-component phase diagram with the representative refractive indices of the different compositions.*

Knowing the densities of the components, convert the data collected in Table 26–1 into weight percentage composition. Plot your composition data thus obtained on a three co-ordinate diagram. Connect the points to form the miscibility curve (dome). Plot the refractive index diagram above the three co-ordinate diagram in a manner similar to that given in Figure 26–4 by drawing vertical lines from the points representing the titration end point compositions. Selecting the proper refractive index scale as the y axis, plot the refractive indices obtained at each titration end point composition.

Once a diagram such as Figure 26–4 is obtained, use the refractive index calibration curve to obtain the tie lines. This is done as follows: Let us assume that after overtitration two phases have been formed in the first bottle. The refractive index of the benzene rich phase (lighter phase) is 1.495. Project a vertical line from the point on the calibration curve corresponding to a refractive index of 1.495 downward until it intersects the miscibility dome on the phase diagram you have drawn. Similarly, assuming that the refractive index of the aqueous phase in the first bottle is 1.3720, project a vertical line from this point on the calibration curve downward until it intersects the miscibility dome. Connect the two points on the miscibility dome to obtain the first tie line. Proceed in a similar manner to establish the rest of the tie lines.

What is the composition at which a plait point (isothermal critical point) may exist in your ternary system?

If you start with a complex mixture of 40 percent benzene, 20 percent DMSO (or DMF, acetone, and so forth), and 40 percent water, what is the estimated composition of the two phases? In what proportion would you obtain these phases?

Do you see any unusual feature(s) on your phase diagram that deserve special attention? Comment.

Material and Equipment. 200 ml. of dimethyl sulfoxide (DMSO) (or dimethylformamide [DMF], acetone, dioxane, and so forth); 200 ml. of benzene; 15 Erlenmeyer flasks of 100 ml. capacity; three 50 ml. burets; thermostat with heater, stirrer and thermoregulator; circulating pump; 15 medicine droppers; refractometer; lens paper.

REFERENCES

1. J. E. Ricci. The Phase Rule. D. Van Nostrand Co., Inc., Princeton, N.J., 1951.
2. S. T. Bowden. The Phase Rule and Phase Reactions. Macmillan & Co., Ltd., London, 1950.

27

PHASE DIAGRAM OF A THREE-COMPONENT AQUEOUS SYSTEM

The separation of pure solids from their aqueous solutions has considerable industrial interest, and it has spurred a large amount of work on the phase diagrams of ternary, quaternary, and quinary aqueous systems. According to the Gibbs phase rule

$$P + \phi = C + 2 \qquad (1)$$

when one deals with three components ($C = 3$) and they form one phase ($P = 1$), there are four degrees of freedom, ϕ, two of which are temperature and pressure. If these two variables are fixed (the experiment is performed at constant temperature and pressure), the two remaining variables are the concentrations of any two of the three components. This is what will determine the system when, let us say, two salts are dissolved in water and form a solution.

When a new phase appears (i.e., when one of the salts crystallizes out), the number of degrees of freedom in such a ternary system will be three. That means that at a given temperature and pressure when a saturated solution is in equilibrium with a pure solid crystal phase, the composition of the saturated solution can be determined if the concentration of one component is known. In a ternary system in which three phases are in equilibrium, the number of degrees of freedom is two (temperature and pressure) and the *composition* of such a system is fixed. The system with no degrees of compositional freedom is called *invariant*.

In order fully to understand ternary systems, experience must be obtained in the use of triple co-ordinates. Let us assume that we have a system composed of two solids, A and B, and water. In order to make the system as simple as possible, we also assume that the two solids, A and B, form only one type of pure crystals each and that they do not have any hydrates (such as $A \cdot nH_2O$). Moreover, in this simplest of the ternary systems we exclude the possibility that solid phases such as $KNaCl_2$ may form and that we have hydrates such as $KCl \cdot H_2O$. (The student may note that, using salts for the two solids, we can keep the system simple by allowing a common ion. If we had used sodium nitrate and potassium chloride, for example, as the starting solids, even with the exclusion of double salts and hydrates, we would have four types of crystals—sodium nitrate, sodium chloride, potassium nitrate, and potassium chloride—rather than only two.)

The behavior of a ternary system at constant temperature and pressure can be presented in a diagram with three co-ordinates (Fig. 27–1). In order to plot the composition of the system let us assume that there is 20 percent component A, 50 percent component B, and 30 percent water. The three corners of the triple co-ordinate graph represent points at which we have either 100 percent component A, 100 percent component B, or 100 percent water. Moving on the line from H_2O to A in Figure 27–1, we trace the positions of a binary system (component A and water) from pure water to then increasing concentrations of A, and finally to pure A. Similarly, base line A-B represents the variable composition of binary solid mixtures of A and B. The line B-H_2O represents the variable composition of the binary system of component B and water.

Now let us return to our ternary mixture. The starting point is the 20 percent mark on the H_2O-A line. This point would indicate a composition of 20 percent A and 80 percent water in a binary system. However, this is not the case in a ternary

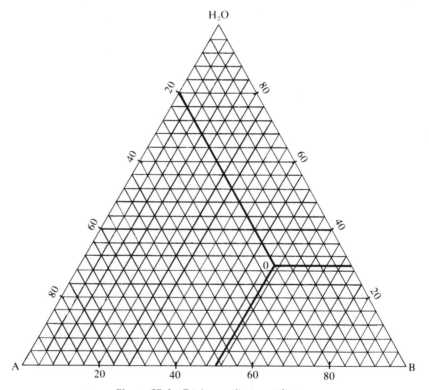

Figure 27–1 *Triple co-ordinate graph paper.*

system. The point corresponding to 20 percent *A* composition in any ternary mixture will be on a line that originates at this point and runs parallel to the H_2O-*B* line (that is, to the side of the triangle opposite the *A* corner).

Similarly, we locate the 50 percent mark on the *A*-*B* base line. Again, this corresponds to a composition of 50 percent *B* and 50 percent *A* in a binary mixture. Other mixtures with 50 percent *B* but variable amounts of *A* and water are indicated by points on a line originating at this point and running parallel to the *A*-H_2O axis (i.e., parallel to the side of the triangle opposite the *B* corner).

Now the line representing 20 percent *A* and that representing 50 percent *B* intersect at point *O*, which indicates the total composition since *A* and *B* are given in percentages and the composition of water is automatically a fixed value. However, to further prove that the *O* is also on the line representing 30 percent water, let us find the point corresponding to 30 percent water on the *B*-H_2O axis. Using this as a starting point, we draw a line parallel to the *A*-*B* axis (the side of the triangle opposite the H_2O corner). We see that this line also intersects the other two lines at point *O*.

Thus, the composition of a ternary system may be described by one point in a triple co-ordinate diagram.

Phase diagrams of ternary systems that are plotted on a triple co-ordinate graph have two important features: an invariant point, *O*, and the solubility curves (Fig. 27–2). In our simple system, the solubilities of *A* and *B* in water are represented by points *C* and *D*, respectively. The line *C*-*O* indicates the composition of the saturated

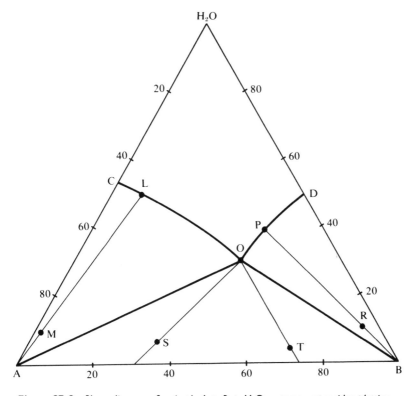

Figure 27–2 *Phase diagram of a simple A + B + H₂O system; wet residue plotting.*

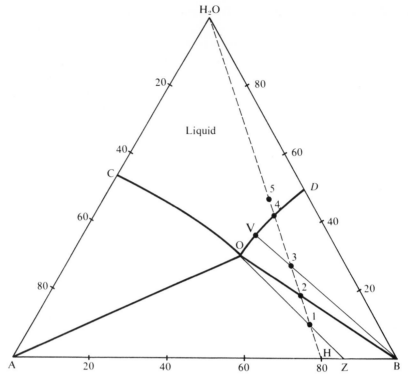

Figure 27–3 *Phase diagram of A + B + H₂O.*

solutions that are in equilibrium with pure solid *A*. The line *O-D* indicates the composition of the saturated solutions that are in equilibrium with pure solid *B*. The points on this solubility line *C-O-D* are obtained by simply analyzing the compositions of saturated solutions and plotting them on a triple co-ordinate graph. The saturated solutions are obtained by starting with different mixtures of *A* and *B* and adding to each enough water to dissolve some, but not all, the solids. After equilibrium has been reached, the saturated solutions are siphoned off and analyzed.

Not only the saturated solutions but also the remainder of the complex (the part remaining after the saturated solution has been siphoned off or decanted) can be analyzed. This remaining portion of the system is called the wet residue because it contains crystals and adhering saturated solutions. Analysis of the wet residue can be performed by weighing it, dissolving it in water to bring it up to volume, and determining the amounts of components *A* and *B*, calculated as a *percentage of wet residue*. These results from the analysis of the wet residues are also plotted on the diagram (*M, S, T, R*). One then connects the points representing the composition of a saturated solution with its wet residue composition (*L-M, P-R, O-S,* and *O-T* in Figure 27–2). These lines, as *L-M* and *P-R* in Figure 27–2, usually converge at either *A* or *B*, indicating simple salts. The lines from the invariant point, *O*, do not converge toward *A* or *B* because the *S* and *T* points represent wet residues in which both types of crystals, *A* and *B*, are present in equilibrium with the saturated solution.

Now the question arises that, once such a phase diagram is obtained, how can we read it for practical use?

In Figure 27–3 the total phase diagram is given for the system investigated in Figure 27–2. The four areas are clearly indicated. The area labeled "liquid" represents unsaturated solutions. The area within C-O-A represents saturated solutions and solid A, the area within points D-O-B represents saturated solutions and pure crystals of B; the area defined by A-O-B indicates saturated solutions and crystals of A and B.

To practice reading such a phase diagram let us follow what will happen if we start with a solid mixture of 20 percent component A and 80 percent component B and add water to the system. The composition of the solid mixture is indicated on the A-B axis at point H. The dilution-solvation process brings us along the line H-H_2O. We can see that this line first enters the area representing both solids A and B in equilibrium with saturated solution of composition O. Let us assume that we proceed by adding water to the complex mixture (solids + saturated solution), until we reach the composition represented by point 1.

A tie line is drawn that connects point 1 to point O. The extension of this tie line intersects the AB axis at Z. We can now describe the system quantitatively. It consists of a saturated solution, the composition of which is indicated by O; the composition of the solids in equilibrium with this saturated solution is indicated by Z.

The tie line, OZ, is divided into two parts—$O1$ and $1Z$. If the length of $OZ = 4$ cm., $O1 = 2.4$ cm., and $1Z = 1.6$ cm. The mixture is, therefore, composed of $(2.4/4.0) \times 100 = 60$ percent solids and $(1.6/4.0) \times 100 = 40$ percent saturated solutions.

Let us now proceed further along the H-H_2O line by adding more water to our complex mixture. We reach a total composition represented by point 2 in Figure 27–3. At this point the mixture consists of a saturated solution with the composition indicated by O and pure solid B. The percentage of the solid in the mixture is represented by $\dfrac{O2}{OB} \times 100$, whereas that of pure B is represented by $\dfrac{2B}{OB} \times 100$.

Upon further dilution we reach a composition of the mixture represented by point 3. The tie line now is the straight line connecting point 3 to point B which, upon extension, intersects the solubility line at point V. The saturated solution now has the composition indicated by V and the solid is pure crystal B. The ratio of the tie line sections again indicates the proportion of solid to solution.

$$\frac{V3}{VB} \times 100 = \text{percentage of solid } B$$

and

$$\frac{3B}{VB} \times 100 = \text{percentage of saturated solution}$$

Proceeding further on the dilution line we come to point 4. Here the solid phase disappears and we have a saturated solution with the concentration represented by point 4. Further dilution with water will give us an unsaturated solution represented by point 5.

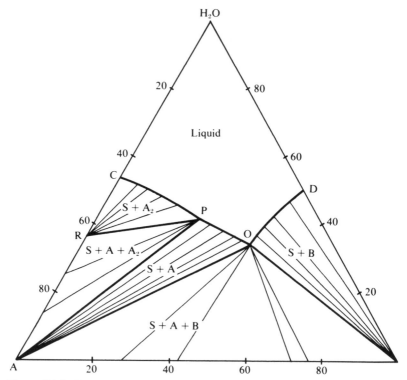

Figure 27–4 *Phase diagram of a three-component system with the presence of a hydrate.*

In practice, one manipulates the concentration of solvent just as in our example we varied the water content. All the preceding considerations refer to a ternary phase diagram with constant temperature and pressure. Under these conditions we change the position of the complex mixture by either evaporation or dilution and move the compositions across phase boundaries.

In practice, one can also manipulate the temperature. At each temperature one obtains a phase diagram similar to that in Figure 27–3. These phase diagrams can be stacked along a temperature co-ordinate, which would be perpendicular to the plane of paper. The variations in temperature move the solubility curve, C-O-D, and the invariant point, O, to different positions. Hence temperature manipulation of a ternary complex mixture does not change the composition by evaporation or dilution but instead the solubility boundaries change with temperature. The same separations can also be accomplished in this fashion. However, we will restrict our consideration to isothermal and isobaric processes.

One of the prime practical examples to which a phase diagram such as that in Figure 27–3 can be applied is the recovery of a salt such as B, from a mixture such as H. To get maximum recovery one should add just enough water to convert the system to the composition represented by point 2. This is the condition at which the ratio of solid B to solution is a maximum. Any further dilution will diminish this ratio.

All these considerations can be applied to a more complex diagram, such as that given in Figure 27–4. In this figure component A can exist either in the anhydrous state or as a hydrate, $A \cdot 2H_2O$, abbreviated as A_2. The composition of this dihydrate is represented by point R. The solubility curves, C-P-O-D, are obtained from analysis of the saturated solutions as before. Similarly, the wet residue compositions are also obtained. The lines connecting the saturated solution and wet residue compositions converge to either R, A, or B, indicating the triangular areas in which the solid is either pure $A \cdot 2H_2O$, A, or B. The areas in which the connecting lines between the saturated solution and the wet residue compositions do not converge indicate that more than one solid phase will appear. Drawing tie lines in a diagram such as Figure 27–4 must be done in a manner similar to that in Figure 27–3, and the process of dilution or concentration will progress along lines such as the H-H_2O in Figure 27–3.

Experimental

The phase diagram of the ternary system of glucose–potassium chloride–water will be determined. (Alternate systems, such as galactose–potassium chloride–water, galactose–sodium chloride–water, or glucose–sodium chloride–water, may be used with a slight modification in calculations.)

1. Prepare seven or eight solid mixtures having different compositions, e.g., 10 per cent potassium chloride and 90 per cent glucose or 20 per cent potassium chloride and 80 per cent glucose. The weight of each component and hence the total weight of these mixtures should be known accurately.

Transfer each mixture into a separate, numbered 50 ml. Erlenmeyer flask and add to each exactly 5 ml. of distilled water. (Make certain that any solid "cake" formed upon the addition of water is broken up to a powdery consistency by use of a spatula.)

The Erlenmeyer flasks should be stoppered and immersed in a 25° C. thermostated bath so that they can be agitated. Various agitators can be used, such as a rotating wheel driven by a motor at a low speed. The wheel is immersed in the thermostat and the Erlenmeyer flasks are tied to different parts of the wheel. Another agitator is a commercial shaker onto which clamps are secured. Each clamp moves with the motion of the shaker and at the same time holds one of the Erlenmeyer flask partly submerged in the thermostat. Allow at least one hour for the system to reach equilibrium. (The experiment may be arranged so that different groups of students perform the experiment at different temperatures, and the isotherms obtained by the different groups may then be combined into a three-dimensional phase diagram.)

2. To determine the concentrations of potassium chloride titrate the solutions with silver nitrate. The following solutions are prepared in advance (*preferably* by the instructional staff):

a. 0.1N silver nitrate solution made up of 8.4935 grams of silver nitrate in 500 ml. and standardized against sodium chloride.

b. 200 ml. of 1 per cent dextrin solution.

   c. 100 ml. of 0.5M sodium acetate buffer (4.1 grams/100 ml.).

   d. Indicator composed of 0.1 gram of dichlorofluorescein dissolved in 70 ml. of ethanol and 30 ml. of water added.

For the determination of glucose, the following solutions should be prepared:

   e. 0.03230N iodine solution made by weighing accurately 1.0250 grams of iodine and adding to it 5 grams of *iodate-free* potassium iodide. The solids are dissolved in distilled water, and diluted to 250 ml.

   f. 5 grams of sodium carbonate, 150 grams of potassium sodium tartrate tetrahydrate, 25 grams of disodium orthophosphate dodecahydrate ($Na_2HPO_4 \cdot 12H_2O$), and 2.49 grams of cupric sulfate pentahydrate ($CuSO_4 \cdot 5H_2O$) all placed in one beaker are dissolved in distilled water, and diluted to 500 ml.

   g. 0.5 gram of soluble starch made into a paste in about 10 ml. of distilled water. Another 60 ml. of water is added and the solution is brought to a quick boil; it is cooled and diluted to 100 ml.

   h. 2 grams of sodium thiosulfate pentahydrate ($Na_2S_2O_3 \cdot 5H_2O$) dissolved in 250 ml. of water. It is standardized by titration against the iodine solution.

   i. Glacial acetic acid.

   j. 1N hydrochloric acid solution.

   3. After equilibrium has been reached in the thermostat, decant the saturated solutions (or use a centrifuge and obtain the supernatant).

   Analysis of the solutions should be as follows: For the glucose analysis, dilute the supernatant so that, for use in the determinations, there will be 10 to 20 mg. of glucose in the aliquot. Transfer the aliquot into an Erlenmeyer flask. Wash it with 20 ml. of water, and add 50 ml. of the copper-containing reagent (item f). Heat the mixture to a boil and let it boil on a low flame for an additional four to five minutes. The solution will turn from blue to brown. Cool the mixture in an ice bath. After cooling, add 1 ml. of glacial acetic acid (item i). Pipet 10 to 30 ml. of the iodine solution (item e) according to the amount of copper reduced. Be sure that excess iodine is present. (A brownish color will indicate this.) Add 15 ml. of 1N hydrochloric acid and titrate with the sodium thiosulfate solution (item h). The end point can be seen clearly by adding a few drops of starch indicator, which in the presence of excess iodine produces a deep bluish purple color and disappears at the end point. When the end point is reached the amount of sodium thiosulfate solution used is noted. Each milliliter of iodine solution used is equivalent to 1 mg. of glucose or galactose but 1.9 mg. of maltose. The calculation is made as follows: Let us assume that 2 grams of the supernatant was diluted to 100 ml. and that 10 ml. of this diluted solution was used as a sample. Then 20 ml. of the iodine solution was added to the mixture and back-titrated with 4.05 ml. of sodium thiosulfate. The standardization of the sodium thiosulfate solution against the iodine solution was

$$\frac{10 \text{ ml. } I_2 \text{ soln.}}{9.2 \text{ ml. } Na_2S_2O_3} = 1.086$$

Therefore, $4.05 \times 1.086 = 4.4$ ml. of 0.323N iodine solution remained of

20 ml. originally added. Hence, $20 - 4.4 = 15.6$ ml. of iodine was used by the reduced copper. Each milliliter of iodine used corresponds to 1 mg. of glucose; hence, a total of 156 mg. of glucose was in 2 grams of sample; the glucose content of the sample was 7.8 percent.

The chlorine determination is done as follows: To a sample containing approximately 30 to 100 mg. of potassium chloride add 10 ml. of distilled water and 1 ml. of sodium acetate buffer (item c). Add 10 ml. of dextrin solution (item b) and 5 drops of indicator (item d). Titrate with 0.1N silver nitrate solution to the end point, indicated by the appearance of a pink color.

The following example may be of help in making calculations on the sample: A sample containing 0.1 gram of supernatant was titrated with 4.15 ml. of silver nitrate solution to the end point.

$$100 \times \frac{4.15 \text{ ml. } 0.1\text{N AgNO}_3 \times 3.55 \text{ mg. Cl}^-}{100 \text{ mg. sample}} \times \frac{75.5 \text{ grams/mole KCl}}{35.5 \text{ grams/mole Cl}^-}$$

$$= 31.33 \text{ percent}$$

TABLE 27–I

Sample No.	Original Composition of the Complex			Composition of the Saturated Solution		
	Glucose (percent)	KCl (percent)	H$_2$O (percent)	Glucose (percent)	KCl (percent)	H$_2$O (percent)
1	7.20	53.06	39.74	9.79	30.02	60.19

4. Since in the preceding procedure the original composition of the complex is accurately known, we can dispense with the analysis of the wet residues.

Tabulate your data on the original composition and the saturated solution composition of each sample as shown in Table 27–1. Plot the results from Table 27–1 on a triple co-ordinate graph. Connect the points representing the composition of each saturated solution with the corresponding points representing the original composition of the ternary system and extend the lines until they intersect the co-ordinate axes. Connect the points indicating the compositions of the saturated solutions to establish a solubility curve. Obtain from the literature, or by experiment, the solubilities of glucose and potassium chloride in water at 25° C. Use these points to complete the solubility curve. On the basis of the converging lines for complex–saturated solution composition determine how many types of solids you may have throughout the whole range of composition.

Draw in the phase boundaries as in Figures 27–3 and 27–4.

Do you have any indication for the presence of hydrate(s)? Do you have any indication for the presence of a double salt? What would be the composition of the different phases if you started with 6 grams of glucose and 2 grams of potassium chloride and added 2 grams of water? In what proportion would you have these phases? What simple solid crystal can you isolate from this original mixture and what quantitative procedure do you have to follow to accomplish it?

Material and Equipment. 100 grams of glucose (or galactose, maltose, or similar substance); 100 grams of potassium chloride (or sodium chloride, lithium chloride, or other compound); 5 grams of silver nitrate; 1 gram of dextrin; 5 grams of sodium acetate; 0.1 gram of dichlorofluorescein indicator; 100 ml. of ethanol; 1.5 grams of iodine; 5 grams of iodate-free potassium iodide; 5 grams of sodium carbonate; 150 grams of potassium sodium tartrate tetrahydrate; 25 grams of disodium orthophosphate dodeca-hydrate ($Na_2HPO_4 \cdot 12H_2O$); 3 grams of cupric sulfate pentahydrate ($CuSO_4 \cdot 5H_2O$); 0.5 gram of soluble starch; 2 grams of sodium thiosulfate pentahydrate ($Na_2S_2O_3 \cdot 5H_2O$); 10 ml. of glacial acetic acid; 200 ml. of 1N hydrochloric acid solution; ten Erlenmeyer flasks of 50 ml. capacity; two 500 ml., one 250 ml., one 200 ml., and two 100 ml. volumetric flasks; two burets of 50 ml. capacity; thermostat with heater, stirrer, and thermo-regulator; commercial shaker or rotating wheel agitator; ice bath; pipets.

REFERENCES

1. F. F. Purdon and V. W. Slater, Aqueous Solution and the Phase Diagram, Edward Arnold Co., London, 1946.
2. S. T. Bowden, The Phase Rule and Phase Reactions, Macmillan & Co., London, 1950.

IX

KINETICS

28

INVERSION OF SUCROSE; KINETICS OF A PSEUDO–FIRST ORDER REACTION DETERMINED BY POLARIMETRY

The hydrolysis of sucrose occurring in aqueous media can be considered a pseudo–first order reaction since one of the components, water, is present in great excess. Thus, the rate of change at any time is proportional to the concentration

$$\text{(1)}$$

of the sucrose. If the initial concentration is a at time t, the sucrose concentration will be $a - x$ (x being the amount of sucrose hydrolyzed) and the rate of reaction will be

$$\frac{dx}{dt} = -\frac{d(a - x)}{dt} = k_1(a - x) \tag{2}$$

The rate constant, k_1, denotes the fractional number of molecules decomposing per unit time.

Integration of equation 2 yields

$$\int_0^x \frac{dx}{(a-x)} = \int_0^t k_1 \, dt$$

$$\ln \frac{a}{(a-x)} = k_1 t \tag{3}$$

and therefore

$$k_1 = \frac{1}{t} \ln \frac{a}{a-x} \tag{4}$$

The half time, $t_{1/2}$, corresponds to the condition in which $x = 0.5a$. Therefore,

$$t_{1/2} = \frac{\ln 2}{k_1} \tag{5}$$

The fractional extent of the chemical change at any time, t, is

$$\frac{x}{a} = 1 - e^{-k_1 t} \tag{6}$$

Whenever possible, the rate of a reaction is studied by physical means, which will not disturb the chemical reaction. In the case of the hydrolysis of sucrose, optical rotation provides such a means.

In general, optical rotation of a substance refers to a process in which the *direction of vibration* of linearly polarized light is changed. (The directions of the vibrations are perpendicular to the direction of propagation of the incident light.)

A linearly polarized light beam can be visualized as the result of two circular rotations, right (deutrorotation) or left (levorotation), designated as *d*- and *l*-components, respectively. These rotations can interfere with each other. In an optically nonactive substance, the velocities of the *d* and *l* components are the same; in an optically active substance they differ. The relative velocities of electromagnetic radiation in different media are related to the refractive indices.

$$v_1/v_2 = n_2/n_1 \tag{7}$$

An optically active substance can be considered a medium that has different refractive indices in the different directions. When linearly polarized light passes through an optically active medium, the planar projection of the two circular components reaches a phase difference, the velocity of one component (either *d* or *l*) being greater than that of the other. The recombination of these components causes a linear vibration that is not in the same direction as the original vibration, but it is rotated at a certain angle (Fig. 28–1).

The d_1, l_1, d_2, l_2, and so forth, are the positions of the two components in the plane of recombination at the same instant of time.

The angle of rotation, α, usually refers to an optical path of 1 decimeter (10 cm.) and is expressed as

$$\alpha = 0.5\phi = (10\pi l/\lambda_0)(n_l - n_d)$$

or

$$\tag{8}$$

$$\alpha^\circ = (1800 l/\lambda_0)(n_l - n_d)$$

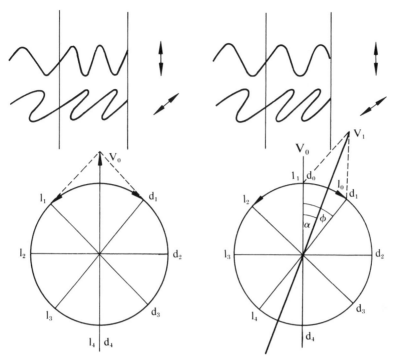

Figure 28-1 *Recombination of dextro- and levorotatory components in A, an optically nonactive medium, and B, an optically active medium.*

where l is the length of the optical path in the medium in decimeters, λ_0 is the wavelength of the electromagnetic radiation in vacuum in centimeters, and $(n_l - n_d)$ is the difference in refractive indices.

Although the physical aspect of optical rotation is relatively easily explained as a difference in refractive indices, one encounters difficulties in explaining how molecules in the liquid state give rise to such anisotropic behavior, i.e., difference in refractive indices.

The usual explanation given in introductory organic chemistry (that an asymmetric center or centers are required for optical rotation) is insufficient. The Pasteur principle, which states that a molecule and its mirror image should not be superimposable, means that the molecule cannot have a center or plane of symmetry, although it may have an axis of symmetry. How this lack of molecular symmetry results in optical activity is explained on the basis of classical theory of Born and Kuhn. According to their theory, the molecule is made up of a number of linear oscillators with fixed mutual orientation. The oscillators are coupled if their orientation is other than 0 or 180 degrees, and maximum coupling is achieved if their orientation is 90 degrees. Electrical coupling between two oscillators means that if a dipole is induced in one and it vibrates with a certain frequency, this will produce a vibration in the second oscillator with the same frequency, although in a different direction. Optical rotation results when a set of coupled oscillators does not have a plane or center of symmetry.

Two other theories of optical rotation explain the phenomenon on the basis of quantum mechanics. Only a qualitative description of these theories will be given here. In the one electron theory of Eyring, optical activity is considered to be the result of the transition of a single electron from its ground state to its excited state. The electron undergoing this transition is localized in one part of the molecule, the chromophoric group. The transition can be $\pi-\pi^*$, $n-\pi^*$, and so forth. The optical rotatory power is related to matrix elements of the magnetic and electric dipole moments. The average magnetic and electric dipole moments are calculated from the perturbed wave functions and averaged over all molecular orientations.

$$\bar{m} = \int \Psi^* m \Psi \, d\tau$$

$$\bar{\mu} = \int \Psi^* \mu \Psi \, d\tau \tag{9}$$

The perturbed wave functions in turn are calculated from unperturbed electronic wave functions by first order perturbation techniques. When the electron (and the electronic transition) is in its local atomic field, no rotatory power exists (unperturbed electronic wave). A dissymmetric field is superimposed upon this chromophoric electron by neighboring group effects in an asymmetric molecule. These group effects may be permanent or induced dipole effects of neighboring groups, electrical field effects of ions, or overlapping of the electronic clouds of the chromophoric group with neighboring groups. Thus, the perturbed wave functions are calculated as a result of such interaction. The one electron theory of Eyring considers interactions of the electron and the neighboring groups one at a time.

The second theory, that of Kirkwood and coworkers, is a quantum mechanical formulation of the Born-Kuhn theory. The asymmetric molecule is considered as a sum of N groups, each of which possesses cylindrical symmetry. The electric and magnetic dipoles of the molecule are expressed as the sum of the contributions from the N groups. The optical activity is considered largely to be a result of interaction of the electric dipole of one group with that of another.

The electric vector of the linearly polarized incident beam induces dipole moments in each of the N molecular groups. These induced dipole moments oscillate; hence, they are the source of secondary radiation that is also linearly polarized but not in phase with the incident radiation. The resulting radiation propagated in the medium is the superimposition of the secondary radiation upon the incident beam. The phase difference between the primary beam and the secondary radiation leads to the rotation of the plane of polarization (see Fig. 28–1). Since the polarizability theory neglects everything but electric dipole-dipole interactions, it cannot be applied to molecules with large magnetic moments.

The optical rotation is measured using a polarimeter. Since optical rotation is both temperature and wavelength dependent, one works with monochromatic radiation and at a specified temperature. The monochromatic light is usually provided by a sodium lamp (D line). The light passes through a polarizer and enters the polarimeter tube as linearly polarized light (Fig. 28–2).

The polarimeter tube is usually 20 cm. long. (Its length must be ascertained for accurate calculations.) The light beam emerging from the polarimeter tube is still

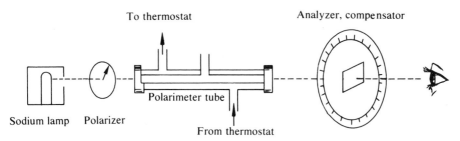

Figure 28–2 *Schematic diagram of a visual polarimeter.*

linearly polarized, but the plane of polarization has been changed (Fig. 28–1). To null the instrument, an analyzer or compensator is employed, the rotation of which indicates the angle of rotation of the light.

A combination of polarizer-analyzer as a set of Nicol prisms is given in Figure 28–3.

One can see that as the second Nicol prism is rotated 90 degrees, the optical axis changes from a perpendicular (crossed) orientation to a parallel orientation.

Instead of an analyzer, a compensator is used in many polarimeters. These are made of quartz wedges that have different rotations, depending upon the thickness of the quartz. Therefore, by lateral positioning of quartz wedges, the same amount of rotation of the sample is produced but in the opposite direction. Such a compensator assembly is given in Figure 28–4.

Since the human eye is more sensitive to differences in light intensities than to measurements of absolute intensities, most visual polarimeters are constructed on the half-shade field principle. Under this condition the intensities of two half-shades are matched at the zero position rather than by setting perpendicular (crossed) prisms to minimum transmittance.

The Nakamura half-shade arrangement is one of many employed in polarimeters, and it is illustrated in Figure 28–5.

When the compensating quartz wedge is turned at an angle of 2ϵ, the fields of the two half-shades are matched, as in Figure 28–6*b*.

The compensator or analyzer is usually fixed in a graduated circle in which the outer circle is marked in degrees and decimal fractions of degrees, and the inner circle is rotated to compensate at zero position. A vernier is also fitted for precision

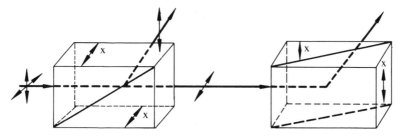

Figure 28–3 As shown here, with crossed Nicol prisms, ▱ ◩ , no light passes through the analyzer; with parallel Nicol prisms, ▱ ▱ , maximum light passes through them. The ◄──►x indicates the optical axis of the Nicol prism.

Movable wedges

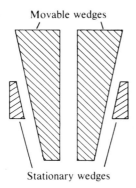

Stationary wedges   **Figure 28–4** *The Schmidt-Haensch quartz wedge compensator.*

work, and most commercial instruments can be read to 0.01 degree. These graduated circles are equipped with a handle for rough adjustment to zero position and also with some device (such as micrometer screw) for fine adjustment.

The optical rotation of the polarimeter, α, for a concentration of c, grams/100 ml. of solution, gives the specific rotation $[\alpha]_D^{25}$ with a sodium D line light source at 25° C. according to the following equation:

$$[\alpha]_D^{25} = 100\alpha/lc \tag{10}$$

where l is the length of the polarimeter tube. The specific rotation, therefore, is the rotation exhibited by 1 gram of an optically active substance in 1 ml. of solution having an optical path length of 10 cm.

The corresponding molecular rotation is given by equation 11.

$$[M] = M\alpha/lc \tag{11}$$

In the inversion of sucrose one simply follows the change in the angle of rotation with respect to time.

The sucrose is dextrorotatory, but the resulting mixture of glucose and fructose is slightly levorotatory because the levorotatory fructose has a greater molecular rotation than does the dextrorotatory glucose. As the sucrose is used up and the

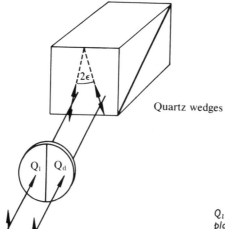

Quartz wedges

Figure 28–5 *Nakamura half-shade. The Q_l and Q_d are thin plates of quartz rotating the plane of polarization left or right by the same amount. The half-shade angle 2ϵ is twice the amount of rotation of either plate.*

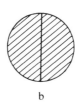

−ϵ   +ϵ

Figure 28–6 *Half-shade fields at rotations of:* a, *less than* 2ϵ, b, *an amount equal to* 2ϵ, *and* c, *more than* 2ϵ.

a   b   c

hydrolysis products are formed, the angle of rotation to the right (as the observer looks into the polarimeter tube) becomes less and less, and finally the light is rotated to the left. The rotation is determined at the beginning (α_0) and at the end of the reaction (α_∞), and the algebraic difference between these two readings is a measure of the original concentration of the sucrose. It is assumed that the reaction goes to completion and practically no sucrose remains at infinite time. At any time, t, a number proportional to the concentration, c, of sucrose is obtained from the difference between the final reading and the reading (α_t) at time t. Thus, the specific reaction-rate constant may be written by analogy to equation 3 as

$$k = \frac{2.303}{t} \log \frac{\alpha_0 - \alpha_\infty}{\alpha_t - \alpha_\infty} \tag{12}$$

The reaction proceeds too slowly to be measured in pure water, but it is catalyzed by hydrogen ions. The water is in such great excess that the reaction follows the equation for a first order reaction, although two molecules are involved in the reaction.

Guggenheim has described a method for calculating the rate constant of a first order reaction without an infinite time value. This method is useful if the reaction does not go to completion in one laboratory period and has the added advantage that each plotted point does not depend upon a single observation of the reading at infinity time.

The Guggenheim method may be applied directly if data are taken at equal time intervals. If the data are taken at unequal time intervals, a plot of concentration, angle of rotation, or other measure of the extent of reaction vs. time (t) is prepared and a smooth curve drawn. The required data may be read from this curve.

The data are arranged in two sets. For each observation (c_1) at time t in the first set, another observation (c_2) is taken at time $t + \Delta t$, where Δt is a fixed time interval. If a plot of $\log (c_1 - c_2)$ vs. t is prepared, the points will be on a straight line of slope $-k/2.303$. The constant time interval Δt may be taken as approximately one-half the duration of the experiment. If Δt is too small, there will be a large error in $c_1 - c_2$.

The equation for this method may be derived as follows: From the integrated form of the first order reaction differential equation,

$$c = c_0 e^{-kt} \tag{13}$$

Thus, the concentrations c_1 and c_2 at two times differing by Δt are

$$c_1 = c_0 e^{-kt} \tag{14}$$

$$c_2 = c_0 e^{-k(t+\Delta t)} \tag{15}$$

Subtracting,

$$c_1 - c_2 = c_0 e^{-kt}(1 - e^{-k\Delta t}) \tag{16}$$

Taking logarithms,

$$\log(c_1 - c_2) = -kt/2.303 + \log[c_0(1 - e^{-k\Delta t})] \tag{17}$$

Thus, the slope of a plot of $\log(c_1 - c_2)$ vs. t is $-k/2.303$.

If, instead of measuring concentration directly, we measure some linear function of the concentration (in our case optical rotation), an equation of the same form as equation 17 applies. For example, if

$$\alpha = ac + b \tag{18}$$

$$c_1 - c_2 = (\alpha_1 - \alpha_2)/a \tag{19}$$

and

$$\log(\alpha_1 - \alpha_2) = -kt/2.303 + \log[ac_0(1 - e^{-k\Delta t})] \tag{20}$$

Experimental

One hundred ml. of 4N hydrochloric acid solution and 100 ml. of 4N monochloroacetic acid solution are prepared.

Twenty grams of pure cane sugar (sucrose) are dissolved in water (filtered, if necessary, to give a clear solution) and diluted to 100 ml.

Two jacketed polarimeter tubes are connected in series with the circulating water from a thermostat at 25° C. A zero reading is taken. Obtain a reading on the sucrose solution in order to calculate the molecular rotation from equation 11. Filling of the polarimeter tube should be accomplished as follows: Remove one of the windows of the polarimeter tube. Holding the tube in a vertical position, pour in the solution to the rim. With the window moving in a lateral sweep, slide the glass over the tube, thus assuring that no air bubble is trapped in the polarimeter tube. Screw on the cap, holding the window in place.

After the sugar solution and the 4N hydrochloric acid solution have been in the thermostat for a few minutes, 25 ml. of each are mixed together thoroughly. Note the time of mixing as the zero time. One of the polarimeter tubes is rinsed out with successive small portions of the solution, and then the tube is filled with the solution and stoppered. The second tube is filled, in a similar manner, with a mixture of exactly 25 ml. each of the sugar solution and 4N monochloroacetic acid solution. The tubes are filled as soon as possible after mixing so that an early reading of the angle of rotation may be obtained.

The time of the first reading is recorded, and polarimeter readings of the hydrochloric acid solutions and the corresponding times are taken as rapidly as convenient (10 minute intervals) over a period of one hour or so. As the reaction rate decreases, the observations should be made every 30 minutes. The observations should extend over a period of three hours or more. The reaction is much slower with the monochloroacetic acid, and the readings are taken less frequently, i.e., at convenient intervals of time when the polarimeter is not being used for readings on the hydrochloric acid solution.

The final readings (α_∞) are taken after the solutions have been in a tightly stoppered flask long enough for the reactions to be completed, at least two days for the hydrochloric acid and a week for the monochloroacetic acid. If it is not convenient to obtain the final reading for the monochloroacetic acid, it may be assumed that α_∞ will be the same as for the hydrochloric acid.

Prepare a plot of $\alpha_t - \alpha_\infty$ vs. time. As stated before, the concentration of sucrose is proportional to $\alpha_t - \alpha_\infty$ where α_t is the angle of rotation at time t and α_∞ is the final angle (negative in this particular case). Plot the logarithms of $(\alpha_t - \alpha_\infty)$ against time. Plotting the logarithms of $(\alpha_t - \alpha_\infty)$ gives the same slope as plotting the logarithms of $(\alpha_t - \alpha_\infty)$ multiplied by a constant. The best straight lines are drawn through the points, one for the hydrochloric acid and one for the monochloroacetic acid. The specific reaction rate constants k are calculated from the slopes of the lines.

In calculating the results of this experiment, treat at least one set of data by the Guggenheim method for comparison with the usual method. If the infinite time reading cannot be obtained conveniently, this method may be used exclusively.

Calculate the specific and molecular rotation of sucrose and compare it with data from the literature.

Material and Equipment. 20 grams of sucrose; 100 ml. of 4N hydrochloric acid solution; 100 ml. of 4N monochloroacetic acid solution; three 250 ml. Erlenmayer flasks; polarimeter with jacketed polarimeter tubes about 20 cm. long; thermostat with heater, stirrer, and thermoregulator; circulating pump.

REFERENCES

1. W. Heller. In Weissberger, A. (ed.). Techniques of Organic Chemistry. 3rd. Edition, Vol. I, Part III, Interscience Publishers, Inc., New York, 1960, Chapter 33.
2. S. F. Mason. Quart. Rev., **17**, 20 (1963).
3. C. Djerassi. Optical Rotatory Dispersion. McGraw-Hill Book Co., New York, 1960.
4. E. A. Guggenheim. Phil. Mag., **2**, 538 (1926).

29

KINETICS OF A GAS PHASE POLYMERIZATION REACTION

Polymerization is a chemical reaction whereby low molecular weight compounds form large molecular weight compounds (macromolecules). In its simplest form, only one type of low molecular weight molecule is used and a long chain molecule is created. The small molecule is known as a monomer; since it is repeated many times in the macromolecule, the latter is called a polymer.

An example is the polymerization of methylacrylate, which will be studied in the present experiment. The reaction is

$$m \ CH_2{=}CH{-}\overset{\overset{\displaystyle O}{\|}}{C}{-}O{-}CH_3 \xrightarrow{h\nu} -\overset{\overset{\displaystyle COOCH_3}{|}}{\underset{\underset{\displaystyle H}{|}}{C}} \left[CH_2{-}\overset{\overset{\displaystyle COOCH_3}{|}}{\underset{\underset{\displaystyle H}{|}}{C}} \right]_n CH_2{-}\overset{\overset{\displaystyle COOCH_3}{|}}{\underset{\underset{\displaystyle H}{|}}{C}}{-}$$

Most organic polymeric materials are obtained from monomers containing double bonds. One of the mechanisms by which such reactions proceed is a chain reaction involving free radicals. (Other mechanisms involve anionic or cationic polymerization using special catalysts.) Chain reactions have three distinct steps:

a. Initiation $\qquad\qquad\qquad M \xrightarrow{k_1} R\cdot \qquad\qquad\qquad (1)$

b. Propagation $\qquad R\cdot + \underset{/}{\overset{\backslash}{C}}{=}\underset{\backslash}{\overset{/}{C}} \longrightarrow R{-}C{-}C\cdot \qquad (2)$

$$R\cdot + M \xrightarrow{k_2} RM\cdot$$
$$RM\cdot + M \xrightarrow{k_2} RM_2\cdot$$
$$RM_2\cdot + M \xrightarrow{k_2} RM_3\cdot$$

$$\cdot$$
$$\cdot$$
$$\cdot$$

$$RM_n\cdot + M \xrightarrow{k_2} RM_{n+1}\cdot$$

c. Termination $\qquad RM_n\cdot + RM_m\cdot \xrightarrow{k_3} RM_{m+n} \qquad (3)$

278

The initiation reaction that forms free radicals can be achieved in many ways. Most often it is accomplished by using a substance called an initiator, such as a peroxide, which, when added in small amounts, steadily produces free radicals during the polymerization. The rate of such initiation, v_i, is usually first order with respect to the initiator (I).

$$\text{Rate} = v_i = k_1(\text{I}) \tag{4}$$

If the initiator interacts with the monomer, M, the rate of initiation is

$$v_i = k_1(\text{I})(\text{M}) \tag{5}$$

If the initiation is achieved photochemically (i.e., by ultraviolet irradiation), the rate is proportional to the intensity of light, ϕ,

$$v_i = k_1\phi \tag{6}$$

and in most cases it is second order with respect to the concentration of monomer.

$$v_i = k_1(\text{M})^2\phi \tag{7}$$

For the propagation steps we may assume that the free radicals produced (R·, RM·, RM$_2$·, and so forth) are unstable and react rapidly. Therefore, after a short initial build up they achieve a steady state whereby their rates of production are equal to their rates of disappearance. Hence,

$$\frac{d(\text{R·})}{dt} = 0 = v_i - k_2(\text{R·})(\text{M}) - k_3(\text{R·})\left[\sum_{n=0}^{\infty}(\text{RM}_n·)\right] \tag{8}$$

$$\frac{d(\text{RM·})}{dt} = 0 = k_2(\text{R·})(\text{M}) - k_2(\text{RM·})(\text{M}) - k_3(\text{RM·})\left[\sum_{n=0}^{\infty}(\text{RM}_n·)\right] \tag{9}$$

.

.

.

and so forth

In these rate equations one assumes that the rate constants of the propagation steps are each equal to k_2. This implies that the reactivity of radical R· or RM$_n$· is independent of its size. Rate constant k_3 refers to the termination step and is also independent of the size of the combining radicals.

If all steady state equations of the type given in equations 8 and 9 are summed, the result is

$$v_i = k_3\left[\sum_{n=0}^{\infty}(\text{RM}_n·)\right]^2 \tag{10}$$

The overall rate of reaction can be determined by monitoring the loss of the monomer or the appearance of the polymer. In a gas phase polymerization reaction, the change in pressure with time can be recorded. Assuming that no degradation side reactions occur that may produce gases, the pressure measured is proportional to the concentration of monomer since the polymer has no appreciable vapor pressure.

$$-\frac{d(\text{M})}{dt} = v = k_2[\text{M}]\left[\sum_{n=0}^{\infty}(\text{RM}_n·)\right] \tag{11}$$

Substituting equation 10 into equation 11,

$$\frac{-d(M)}{dt} = v = (k_2/k_3^{1/2})v_i^{1/2}(M) \tag{12}$$

This indicates that the order of the overall reaction depends on the rate of initiation, v_i.

If we assume that a photochemically initiated reaction proceeds at the rate given in equation 7, then

$$-\frac{d(M)}{dt} = \frac{dP}{dt} = v = k_2(k_1/k_3)^{1/2}(M)^2\phi^{1/2} \tag{13}$$

and the overall reaction is second order with respect to monomer concentration.

On the other hand, if the initiation rate is the one indicated in equation 4, the overall reaction will be first order with respect to monomer concentration.

If the initiation reaction is that represented by equation 5, the overall order will be 1.5.

One way to determine which initiation reaction took place is to examine the data in terms of the integrated first and second order rate equations to see which equation is more in agreement with the data throughout the experiment. For example, polymerization of methylacrylate is monitored by observing the decrease in pressure with time. At low vapor pressures (less than 1 atm.) the ideal gas law may be used; thus, the pressure in a closed vessel is proportional to the molar monomer concentration, C.

$$P = \frac{n}{V}RT = CRT \tag{14}$$

Since M in equation 13 and all other rate equations is the molar monomer concentration, and for a gaseous reaction is directly proportional to the vapor pressure of the monomer, for a first order reaction

$$-\frac{dP}{dt} = kP \tag{15}$$

$$\ln P/P_0 = kt \tag{16}$$

Hence, $(\ln P/P_0)/t$ should be constant or the plot of $\ln P/P_0$ vs. t should yield a straight line if the reaction is first order. The P_0 is the initial pressure, and P is the pressure at any time, t.

For a second order reaction,

$$-\frac{dP}{dt} = kP^2 \tag{17}$$

$$\frac{1}{P} - \frac{1}{P_0} = kt \tag{18}$$

Hence, a plot of $1/P - 1/P_0$ vs. time should yield a straight line.

Once the order of reaction is established and the rate constant is obtained for different initial pressures (monomer concentration), the temperature dependence of the reaction may be studied.

The dependence of the rate constant on the absolute temperature is given by the Arrhenius equation

$$k = Ae^{-\Delta E_a/RT} \tag{19}$$

where ΔE_a is the energy of activation and A is a preexponential factor.

If the Arrhenius equation is operative, rate constants obtained at two different temperatures are sufficient to determine ΔE_a and A. Rate constants usually are obtained at several different temperatures and $\ln k$ is plotted against $1/T$. A straight line indicates that the Arrhenius relationship is applicable to the reaction investigated. Then from the slope the ΔE_a and from the intercept the preexponential factor may be obtained.

Some of the theories of chemical kinetics can be applied to gain more insight into the reaction mechanism.

Since methylacrylate polymerization is a gas phase reaction, the collision theory may be applicable. This theory gives the rate constant as

$$k = \kappa\pi\, d^2 \left[\frac{8k'T}{\pi m} \right]^{1/2} p^{-\Delta E_a/RT} \tag{20}$$

In equation 20, κ is a collision symmetry number, which is usually considered to be $\frac{1}{2}$ for collisions between identical molecules, d is the molecular diameter, which can be obtained from second virial coefficients, from the van der Waals constants, or from transport phenomena such as gas viscosity or diffusion, k' is the Boltzmann constant, T is the absolute temperature, m is the mass of a gas molecule, π equals 3.14, and R is the gas constant. The quantity p is an arbitrary steric factor indicating the necessary orientation requirements for molecules of sufficient energy to react. The smaller the value of p the more stringent are the orientation requirements.

According to equation 20, if $\ln (k/T^{1/2})$ is plotted against $1/T$ a straight line is obtained; the slope equals $-\Delta E_a/R$ and the intercept equals

$$\left(\ln \kappa\pi\, d^2 \left[\frac{8k'}{\pi m} \right]^{1/2} p \right).$$

Therefore, if the molecular diameter is known, the steric factor can be evaluated. A steric factor of unity would mean that each molecular collision having energy of ΔE_a or greater will result in reaction. A steric factor of 10^{-2} indicates that only one of every 100 energetically sufficient collisions will produce a reaction because of orientation requirements.

Alternatively, one may apply the transition state theory for a bimolecular gas phase reaction at constant volume.

$$k = \frac{k'T}{h} e^{-\Delta G^*/RT} = \frac{k'T}{h} \exp 2 \cdot \exp\left(-\Delta E^*/RT\right) \exp\left(\Delta S^*/R\right) \tag{21}$$

In the preceding equation k' is the Boltzmann constant, h is Planck's constant, R is the gas constant, T is the absolute temperature, and ΔG^*, ΔE^*, and ΔS^* are the Gibbs free energy, internal energy, and entropy of activation, respectively. The term $\exp 2$ enters into equation 21 because the reaction is bimolecular.

A plot of $\ln (k/T)$ vs. $1/T$ will yield a slope that is equal to $\Delta E^*/R$ and then, from equation 21, the ΔS^* at the appropriate temperature can be calculated.

A negative entropy of activation indicates that the transition state complex has fewer degrees of freedom than the reactants, whereas a positive entropy of activation implies the opposite. The magnitude of the entropy of activation is indicative of the configuration of the activated complex.

Experimental

   CAUTION: Although no accident or mishap has occurred in five years of laboratory experimentation, the following experiment must be carried out under strict safety rules.
   Methylacrylate is a flammable liquid. Its vapor forms explosive mixtures with air. Since the vapor is denser than air it may travel a considerable distance to a source of ignition and flash back.
   It is essential, therefore, that no cigarette smoking, burning Bunsen burners, or any other ignition source should be in the vicinity of the experimental setup.
   Since the distillation and polymerization is done in vacuum, no explosive mixture of methylacrylate vapor with air is formed.
   In case of fire, water is ineffective and carbon dioxide or foam extinguishers should be used.
   Methylacrylate vapor is also a respiratory irritant, and prolonged contact with the skin or eyes may cause damage. Therefore, if methylacrylate spills, a self-contained breathing apparatus should be used when mopping up.
   Methylacrylate vapor polymerizes extremely rapidly upon irradiation with light of 2500 Å or less, with a dense white cloud of polymer forming within the reaction vessel. The reaction is so rapid that the polymer particles must be aggregates of many polymer molecules. The quantum yield for the reaction is high, and the polymer is insoluble in most solvents—the product simply swells slightly. During polymerization there is an additional complicatory reaction that produces hydrogen. This is likely to be dehydrogenation of the monomer to yield a residue, which might be methylacrylate radical ($CH\!=\!C\cdot COOCH_3$), although no definite evidence of the formation of this compound has been presented.
   This decomposition process is due to the absorption of high energy radiation, which results in instability of the molecule, causing scission or some other decomposition reaction. This is amply demonstrated by the fact that the use of light of shorter wavelength and consequently of higher energy increases the decomposition process at the expense of the polymerization reaction. As a consequence of the production of hydrogen, the pressure will never decrease to zero at the end of the reaction.
   Initially, the rate of decrease of pressure because of polymerization is about 2000 times the rate of hydrogen production which, although remaining less than the polymerization rate, increases as the reaction proceeds.

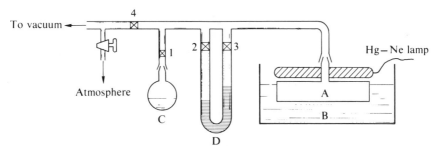

Figure 29–1 *Vacuum apparatus.*

The vacuum apparatus necessary for study of the polymerization of methylacrylate vapors is shown in Figure 29–1. It consists of a cylindrical silica reaction vessel, *A*, of about 100 cc. capacity. The reaction vessel is thermostated at the desired temperature in *B*. However, the wall of the reaction vessel that will be illuminated by a mercury-neon lamp should be out of the thermostat so that the water will not absorb the radiation, which initiates the reaction. The initiation is accomplished by a low pressure mercury-neon or mercury-argon discharge tube. These are available from several commercial sources, such as Life-lite or Fluorolite lamps by Ultraviolet Technical Laboratory, Los Angeles, or S.C. 2537 by Hanovia Chemical Co., Newark, New Jersey.

The input of the lamp is about 60 watts and stabilization is reached within a few minutes. The output of these lamps is usually 10^{16} to 10^{17} quanta sec.$^{-1}$ cm.$^{-1}$ length of lamp at 2537 Å. The operating voltage for a mercury-neon lamp is 10 volts/centimeter and for a mercury-argon lamp 15 volts/centimeter. About ten times the operating voltage is needed to start these lamps. The intensity of the second resonance mercury line is diminished by the air between the discharge lamp and the reaction vessel, (2 to 3 cm.) and by the Vycor (fused silica).

The remaining part of the system consists of a vacuum pump with liquid nitrogen traps, a sample holder, *C*, and a mercury manometer, *D*. Since the methylacrylate vapor is soluble in most conventional greases, either greaseless glass stopcocks or mercury cutoffs (Fig. 29–2) should be used at positions 1, 2, 3, and 4. Use safety glasses to avoid any possible damage to the eye by ultraviolet radiation.

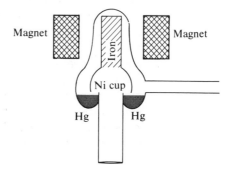

Figure 29–2 *Mercury cutoff.*

Distill commercial methylacrylate in vacuum twice to remove any stabilizers, such as hydroquinones. Place the distilled methylacrylate in the sample chamber and evaluate the whole system. Take care to remove the last traces of oxygen since it acts as an inhibitor. To do so, pump on the whole system with valves 1, 2, 3, and 4 open until the methylacrylate comes to a quick boil. After one minute of boiling, close cutoff valve 1 and continue to evacuate the system for another five minutes. Then close cutoff valve 3, establishing the right arm of the manometer as the reference arm. Isolate the system from the vacuum pump by closing cutoff valve 4.

Set the thermostat at 20° C., and after this temperature has been established, open cutoff valve 1 of the sample chamber and admit about 45 torrs (mm. of mercury) methylacrylate vapor pressure to the reaction vessel. Close cutoff valve 1. Switch on the mercury-neon discharge lamp and place a protecting dark filter or thin metal plate between the lamp and the reaction vessel. Record accurately the initial pressure of the methylacrylate vapor. A few minutes after the mercury-neon lamp becomes stabilized, suddenly remove the filter or metal plate and simultaneously start a stopwatch. Record the decrease in pressure *every 30 seconds* for about ten minutes.

(If, instead of a mercury-neon lamp, a weaker hydrogen lamp is used that has an output of 0.8×10^{14} quanta sec.$^{-1}$ cm.$^{-1}$ at 2000 to 2300 Å, the reaction must proceed for 40 to 50 minutes.)

After the first experiment, replace the filter between the lamp and the plate. Open cutoff valves 3 and 4 (cutoff valve 2 is already open) to allow atmospheric pressure to be reached in the system. Remove the reaction vessel, and wash it thoroughly to remove all the polymethylacrylate. Dry the reaction vessel and put it back into the vacuum system. Pump on the system for five minutes and repeat the experiments with 35 and finally 25 torrs initial methylacrylate vapor pressure.

After conducting the three experiments at 20° C., repeat the polymerization reaction at 30° C. with 55 and 45 torrs initial vapor pressure. All five experiments can be done comfortably within a three-hour period.

Using equations 16 and 18 establish the *order* of reaction with respect to methylacrylate pressure at 20° C.

Calculate the appropriate rate constants at 20 and 30° C. Assuming that the Arrhenius relationship is operative:

a. Calculate the energy of activation and the steric factor of the collision theory from equation 20. The diameter of the methylacrylate molecule can be considered to be 4 Å.

b. Calculate the energy and the entropy of activation in terms of transition state theory as given by equation 21.

What does the sign and the magnitude of the energy of activation indicate about this reaction? What conclusions can you draw from the magnitude of the steric factor? What does the entropy of activation indicate about the transition state complex? What is the relationship between the steric factor and the entropy of activation?

Material and Equipment. 100 grams of methylacrylate; vacuum distillation apparatus with greaseless joints; vacuum reaction apparatus with Vycor reaction vessel; mercury manometer, sample holder, and liquid nitrogen trap all joined by greaseless joints and stopcocks or mercury cutoff valves; low pressure mercury-neon or mercury-argon discharge tube (or hydrogen lamp); thermostat with heater, stirrer, and thermoregulator; safety glasses.

REFERENCES

1. H. W. Melville. Proc. Roy. Soc. A., **167**, 99 (1938).
2. C. Tanford. Physical Chemistry of Macromolecules. Wiley & Sons, Inc., New York, 1961, Chapter 9.
3. W. A. Noyes, Jr., and P. A. Leighton. Photochemistry of Gases. Reinhold Publishing Corp., New York, 1941.
4. M. Pirani and J. Yarwood. Principles of Vacuum Engineering. Reinhold Publishing Corp., New York, 1963.

30

EFFECT OF SOLVENT ON THE KINETICS OF ACETAL HYDROLYSIS BY DILATOMETRY*

The immediate environment of molecules (the solvent) has a very important influence on the rate of chemical reaction. The effects are manifold, and the most important among them are the following:

a. The general solvating power of the solvent affects the rate of reaction by simply increasing the possible concentrations of the reactants and, thereby, increasing the possible rates. This, however, does not affect the rate constant.

b. Selective solvating power of the initial states and the transition states affects the rate constants. For example, if the solvent has greater solvating power for the transition state than for the initial state, the energy of activation may be decreased, which results in a high rate constant. This aspect is especially important when the solvent is a binary mixture. Under such conditions, special solvent structures may be formed around both the initial states and the transition states, neither of which may be the same as that found in the bulk of the binary solvent mixture.

c. The solvent may not only stabilize the transition state but also help its polarization or ionization and thereby assist in reaching the final state (product).

All these effects are described here only qualitatively, but they can also be formulated quantitatively. For example, the dielectric constant of the medium in a reaction between spherical polar molecules having dipole moments of μ_i and an average radius of r_i affects the rate constant as follows:

$$\ln k = \ln k_0 + \frac{1}{kT} \frac{\epsilon - 1}{2\epsilon + 2} \left[\frac{\mu_*^2}{r_*^3} - \frac{\mu_A^2}{r_A^3} - \frac{\mu_B^2}{r_B^3} \right] \tag{1}$$

where k is the rate constant in a medium with dielectric constant ϵ, and k_0 would be the rate constant in a medium with unit dielectric constant ($\epsilon = 1$). The subscripts A and B refer to the reactants, and the $*$ to the transition state.

Equation 1 predicts (as was discussed under point b) that the rate constant increases with increasing dielectric constant if the transition state (activated complex) is more polar than the reactants. However, equation 1 is of limited accuracy. One reason is that the macroscopic dielectric constant may not represent the microscopic dielectric constant of the solvent cages formed around the reactants and transition states. This is especially the case with solvents made of binary mixtures if the microscopic structure of the solvent changes as the composition and, hence, the dielectric constant of the medium changes.

Whether such special solvent structure effects are in play can then be proved by plotting $\ln k$ vs. $(\epsilon - 1)/(2\epsilon + 2)$ over a variety of compositions of binary mixtures. A deviation from linearity may indicate special solvent structure formations at different compositions.

In the present experiment the effect of the solvent on the rate of acid-catalyzed hydrolysis of acetal will be studied. The solvent will be the dimethyl sulfoxide (DMSO)–water system (or dimethylformamide [DMF]–water, acetone–water, or other systems).

Diethylacetal hydrolysis is catalyzed by hydrogen ions and the reaction is

$$CH_3CH(OC_2H_5)_2 + H_2O \xrightarrow{H^+} CH_3CHO + 2C_2H_5OH \qquad (2)$$

This reaction has been studied to a great extent (Long and Paul; Kreevoy and Taft). It is accepted that the first step is very fast equilibration of the acetal with the hydrogen ions.

$$CH_3CH(OC_2H_5)_2 + H^+ \rightleftharpoons [CH_3CH(OC_2H_5)_2]H^+ \qquad (3)$$
$$(A) \qquad\qquad\qquad (AH^+)$$

This is followed by

$$[CH_3CH(OC_2H_5)_2]H^+ \rightleftharpoons CH_3CHOC_2H_5$$
$$\overset{|}{H}OC_2H_5 \qquad (4)$$
$$(I)$$
$$\downarrow k$$
$$(CH_3CHOC_2H_5)^+ + HOC_2H_5$$

$$CH_3CH(OC_2H_5)^+ + H_2O \rightarrow CH_3CHO + C_2H_5OH + H^+ \qquad (5)$$

The rate-determining step is represented by equation 4, which is a unimolecular and a first order reaction.

$$-\frac{dc_A}{dt} = kc_I \qquad (6)$$

However, the transition species (I) is formed by the step indicated by equation 3; hence it can be given as

$$c_I = K^* c_{H^+} c_A \qquad (7)$$

Substituting equation 7 into equation 6 we obtain

$$-\frac{dc_A}{dt} = kK^* c_{H^+} c_A \tag{8}$$

The rate is proportional to the concentration rather than to the activities of the acetal and hydrogen ions because the equilibrium constant, K^*, contains the activity coefficients

$$K^* = \frac{c_{AH^+}}{c_A c_{H^+}} \frac{\gamma_{AH^+}}{\gamma_A \gamma_{H^+}} \tag{9}$$

and, therefore, the activity coefficients of the acetal and hydrogen ions are canceled out in equation 8.

Since the hydrogen ion acts as a catalyst and is regenerated in the final step of the reaction (equation 5), the rate of reaction can be considered to be pseudo–first order. In order to calculate the rate constant the Guggenheim method (reference 5) will be employed, which is useful when the reaction does not go to completion within the experimental time allotted.

For the measurement of the rate, a physical quantity that is a linear function of the concentration can be used. In the present experiment we use a dilatometer in which changes in volume are observed. The dilatometer is a thermostated vessel terminating in a graduated capillary. The capillary has a small radius and the changes in the total volume of the solution can be observed by the rise of the meniscus in the capillary. Hence, the experimentally observed meniscus height is a linear function of the total volume. On the other hand, the total volume of the solution, V, is a linear function of the number of moles, n_i, of component species i,

$$V = \int_i n_i \bar{V}_i \tag{10}$$

because the partial molal volumes, $\bar{V}_i$, of the solvent and reactants can be considered to be constant in dilute solution. Finally, the number of moles, n_i, is a linear function of the disappearing acetal concentration. Thus the meniscus height, h, in arbitrary units is related to the acetal concentration.

$$h = a - bc_A \tag{11}$$

where a and b are constants.

The Guggenheim method may be applied directly if data are taken at equal time intervals. If the data are taken at unequal time intervals, a plot of concentration, meniscus height, or other measure of the extent of reaction vs. time, t, is prepared and a smooth curve drawn. The required data may be read from this curve.

The data are arranged in two sets. For each observation (c_1) at time t in the first set, another observation (c_2) is made at time $t + \Delta t$, where Δt is a fixed time interval. If a plot of log $(c_1 - c_2)$ vs. t is prepared, the points will be on a straight line of slope $-k/2.303$. The constant time interval, Δt, may be considered to be approximately one-half the duration of the experiment. If Δt is too small, there will be a large error in $c_1 - c_2$.

The equation for this method may be derived as follows: From the integrated form of the differential equation for the first order reaction,

$$c = c_0 e^{-kt} \tag{12}$$

Thus, the concentrations c_1 and c_2 at two times differing by Δt are

$$c_1 = c_0 e^{-kt} \tag{13}$$

$$c_2 = c_0 e^{-k(t+\Delta t)} \tag{14}$$

Subtracting,

$$c_1 - c_2 = c_0 e^{-kt}(1 - e^{-k\Delta t}) \tag{15}$$

Taking logarithms,

$$\log (c_1 - c_2) = -kt/2.303 + \log [c_0(1 - e^{-k\Delta t})] \tag{16}$$

Thus, the slope of a plot of $\log (c_1 - c_2)$ vs. t is $-k/2.303$.

If instead of measuring concentration directly, some linear function of the concentration (in our case meniscus height) is measured, an equation of the same form as equation 16 applies. For example, using equation 11,

$$c_1 - c_2 = (h_1 - h_2)/a \tag{17}$$

and

$$\log (h_1 - h_2) = -kt/2.303 + \log [ac_0(1 - e^{-k\Delta t})] \tag{18}$$

Experimental

Set up a constant temperature bath at 25° C. and maintain the temperature within $\pm 0.001°$ C. The importance of this constant temperature can be understood by considering that small fluctuations in the temperature of the bath can cause changes in the height of the meniscus that are of the same order of magnitude as that caused by the chemical reaction. Provide a Beckmann thermometer so that the constancy of the bath can be recorded during the experiment.

The dilatometer is shown in Figure 30–1. Essentially, it consists of a 100 cc. Pyrex tube, A, that is connected by a stopcock to a U tube, B. The left arm of the U tube is a Pyrex tube of 2 mm. inside diameter and 8 cm. long, and the right arm is a Pyrex tube of 2 cm. inside diameter and about 10 cm. long, containing about 30 to 50 cc. of solution. The right arm of the U tube is narrowed and ends in a graduated capillary, C, of 0.5 mm. inside diameter and about 15 cm. long.

Thoroughly clean the dilatometer by removing the stopcock and inserting two corks in its place. Fit a rubber stopper with a glass tube and insert it into the end of tube A, and then connect the dilatometer to an aspirator. Inverting the dilatometer and dipping its capillary end into a detergent solution in a beaker, wash the dilatometer and rinse it with water, then

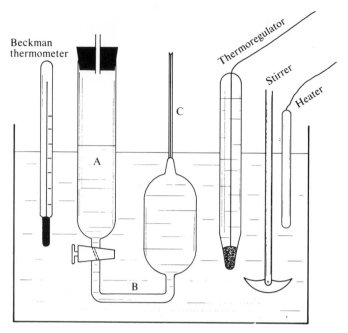

Figure 30–I *Dilatometer.*

with distilled water and finally with acetone. With continued use of the aspirator, dry the dilatometer by allowing air to pass through it. Carefully grease the stopcock with silicone grease. (Care must be taken not to use regular stopcock grease because DMSO, DMF, and so forth, may dissolve some grease.) Then immerse the dilatometer in the constant temperature bath as indicated in Figure 30–1.

Prepare six 250 ml. solutions of the DMSO-water solvent system (or DMF-water, acetone-water, and so forth). The range of composition should vary from 0 to 0.9 mole fraction of DMSO (DMF, acetone, and so forth). Make each solution 5×10^{-3}M with respect to hydrochloric acid by using the concentrated acid.

Thermostat these solutions in the bath at least 15 minutes before the start of the experiment.

The first experiment will be made in water and successive experiments will be done in the other five DMSO-water solvents.

Pipet 1 ml. of diethylacetal (dry and of high purity) into a 100 ml. volumetric flask. Add the aqueous hydrochloric acid solution from the thermostated flask and swirl it to dissolve the acetal. Note the time of mixing as t_0. Bring the mixture to volume.

With the stopcock of the dilatometer closed, transfer approximately 90 ml. of the reaction mixture into chamber A of the dilatometer (Fig. 30–1). Insert the rubber stopper with the tube at the top of chamber A and apply slight suction to dislodge any air bubbles trapped during the transfer.

Open the stopcock and allow the U tube, B, to be partly filled. Again apply suction at the tube affixed to the rubber stopper to dislodge any air bubbles that may be trapped below the stopcock. Close the stopcock and raise the dilatometer from the bath to see whether any bubbles are visible. If there are none, lower the dilatometer in the bath and allow chamber B to be filled. Do this very slowly so that the stopcock may be closed immediately after the solution has entered the capillary.

With a cathetometer measure the height of the meniscus at one minute intervals for 15 to 30 minutes. The data should be tabulated as t_1, $h = f(t)$, Δh, and log Δh.

At the end of the procedure clean the dilatometer and repeat the experiment with the next DMSO-water system. When the concentration of DMSO reaches 40 per cent and above, use only *0.5 ml. of acetal* as the initial concentration.

Plot the log Δh against t for each of the six solvent systems. From the best straight line, obtain the rate constant for each system.

Plot the log k against the mole fraction of DMSO.

Use your previous data on the dielectric constants of DMSO-water mixtures (Experiment 15) and plot log k against $(\epsilon + 1)/(2\epsilon + 2)$.

What general conclusions can you draw regarding the solvent effect upon the rate of reaction? What specific conclusion can you draw from your graph of log k vs. $(\epsilon + 1)/(2\epsilon + 2)$ in view of the relationship given in equation 1? Your comments should take into account what you have already learned of your solvent system (DMSO-water, DMF-water, acetone-water, and so forth) from previous experiments.

Material and Equipment. 10 ml. of diethylacetal; 500 ml. of dimethyl sulfoxide (DMSO) (or dimethylformamide [DMF], acetone, dioxane, and so forth); 100 ml. of concentrated hydrochloric acid; 100 ml. volumetric flask; silicone stopcock grease; dilatometer of 100 ml. capacity; thermostat to be maintained at $25 \pm 0.001°$ C.; Beckmann thermometer; pipets.

REFERENCES

1. F. A. Long and M. A. Paul. Chem. Rev., **57**, 935 (1957).
2. M. M. Kreevoy and R. W. Taft. J. Am. Chem. Soc., **77**, 3146 (1955).
3. R. K. Wolford and R. G. Bates. J. Phys. Chem., **66**, 1496 (1962).
4. R. K. Wolford. J. Phys. Chem., **68**, 3392 (1964).
5. E. A. Guggenheim. Phil. Mag., **2**, 538 (1926).

31

PRIMARY SALT EFFECT IN KINETICS OF IONIC REACTIONS

Reactions in liquid phase differ considerably from those in gas phase. The velocity of the reaction is influenced largely by the solvent since in most cases the solvent interacts strongly with the reactants. This is especially the case for ionic reactions in aqueous systems, because it is the water that allows the dissociation of a salt into its ionic species because of the energy released in the solvation process. The effect of the solvent and the behavior of ions in aqueous solution are demonstrated in the activity coefficient of the reacting species.

If we consider the transition state theory explaining the velocity of the reactions, the rate constant, k, is written as

$$k = \frac{k'T}{h} K^* \frac{\gamma_A^a \gamma_B^b}{\gamma_{M^*}} \tag{1}$$

where k' is the Boltzmann constant, T is the absolute temperature, h is Planck's constant, K^* is the equilibrium constant between activated complex and reactants given in terms of concentrations, and the γ's are the activity coefficients.

One visualizes that the ionic reaction in aqueous systems deviates from some ideal reference state because of solute-solute and solvent-solute interaction. This deviation is expressed in terms of free energy change, ΔG_i^n, involved in the transition from the reference state to the actual state. The ΔG_i^n is related to the activity coefficient, γ_i, of the ion i.

$$\Delta G_i^n = k'T \ln \gamma_i \tag{2}$$

Using Debye-Hückel potentials, the activity coefficient is given as

$$k'T \ln \gamma_i = \frac{z_i^2 e^2}{2\epsilon a} - \frac{z_i^2 e^2}{2\epsilon(1 + \beta a)} \tag{3}$$

where the z_i's are the ionic valences of the *reacting* species, ϵ is the macroscopic dielectric constant, e is the electronic charge, β is the ionic strength, and a is the closest approach between two ions.

The change from a reference state to the actual state at constant temperature and pressure can be expressed in terms of the equilibrium constants of those two states.

$$\Delta G^n = RT \ln \frac{K_{actual}}{K_{ref}} \tag{4}$$

Using the transition state terminology given in equation 1, the rate constant of a reaction in the actual state is

$$\ln k = n\, k_0 - \frac{z_1 z_2 e^2}{\epsilon a k'T} + \frac{z_1 z_2 e^2}{\epsilon k'T(1 + \beta a)}, \tag{5}$$

where k_0 is the rate constant in the reference state. For ionic reactions, this reference state is a hypothetical state where $\epsilon \to \infty$ and $\beta \to 0$. Under these conditions

$$\frac{z_1 z_2 e^2}{\epsilon a k'T} \to 0 \text{ since } \epsilon \to \infty \quad \text{and} \quad \frac{z_1 z_2 e^2 \beta}{\epsilon k'T(1 + \beta a)} \to 0 \text{ since } \beta \to 0$$

Hence the equality $k_0 = \dfrac{k'T}{h} K^*$ refers to such a hypothetical situation.

For dilute aqueous solutions at $25°$ C., equation 5 reduces to

$$\log k/k_0 = 1.02 z_1 z_2 \beta^{1/2} \tag{6}$$

where

$$\beta = \tfrac{1}{2} \sum c_i z_i^2 \tag{7}$$

Equation 6, which is known as the Brønsted relation, predicts that the rate constant should increase with increasing ionic strength, β, when the reacting ions have the same charge and should decrease with increasing ionic strength when the reacting ions are oppositely charged. This effect is called the primary salt effect.

In calculations of ionic strength, β, (equation 7) all ionic species are summed up whether they participate in the reaction or are present as neutral electrolytes. Equation 6 would predict that a straight line is obtained when $\log k$ is plotted against $\beta^{1/2}$ (Figure 31–1). This has been observed in many ionic reactions at low ionic strength, but deviation from such a straight line plot is expected when the ionic strength increases and the Debye-Hückel limiting law is no longer applicable.

In the present experiment, a simple ionic reaction will be studied in which the iodide ion is oxidized to iodine by persulfate ion. The stoichiometry of the reaction is given in equation 8.

$$2I^- + S_2O_8^= \to I_2 + 2SO_4^= \tag{8}$$

One can monitor the rate of appearance of iodine either by titrating with sodium thiosulfate at set intervals or by using polarized electrode techniques (Indelli and Prue).

A simple modification of this reaction is called the iodine clock reaction, in which, rather than measuring the rate of production of iodine, we measure the time required

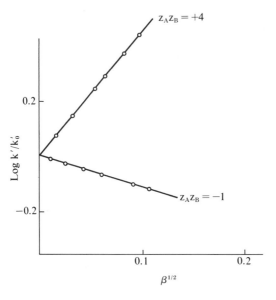

Figure 31-1 *Test of Brønsted relation.*

for free iodine to appear when an initial amount of sodium thiosulfate has been added to the solution.

In the iodine clock reaction the following reaction occurs simultaneously with that given in equation 8.

$$2S_2O_3^= + I_2 \rightarrow 2I^- + S_4O_6^= \tag{9}$$

The reduction of free iodine to the iodide ion by the thiosulfate ion (equation 9) is a fast reaction compared to the oxidation of the iodide ion by the persulfate ion. Therefore, equation 8 is the rate-determining step.

When persulfate ion is added to a reaction mixture containing the iodide ion and a small amount of thiosulfate ion, a slow oxidation process ensues. However, the liberation of free iodine is not indicated by the appearance of a blue color of the starch-iodine indicator because the liberated iodine is reduced immediately back to the iodide ion. This process continues until the added thiosulfate ion is exhausted (i.e., in reaction 8 the amount of persulfate ions used up equals the added moles of thiosulfate ions). At this point the liberation of iodine commences and is indicated by the appearance of the blue color of the starch-iodine complex. Before the blue color appears a minimal amount of free iodine must be liberated because the blue color is caused by polyiodine chain buildup inside the amylose helix of the starch indicator. This means that the color appears later than the first iodine molecule is liberated. In order to make this induction period as uniform as possible, a small amount of acid (0.0005M hydrochloric acid) and 5×10^{-6}M ethylenediamine tetra-acetate (EDTA) is added to the test solutions.

Since in the scheme equation 8 is the rate-determining step, the reaction should be first order with respect to both iodide and persulfate ions. When the initial amount of iodide ion exceeds the persulfate ion concentration (a tenfold difference will do) and sodium thiosulfate is added to allow the fast reaction of equation 9 to occur, there is hardly any change in the concentration of iodide ion up to the point at which the blue color appears. Therefore, the reaction is a pseudo–first order reaction under this condition

$$-\frac{d(S_2O_8^=)}{dt} = k_{app}(S_2O_8^=) \tag{10}$$

and the apparent rate constant, k_{app}, includes the constant iodide ion concentration.

$$k_{app} = k(I^-) \tag{11}$$

We assume that the rate of reaction is

$$-\frac{d(S_2O_8^=)}{dt} = \frac{\Delta x}{\Delta t} \tag{12}$$

where Δx is the number of moles of thiosulfate ion added to the solution, and Δt is the time elapsed from the moment of mixing of the persulfate with the reagents until the blue color appears. Thus, with this assumption, k_{app} can be calculated from equation 10 by varying the initial persulfate ion concentration. Knowing also the initial iodide ion concentration, the rate constant can be obtained (equation 11).

Furthermore, by varying the ionic strength of the medium through the addition of neutral electrolyte (potassium nitrate), one can test whether the Brønsted relationship is obeyed.

Experimental

Prepare the following stock solutions in water: 250 ml. of 0.1M potassium iodide; 150 ml. of 10^{-3}M sodium thiosulfate; 300 ml. of 10^{-2}M potassium persulfate solution. Prepare 1 liter of a solution that is 0.001N in hydrochloric acid and 1×10^{-5}M in EDTA. This solution will be called the "solvent." Using this solvent prepare 200 ml. of 1M potassium nitrate solution. Thermostat all these stock solutions at 25° C.

Prepare starch indicator by dispersing 1 gram of starch in about 20 ml. of cold water, making a fine paste. Add 80 ml. of cold water, stir the solution, and bring it to a quick boil. Cool the starch solution to room temperature and test a 5 ml. portion of it by adding iodine solution. The appearance of an intense blue color shows that the starch is suitable as an indicator. (Many students' experiments have to be repeated when this precaution of testing has not been performed because some starch preparations in stockrooms retrograde during storage and do not give suitable color for use as indicators.)

Set up ten 250 ml. Erlenmeyer flasks in a thermostat at 25° C. and number them. Add a lead collar or sinker to keep them firmly in the thermostat. Add the reagents to the flasks according to the scheme given in Table 31–1.

BE SURE THAT THE POTASSIUM PERSULFATE SOLUTION IS ADDED *LAST* TO EACH FLASK AND THAT THE START OF THE REACTION IS NOTED AS THE TIME OF MIXING THE POTASSIUM PERSULFATE WITH THE MIXTURE.

TABLE 31-1. Volume of Reagents Added (in Milliliters)

Flask No.	10^{-1}M KI	10^{-3}M $Na_2S_2O_3$	1M KNO_3	Solvent	Starch	10^{-2}M $K_2S_2O_8$
1	20	10	0	59	1	10
2	20	10	0	44	1	25
3	20	10	0	34	1	35
4	20	10	1	43	1	25
5	20	10	3	40	1	25
6	20	10	5	38	1	25
7	20	10	10	33˙	1	25
8	20	10	20	23	1	25
9	20	10	25	18	1	25
10	20	10	35	8	1	25

Occasional mixing of the reaction flasks is in order. The end of the reaction is noted as the time at which the blue color of the starch-iodine complex appears in each flask. In order to avoid confusion, it is preferable to start the reaction in the first three flasks only (in the absence of neutral electrolyte), and to start the reactions in the other flasks after these reactions have been completed.

Calculate the apparent and the true rate constants for the reaction without added electrolytes from the first three flasks.

Comment on the agreement obtained among the three values.

Plot the log k values obtained vs. $\beta^{1/2}$.

Is the Brønsted relation obeyed? Over what concentration range? In what direction does the deviation occur? What is your explanation for the deviation?

Material and Equipment. 5 grams of potassium iodide; 1 gram of potassium persulfate; 25 grams of potassium nitrate; 1 gram of sodium thiosulfate pentahydrate; 1 gram of ethylenediamine tetraacetate (EDTA); 10 ml. of concentrated hydrochloric acid; 1 gram of soluble starch indicator; 0.1 gram of iodine; ten 250 ml. Erlenmeyer flasks; lead collar or sinker; five 50 ml. burets; one 10 ml. graduated pipet; thermostat with heater, stirrer, and thermoregulator.

REFERENCES

1. I. Amdur and G. G. Hammes. Chemical Kinetics. McGraw-Hill Book Co., New York, 1966.
2. O. M. Knudson and C. V. King. J. Am. Chem. Soc., **60,** 687 (1938).
3. A. Indelli and J. E. Prue. J. Chem. Soc. (London), **1959,** 107.

32

KINETICS OF ENZYME
REACTIONS; ACTIVITY
OF TYROSINASE

Many chemical reactions, and most biochemical reactions, proceed slowly at normal temperatures when only reactants and products are present. These same reactions, however, may occur rapidly in the presence of certain other compounds that are not used up in the reaction. These additional compounds, or catalysts, make many reactions possible at room temperature. When the catalyst and the reaction mixture form a single phase, homogeneous catalysis is said to occur. When the catalyst forms a separate phase [for example, solid triethyl aluminum and titanium chloride in the polymerization of olefins] heterogeneous catalysis is said to occur.

Most biochemical reactions are catalyzed by complicated macromolecules (proteins or glycoproteins) called enzymes. Since a biochemical reaction usually takes place in an aqueous system and the enzymes are soluble in these systems, this is an example of homogeneous catalysis. However, since the enzymes are high molecular weight compounds, their large size results in the formation of colloidal solutions in which the surface properties of the enzymes are of primary importance. From this point of view, enzyme catalyzed reactions have the characteristics of heterogeneous catalysis in which the important quantity is the molecular surface available.

The conformation of the active enzyme, i.e., the specific three-dimensional structure into which the protein chains are coiled or twisted, is labile. Changes in temperature, pH of the medium, or addition of organic compounds, urea and so forth, change the specific active conformation. This process is called denaturation and can be either reversible or irreversible. Since enzyme activity can be decreased or completely lost by such conformational changes, it is clear that, in addition to molar concentration, the specific conformation is of great importance.

The mechanism of an enzyme reaction is visualized as follows: The reacting small molecules approach the enzyme and are temporarily adsorbed. This adsorption is very specific in most cases. We distinguish between the "active site" of the enzyme (the site at which the reaction will occur) and the rest of the molecule. For example, in chymotrypsin the active site is serine, which is part of one protein chain. However, two histidine residues, parts of another protein chain, are near the serine, and by their specific stereochemistry they aid the catalysis. The enzyme papain is composed of a single polypeptide chain with 211 amino acid residues. The active site is a cysteine that is the twenty-fifth amino acid of the sequence. Other amino acid residues in the active site, which is a cleft on the enzyme surface, are: histidine (106), trypto- phan (128), glycine (19), and aspartic acid (105 and 160). (The numbers in paren- theses indicate the positions of the respective amino acids in the polypeptide chain.) One can see that the twisting and folding of the chain brings parts of the chain close enough together for them to act catalytically, whereas they would otherwise be quite far apart in the sequence.

Since the active site is only a small portion of the enzyme molecule, presumably the rest of the molecule is necessary to bring and hold together certain amino acids at the exact spatial arrangement necessary for catalytic action. The rest of the mole- cule may also perform a selective screening, allowing only certain specific molecules to be adsorbed. For instance, in Figure 32–1 substrate 1 must have very special configurational properties to be absorbed in the enzyme crevice at the exact position indicated.

Once such specific adsorption has occurred, the reaction proceeds rapidly and the products diffuse from the surface.

In our experiment we shall study the action of tyrosinase on *l*-tyrosine. Tyro- sinase, or polyphenoloxidase, is an enzyme present in various *l*-plant and animal tissues. It is a copper-containing enzyme, and it oxidizes tyrosine through a number of intermediates to a dark colored polyphenolic compound, melanin. The reaction is as follows (see Dawson and Tarpley):

Figure 32–1 *Schematic diagram of enzyme-substrate interaction. The 1 and 2 are the reacting substrates. The light and dark parts represent two protein chains of a hypothetical enzyme.*

The hypothesis is that tyrosinase catalyzes the first two steps through a change in the valence of the copper.

$$2Cu^{++}\text{ enzyme} + \text{tyrosine} \rightleftharpoons 2Cu^{+}\text{ enzyme} + \text{DOPA}$$

$$2Cu^{+}\text{ enzyme} + \tfrac{1}{2}O_2 \rightleftharpoons 2Cu^{++}\text{ enzyme} + O^{=}$$

and

$$2Cu^{++}\text{ enzyme} + \text{DOPA} \rightleftharpoons 2Cu^{+}\text{ enzyme} + o\text{-quinone} + 2H^{+}$$

$$\tfrac{1}{2}O_2 + 2Cu^{+}\text{ enzyme} + 2H^{+} \rightleftharpoons 2Cu^{++}\text{ enzyme} + H_2O$$

This hypothesis (see Singer and Kearney) is disputed by some authors and, therefore, should not be considered an established mechanism.

The rate of reaction can be monitored manometrically by determining the oxygen uptake. Alternatively, the change in the absorbency of the ultraviolet spectrum can be observed. The tyrosine itself absorbs light in the ultraviolet region, and as soon as the reaction begins an increase in the absorbency at 280 mμ can be observed because of the appearance of intermediates and products. The tyrosinase activity is given in terms of units/milligram. One unit increases the absorbency at 280 mμ by 0.001 per minute at 25° C. Commercial preparations of mushroom tyrosinase are sold with 500 to 700 units/milligram.

For a kinetic study it is better to monitor the absorbency of DOPA at 310 mμ or that of hallachrome at 485 mμ because the unreacted tyrosine does not interfere in these regions.

The effect of reactant (substrate) concentration on the velocity of simple enzyme reactions is usually given by a curve such as in Figure 32–2.

The analysis of such a curve was first described by Michaelis and Menten and later extended by Briggs and Haldane. Assume the following mechanism:

$$E + S \underset{k_2}{\overset{k_1}{\rightleftharpoons}} ES^* \xrightarrow{k_3} P + E \tag{1}$$

where E is free enzyme, S is substrate, ES^* is the activated enzyme substrate complex, and P is the product.

The concentration of free enzyme at any time will be $(E_0 - ES^*)$, that is, the difference between the initial enzyme concentration and that tied up in the enzyme

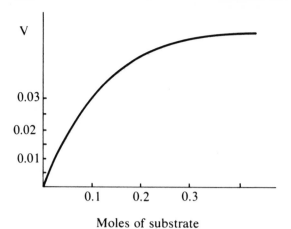

Moles of substrate

Figure 32–2 *Substrate concentra-tion–velocity curve.*

substrate complex. The concentration of the latter (ES^*) is assumed to be very small and in steady state.

Therefore, the following rate equations can be written:

$$\text{Rate} = V = -\frac{d(S)}{dt} = k_1(E)(S) - k_2(ES^*) \tag{2}$$

$$\text{Rate} = V = \frac{d(P)}{dt} = k_3(ES^*) \tag{3}$$

$$\text{Steady rate} = \frac{d(ES^*)}{dt} = k_1(E)(S) - (k_2 + k_3)(ES^*) = 0 \tag{4}$$

Therefore,

$$(ES^*) = \frac{k_1}{k_2 + k_3}(E)(S) = \frac{k_1}{k_2 + k_3}(S)[(E_0) - (ES^*)] \tag{5}$$

$$\frac{(E_0) - (ES^*)}{(ES^*)} = \frac{(E_0)}{(ES^*)} - 1 = \frac{(k_2 + k_3)}{k_1} \cdot \frac{1}{S} \tag{6}$$

$$(ES^*) = \frac{k_1(S)(E_0)}{(k_2 + k_3) + k_1(S)} \tag{7}$$

and from equation 3

$$\text{Rate} = V = \frac{k_3(S)(E_0)}{K_m + (S)} \tag{8}$$

where

$$K_m = \frac{k_2 + k_3}{k_1}$$

The maximum rate, V_{max}, is attained when the concentration of substrate is so high that all the enzyme, (E_0), is in the enzyme-substrate complex (ES^*).

$$(E_0) = (ES)^*$$

in which case equation 3 becomes

$$V_{max} = k_3(E_0) \tag{9}$$

Therefore equation 8 can be rewritten in terms of equation 9.

$$\text{Rate} = V = \frac{V_{max}(S)}{K_m + (S)} \tag{10}$$

Taking the reciprocal of equation 10,

$$\frac{1}{V} = \frac{K_m}{V_{max}} \frac{1}{(S)} + \frac{1}{V_{max}} \tag{11}$$

This is plotted in Figure 32–3.

Therefore, if an enzyme reaction is studied isothermally, by varying the substrate concentration but leaving the enzyme concentration fixed, the data obtained can be plotted as in Figure 32–3. The two constants obtainable from such a plot are K_m and k_3. The latter is obtained from V_{max} via equation 9. The rate constant, k_3, refers to the decomposition of the enzyme-substrate complex.

The K_m, however, is a complex quantity. If *it can be assumed that the* $k_3 \ll k_2$, the K_m can be interpreted as an equilibrium constant indicating the affinity of the enzyme for the substrate. Under this condition, if K_m is large there is little tendency for a strong enzyme-substrate complex formation; on the other hand, small values of K_m would indicate great affinity of a substance for the specific enzyme surface.

In the other extreme case in which $k_2 \ll k_3$, the K_m is simply the ratio of k_3/k_1, and since k_3 is known from the value of V_{max}, the k_1 can be determined.

In many cases, however, k_2 and k_3 are of the same magnitude and, therefore, neither of them can be neglected in favor of the other. Under these circumstances K_m has no simple interpretation.

The literature is full of K_m values interpreted as enzyme-substrate dissociation constants $(k_3 \ll k_2)$ without full justification of such assumption. For example, different values for the rate constants k_1, k_2, and k_3 obtained for cytochrome C reductase in separate experiments were found to be $k_1 = 85 \times 10^6$, $k_2 = 53 \times 10^8$, and $k_3 = 8 \times 10^3$. Under this condition the interpretation of K_m would be justly that of a dissociation constant. For another enzyme, the so-called "old yellow enzyme," catalyzing the same reaction, the values for the same rate constants were

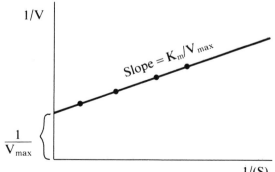

Figure 32–3 *Lineweaver-Burk plot.*

found to be $k_1 = 6 \times 10^6$, $k_2 = 3 \times 10^4$, and $k_3 = 10 \times 10^4$. In the latter case K_m is obviously not a dissociation constant. The units in all cases are $mole^{-1} sec^{-1}$. These data illustrate that when separate determinations are not done to establish the individual rate constants, the identification of K_m as a dissociation constant must be made with care.

It is interesting to compare the Michaelis-Menten analysis of enzyme reaction to that of a Langmuir isotherm in surface adsorption of monolayer. The comparison is more intriguing since our model indicated specific active sites that can be saturated at V_{max}, just as a solid surface can be covered with monolayer. In effect, the Langmuir isotherm has the same mathematical form as equations 8 and 10.

$$y = \frac{y_m P}{a + P} \tag{12}$$

In this comparison the rate of reaction, V, is compared to the amount of substrate adsorbed on a monolayer, y. The variable in equation 12 is the pressure, P, which through gas laws can be related to a concentration term, $P/RT = n/V$. Hence, P can be the counterpart of S, the concentration of substrate in equation 8. With this kind of identification the equality $V_{max} = k_3 E_0$ in equation 9 becomes the amount of material in a completed monolayer, y_m, in equation 12 and constant K_m equals a, which is an equilibrium constant, the ratio of the rate constant of evaporation to that of condensation on the surface. Hence, in this kind of identification K_m is again a dissociation constant.

The purpose of this analogy is to show that phenomenologically the enzyme catalyzed reaction can be related to a surface catalyzed reaction (heterogeneous catalysis).

The purpose of this experiment is to determine whether tyrosinase action on *l*-tyrosine follows the Michaelis-Menten-Briggs-Haldane mechanism. In the affirmative case one also would like to obtain the rate constants k_3 and K_m.

Experimental

The rate of reaction of tyrosinase upon tyrosine will be followed spectrophotometrically. The increase in absorbency is to be monitored at 280, 310, and 485 $m\mu$. If a recording spectrophotometer is available a spectrum around these wavelengths has to be obtained every five minutes for at least an hour. If no recording spectrophotometer is available, the change in absorbency at these wavelengths has to be monitored every five minutes.

Detailed rules for the operation of the particular instrument you will be using are provided in the manual furnished by the manufacturer.

A stock solution of 70 mg. of tyrosine in 500 ml. of distilled water should be prepared.

Mushroom tyrosinase or polyphenoloxidase is sold as lyophilized preparations having an activity of 500 to 700 units/milligram. *CAUTION:*

The instructor should check in advance whether the activity of the enzyme meets the requirements. Good preparations have been obtained from firms such as Mann Research Laboratories, New York, and Schwarz BioResearch Laboratories, Orangeburg, New York. If the enzyme has an activity of 500 unit/milligram, a *stock solution of 12.5 mg/500 ml.* should be prepared. Since lyophilized material tends to clump in distilled water, it is necessary to ensure that all the enzyme solution will be completely dissolved in water by dispersing the clumps mechanically with a spatula.

Such a stock solution when mixed with tyrosine solution in the proportions given in the following paragraphs will have an activity producing an increase in the absorbance of 0.01/minute at 25° C. at 280 mμ. Therefore, sufficient change can be monitored conveniently at five minute intervals for an hour.

Prepare a 25° C. thermostated bath for the samples and thermostat the stock solutions and the samples in this bath.

First run: Obtain the ultraviolet absorption spectrum of a solution made of 8 ml. of tyrosine stock solution and 6 ml. of distilled water run against a distilled water blank. Be sure that silica cuvettes are used. Record the entire ultraviolet spectrum from 260 to 390 mμ or the absorbancies at 275, 280, and 310 mμ.

Second run: Prepare a *blank* made of 8 ml. of distilled water and 6 ml. of enzyme stock solution.

Prepare a sample by mixing 8 ml. of tyrosine stock solution with 6 ml. of enzyme stock solution. Note zero time as the time of mixing.

Leave the reaction mixture in the thermostat in an *open vessel* (Erlenmeyer flask) so that the oxygen necessary for the reaction will be available. *Occasional swirling of the mixture is advisable.* After about three minutes elapse, fill the sample cuvette and obtain first the ultraviolet spectrum (scanning from 260 to 390) and later the visible spectrum (from 400 to 500 mμ). Note the reaction time at each peak (at 275, 280, 310, and 485 mμ). *Return the reaction mixture from the cuvette into the open Erlenmeyer flask.*

Repeat the process every five minutes at the beginning, and every ten minutes during the hour.

Third run: Obtain the ultraviolet absorption spectrum of a solution of 6 ml. of tyrosine stock solution and 8 ml. of distilled water run against a distilled water blank. (This run is similar to the first run but with a different concentration of tyrosine.)

Fourth run: Determine the rate of the reaction, as in the second run, of a solution made up of 6 ml. of tyrosine stock solution, 6 ml. of enzyme stock solution, and 2 ml. of distilled water run against a blank of 6 ml. of enzyme stock solution and 8 ml. of distilled water (the same blank as used in the second run).

Fifth run: Obtain the ultraviolet spectrum of a solution of 4 ml. of tyrosine stock solution and 10 ml. of distilled water run against a distilled water blank.

Sixth run: Determine the rate of reaction, as in the second run, of a solution made of 4 ml. of tyrosine stock solution, 6 ml. of enzyme stock solution, and 4 ml. of distilled water run against a blank of 6 ml. of enzyme stock solution and 8 ml. of distilled water (the same blank as used in the second and fourth runs).

From the spectra obtained in the first and second runs, plot the change in absorbance as a function of time at 275, 280, 310, and 485 mμ. Obtain similar plots for the pairs of runs 3–4 and 5–6. (Note that in these pairs of runs the first, third, and fifth runs give the spectra of different initial concentrations at zero time. Their pairs, 2, 4, and 6, respectively, give the spectra as a function of time. Therefore, the change in absorbency is the difference between the spectra of runs 2 and 1, 4 and 3, and 6 and 5.)

From the initial slopes of the curves of change in absorbency against time, obtain the *rates of reaction* at 275, 280, 310, and 485 mμ for the different concentrations of tyrosine. The units of rate will be absorbency/minute. Make a Lineweaver-Burk plot for each wavelength, 310 and 485 mμ.

Obtain the k_3 and K_m values at each of these wavelengths.

Interpret the data at the two wavelengths with the assumption that K_m is a dissociation constant, and comment on whether this assumption is reasonable.

Material and Equipment. 100 mg. of *l*-tyrosine; 20 mg. of mushroom tyrosinase (polyphenoloxidase) of 500 to 700 units/milligram; thermostat with heater, stirrer, and thermoregulator; ten 100 ml. Erlenmeyer flasks; spectrophotometer (possibly recording) covering both ultraviolet and visible ranges; matched 1 cm. silica cuvettes.

REFERENCES

1. C. R. Dawson and W. B. Tarpley. In Summer, J. B., and Myrbäck, K. (eds.). The Enzymes. Vol. II, Part 1, Chapter 57, Academic Press, New York, 1951.
2. T. P. Singer and E. B. Kearney. In H. Neurath, H., and Bailey, K. (eds.). The Proteins Vol. 2, Academic Press, New York, 1954, p. 135.
3. L. Michaelis and M. L. Menten. Biochem. Zschr., **49**, 333 (1913).
4. G. E. Briggs and J. B. S. Haldane. Biochem. J., **19**, 338 (1925).
5. H. Lineweaver and D. Burk. J. Am. Chem. Soc., **56**, 658 (1934).
6. J. Drenth, J. N. Jansonius, R. Koekoek, H. M. Swen and B. G. Wolthers. Nature **218**, 929 (1968).

X

RADIOACTIVITY

33

RADIOACTIVE DECAY

The disintegration of a radioactive nuclide is a typical first order reaction. For example, the decay of ^{131}I, which emits beta rays, can be written as a stoichiometric equation:

$$^{131}\text{I} \rightarrow {}^{131}\text{Xe} + \beta^- \tag{1}$$

Although equation 1 is sufficient for the discussion of kinetics, one should be aware that even with a single radionuclide there are multiple pathways between reactant and products. This is illustrated in Figure 33–1 for ^{131}I. Thus, ^{131}I emits four beta rays within the range of 0.250 to 0.185 Mev. The most frequent is the 0.608 Mev. emission (87.2 percent of the time). These yield different excited ^{131}Xe species, which return to the ground state by gamma ray emission. In accordance with Figure 33–1, six gamma rays are emitted within the range of 0.080 to 0.722 Mev. (Mev. $= 10^6$ electron volts).

In spite of the multiple pathways of ^{131}I decay, it is a simple process because it can be represented by one stoichiometric equation (1). It should not be confused with multiple pathway processes for which different stoichiometric equations must be written. For example, ^{112}In decays 44 percent of the time by positron emission, 24 percent of the time by beta ray emission, and 32 percent of the time by electron capture.

As a first order reaction, the number of atoms disintegrating per unit time is proportional to the number of radioactive nuclides present, N,

$$-\frac{dN}{dt} = \lambda N \tag{2}$$

where λ is the disintegration rate constant, expressed in reciprocal seconds (sec.$^{-1}$).

The integrated form of equation 2 is

$$\ln \frac{N}{N_0} = -\lambda t \tag{3}$$

where N_0 is the number of radionuclides present at zero reference time.

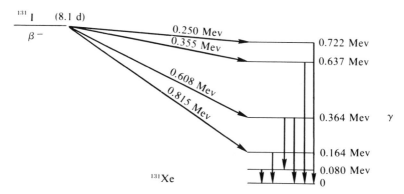

Figure 33–I *Decay scheme of* ^{131}I. *(From Lederer, Hollander, and Perlman.)*

In general, in radioactive decay, the counting rate of a sample rather than the number of atoms disintegrating is measured. The counting rate is proportional to the disintegration rate or absolute activity. The latter, in turn, is proportional to the number of radionuclides present. Hence,

$$-\frac{dN}{dt} = A = N\lambda \tag{4}$$

and

$$\frac{dC}{dt} = R = bA \tag{5}$$

where A is the disintegration rate, R is the counting rate, and b represents the respective proportionality constants.

Therefore, equation 3 becomes

$$\ln \frac{N}{N_0} = \ln \frac{A}{A_0} = \ln \frac{R}{R_0} = -\lambda t \tag{6}$$

The disintegration rates, A, usually are given in disintegration per second (dps). The unit of absolute activity is the curie, which is defined as 3.700×10^{10} dps. Most tracer level work is performed by using only microcuries of activity ($\mu c = 10^{-6}$ curie). The counting rates are usually reported as counts per second (cps) or counts per minute (cpm).

The decay process is characterized by the disintegration rate constant, λ. Another convenient characterization is the half-life, $t_{1/2}$, which is the time required to reduce the activity of the radionuclide by 50 percent. Thus, according to equation 6,

$$\ln \frac{R}{R_0} = \ln \left(\tfrac{1}{2}\right) = -\lambda t_{1/2}; \qquad t_{1/2} = \frac{0.693}{\lambda} \tag{7}$$

A third characteristic constant of radioactive decay is the mean life, τ. The actual life-time, t, of a single radionuclide can be any value between zero and ∞.

The average (mean) value of t is

$$\tau = \frac{\displaystyle\int_0^{N_0} t \, dN}{\displaystyle\int_0^{N_0} dN} = \frac{1}{N_0} \int_0^{N_0} t \, dN \tag{8}$$

But according to equations 2 and 3,

$$dN = -\lambda N_0 e^{-\lambda t} \, dt \tag{9}$$

Therefore,

$$\tau = \frac{1}{N_0} \int_0^{N_0} t \, dN = \frac{1}{N_0} \int_0^{\infty} t\lambda N_0 e^{-\lambda t} \, dt \tag{10}$$

From the table of integrals this is equal to

$$\tau = \frac{1}{\lambda} = \frac{t_{1/2}}{0.693} \tag{11}$$

The mean time or relaxation time is then the time required to reduce the disintegration rate by a factor of $1/e$.

The mean time, τ, is useful in the calculation of the total number of particles emitted. For example, for a 1 μc sample of ^{131}I (8.1 days half-life) complete decay would be the number of disintegrations/second multiplied by the mean life of the ^{131}I. One μc equals 3.7×10^4 dps, and the half-life of ^{131}I is 7×10^5 seconds; the corresponding mean life from equation 11 is 1.01×10^6 seconds. For complete decay of 1 μc of ^{131}I one would get 3.7×10^{10} disintegrations and 3.7×10^{10} beta particles.

If the average energy of a beta particle is considered to be 0.608 Mev. (see Fig. 33–1) and the average energy of the accompanying gamma rays is considered to be 0.364 Mev., the total energy from a 1 μc sample of ^{131}I would be

$$3.7 \times 10^{10}(0.608 + 0.364) = 3.8 \times 10^{10} \text{ Mev.}$$

Such calculations are of primary importance when short-lived radionuclides are used for medical research and therapy. Although there are different standards for radiation exposure of different parts of the body, the permissible level for body exposure is about 0.1 rem. (roentgen equivalent men) over a 24-hour period. For beta radiation this is equivalent to absorbing 10 ergs per gram of body weight.

In order to determine the λ, $t_{1/2}$, or τ of a radioactive decay process, the radiation must be detected and measured. The most commonly used device is the Geiger-Müller counter. Basically, it consists of a pair of electrodes surrounded by a rare gas that is easily ionized by the radiation. Geiger-Müller tubes come in many different forms and shapes, but the most common one is a bell-type tube, as shown in Figure 33–2.

The tube is made of either a metal or a glass cylindrical container. If it is metal it may serve directly as the cathode. If it is glass its inner surface is coated with a conductive material to serve as the cathode.

In the center, coaxially aligned, is the collector electrode (anode), which is a fine wire (often tungsten), three- or four-thousandths of an inch in diameter. One

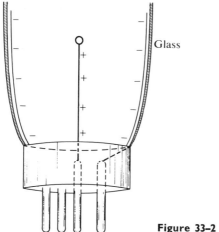

Figure 33–2 *Geiger-Müller tube.*

end of the collector may be attached to the insulator and the free end may be covered by a glass bead.

It is charged positively to about 1200 volts with respect to the cathode. The space between the two electrodes is filled with helium or argon at a pressure of 70 to 200 torr. The end window is a thin plate of mica or Mylar, which keeps the gas in the tube and at the same time allows beta particles to pass through the plate. In addition to the rare gas, a small quantity of quencher gas is also present. This may be an organic quencher, such as ethyl alcohol (i.e., 90 torr argon and 10 torr ethyl alcohol) or ethyl formate, or an inorganic quencher, which can contain about 0.1 percent chlorine.

The beta particle entering the counter ionizes the rare gas, producing the helium ion and an electron. The electron is collected on the anode and the helium ion travels to the cathode. However, some of the electrons thus produced may acquire enough energy so that in collisions with helium atoms they may produce additional ion pairs. This process would result in an avalanche of electrons and helium ions once the ionization is initiated, and the tube would discharge continuously.

The quencher gas prevents this. With an organic quencher the molecules decompose irreversibly and such tubes have a useful life of about 10^8 counts. Inorganic

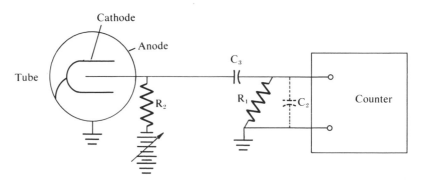

High voltage power supply

Figure 33–3 *Geiger-Müller tube and scaler circuit.*

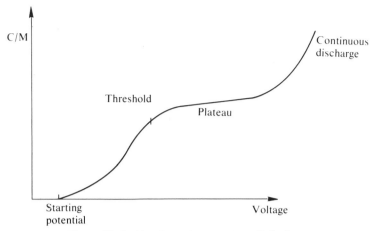

Figure 33–4 *Plot of counting rate vs. applied voltage.*

halogen quenchers are longer lived because the chlorine atoms formed by the collision between the electrons and the free chlorine will recombine, but the tubes are less sensitive than those with organic quenchers.

The electrons collected on the anode of the Geiger-Müller tube cause a discharge in the form of a voltage pulse. The counter (scaler) registers a count each time the voltage pulse exceeds the level set by the discriminator of the counter.

A common circuit that incorporates the Geiger-Müller tube is given in Figure 33–3. A more detailed account of the electronic components of the apparatus may be found in Part XIV of this book.

In Figure 33–3 the cathode is grounded and the anode has high positive potential relative to ground. The C_3 capacitor blocks the D.C. potential, preventing it from reaching the counter. The counter input capacity is represented by C_2. Resistor R_2 isolates the power supply from the collector anode, thus allowing the voltage to drop at this point upon the collection of electrons. Resistors R_2 and R_1 allow the equilibrium voltage to be reestablished after each discharge.

When a sample is to be counted, the operating potential has to be established. Below a certain potential no counting is observed (Fig. 33–4).

The level of this starting voltage is set by the discriminator of the scaler. When the voltage is increased, the counting rate increases, reaching a plateau at which the quencher operates most successfully. Beyond this potential a rapid increase in the counting rate occurs, at which point the quencher cannot perform its function and continuous discharge takes place. Such a process would ruin the tube. Geiger-Müller tubes should be operated at a voltage corresponding to the plateau, preferably about 50 volts above the threshold value.

Experimental

The proper choice of a radioactive nuclide is important in this experiment. Since the half-life of a radioisotope is to be determined, it should be a short-lived beta-emitting material so that the experiment can be performed in one afternoon. We have found that ^{132}I is well suited for this purpose. It is the thermal neutron fission product of ^{235}U and it may be available to laboratories within a 100-mile radius of a reactor. Otherwise, ^{24}Na is recommended and can be obtained from Nuclear-Chicago Corporation, Chicago, Ill., New England Nuclear Corp., Boston, Mass., and other sources.

Procedures for preparing both of these radioisotopes will be given here. However, it must be emphasized that in handling radioisotopes special precautions have to be enforced; moreover, the instructor should have the proper facilities, especially for the disposal of radioactive wastes. All the preparative work and the equipment (dishes, pipets, and so forth) should be confined to one corner of the laboratory (or in a separate laboratory) and the measurement equipment kept in a different area. In the working area the tabletop and working trays should be covered with thick absorbent paper that has a water-resistant underlayer (Kimberly-Clark Corp., Neenah, Wis., Type 201 Kimpak, Cromwell Kraft X 37). If radioactive solution is accidentally spilled, the underlayer prevents the liquid from soaking through the paper. Such absorbent papers and other wastes can be disposed of in special containers labeled as such.

In handling radioactive nuclides rubber gloves should be worn. At the end of the preparative part of the experiment the rubber gloves should be washed before they are removed from the hands. The gloves then should be checked for radiation. If no contamination is found the gloves can be removed and stored and then the hands washed thoroughly.

Eating, drinking, smoking, and using cosmetics in the preparative area should be strictly forbidden. Pipetting should *never* be performed by mouth suction.

All labeled material, plus washings from pipets, and so forth, should be collected and the instructor should dispose of them according to AEC regulations.

PREPARING THE SAMPLE OF ^{132}I

The ^{132}I is received in an alumina-column generator that is mounted in a hood. The generator is "milked" by pouring 25 ml. of 0.01 M ammonium hydroxide into the top of the generator and collecting the product solution in a volumetric flask as it drips from the bottom. Two or three drops of product near the end are caught in a planchet; the rest of the product is discarded. The solution in the planchet is dried carefully under the heat lamp in the hood. In essence, the material provided in the generator is ^{132}Te, which is a weak beta-emitting radioisotope with a half-life of 78 hours. The generator can be used for a maximum of three weeks.

Before the first milking and when the unit has not been milked for several days, the generator is washed with 50 to 100 ml. of 0.1 M ammonium hydroxide and the product solution is discarded. This prevents contamination of the ^{132}I. Half an hour must then be allowed after washing for more ^{132}I to accumulate. The column is then hot-milked as already described.

PREPARING THE SAMPLE OF ^{24}Na

A radioactive ^{24}Na solution containing about 2 to 3 mc/liter should be prepared by diluting a commercial sample having known activity (e.g., radioactive sodium chloride, pure or mixed with regular sodium chloride). Then 0.1 ml. of the 2 mc ^{24}NaCl solution is transferred into a planchet with a micropipet (Fig. 33–5).

The planchet is dried with a heat (infrared) lamp under the hood. The sample is now ready for counting.

CONDUCTING THE COUNTING PROCEDURE

In the first counting experiment the threshold voltage or operative voltage of the Geiger-Müller tube will be established.

1. Before connecting the scaler to a 110 volt 60 cycle outlet, be sure that the master switch of the scaler is turned off. Also, the high voltage switch should be off and the high voltage adjust knob turned to the extreme counterclockwise position. Check that the Geiger-Müller tube is connected to the Geiger-Müller input jack.

2. Plug the scaler into the 110 volt line. Turn on the master switch. Allow two minutes for warm-up. The pilot lamp should indicate that power is being supplied to the instrument.

3. Place the sample in the planchet near the mica window of the Geiger-Müller tube. Turn the count switch to the count position and turn on the high voltage switch. Increase the voltage by turning gradually the high voltage adjust knob. The scaler starts to count when the starting voltage is reached.

4. Continue to increase the voltage slowly and note the counting rate (counts per minute [c/m]) at each voltage reading. The operating plateau has been reached when the counting rate increases only very little with increased voltage.

Continue to increase the voltage carefully but avoid reaching the point at which continuous discharge takes place (see Fig. 33–4).

5. Set the operating voltage back to about 50 volts above the threshold voltage established in the third and fourth steps (see Fig. 33–4).

6. Turn the count switch to stop position and push the reset switch. The scaler is now ready for operation.

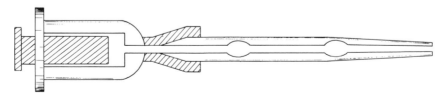

Figure 33–5 *Micropipet connected to a syringe.*

In the second part of the experiment establish the half-life of your radioactive nuclide.

With the sample in position near the window of the Geiger-Müller tube, count the disintegrations by repeating steps 3 through 6. If your sample is ^{132}I, count for two 3-minute intervals every 20 minutes for three hours. The average c/m of the two 3-minute counts should be plotted vs. time on semilog paper.

If your sample is ^{24}Na, note the time necessary to reach 5000 counts. Reset the counter. Repeat the count every hour and during the periods approximately 16 to 24 and 48 hours from the start of the experiment.

Plot the counting rate vs. time on semilog paper. Establish the disintegration rate constant, λ, the half-life, $t_{1/2}$, and the mean time, τ, by using equations 6, 7, and 11.

Compare the half-life value you obtained to the value given in the literature.

Assuming that the average energy of the total beta and gamma rays per disintegration is 1 Mev., what is the permissible activity (millicuries or microcuries) of your radioactive nuclide for medical administration so that the total radiation should not exceed 1×10^6 Mev?

Material and Equipment. ^{132}Te in an alumina-column generator; 200 ml. of 0.01M ammonium hydroxide (or a sample of ^{24}NaCl); absorbent filter paper with a water-resistant underlayer; micropipet graduated to 0.01 ml.; syringe; drying planchet; infrared lamp; Geiger-Müller counter; stopwatch.

REFERENCES

1. C. M. Lederer, J. M. Hollander and I. Perlman. Table of Isotopes, 6th Edition, Wiley & Sons, Inc., New York, 1967.
2. R. T. Overman and H. M. Clark. Radioisotope Techniques. McGraw-Hill Book Co., New York, 1960.
3. G. D. Chase. Principles of Radioisotope Methodology. Burgess Publishing Co., Minneapolis, 1959.
4. W. J. Price. Nuclear Radiation Detection. McGraw-Hill Book Co., New York, 1964.
5. G. R. Choppin. Experimental Nuclear Chemistry. Prentice-Hall, Inc., Englewood Cliffs, N.J., 1961.

34

RADIOISOTOPE DILUTION TECHNIQUE

The advances in radioisotope techniques and instrumentation in recent years have made it possible to measure accurately the specific activities of many radioisotopes. As a result, the analysis of a number of materials, which are difficult to isolate quantitatively, is now possible.

In general, classic gravimetric analysis relies upon the fact that the substance sought can be isolated in its purest form without any loss during the separation steps and then weighed. Volumetric analysis is based upon the assumption that the titrate is specific to the material sought and no other component of the mixture will interfere with the titrations. In many cases these conditions are not fulfilled. Isotope dilution techniques are helpful in these instances.

The direct isotope dilution technique is used when one wants to determine the concentration of a specific compound (organic or inorganic) that would be hard and tedious to isolate quantitatively in its pure form from a mixture. Under these conditions a certain amount of tagged (radioisotopic) species of the same compound is added to the mixture. After thorough mixing, the mixed compound (labeled and nonlabeled) is isolated in its purest form. However, the extraction does not have to be quantitative; the only criterion is that the isolated material should be pure. The activity of the pure labeled compound and that of the isolated mixture is determined. From the weight of the spike (the added labeled compound) and from the weight of the sample mixture, the amount of unknown in the original mixture can be calculated. An example is the determination of sulfate ion ($SO_4^=$) in a biological fluid that contains colloidal materials. The precipitation of the sulfate ion by the barium ion is the conventional technique. Gravimetric analysis may be difficult because barium sulfate forms a very fine precipitate, which may pass through filters. Furthermore, in the presence of colloidal particles, the barium sulfate aggregates may not grow to the sizes at which precipitation occurs; the protective colloid may keep them in solution.

315

Let us assume we use a spike (radioactive sulfate) that, after precipitation as a barium salt, yields the following data: A sample of 0.1 gram of spike has an activity of 1500 counts per second. The specific activity, SA^*, of the spike then will be

$$SA^* = 1500/0.1 = 15,000 \text{ c/s/gram}$$

The compound isolated from the biological mixture as a barium salt weighed 0.5 gram and had an activity of 100 c/s.

The specific activity of the spike is given as

$$SA^* = A^*/W^* \tag{1}$$

where A^* is the activity and W^* is the weight of the labeled spike.

On the other hand, the mixture of labeled and nonlabeled material has a specific activity, SA,

$$SA = A/W + W^* \tag{2}$$

where W is the weight of the unlabeled compound and A is the activity of the isolated mixture. Combining equations 1 and 2 and solving for W, one obtains

$$W = W^*(SA^*/SA - 1) \tag{3}$$

Substituting now the values obtained in the experiment into equation 3, the weight of unknown nonlabeled barium sulfate is

$$W = 0.1(1500/100 - 1) = 1.4 \text{ grams}$$

It is obvious that only one-third of the barium sulfate was isolated in the experiment. Under these circumstances the direct isotope dilution technique is useful for analysis. This method has become important in biochemical studies, especially in hormone analysis (Hales and Randle) and in enzyme-metabolic product assay (Newsholme and Taylor).

In order to detect radiation and, therefore, the activity of a sample, a counting device such as the Geiger-Müller counter is used. The description and operation of such a device was given in Experiment 33.

In quantitative measurements of beta-emitting substances a special precaution must be taken. Beta rays are easily absorbed by materials and, therefore, the thickness of the sample determines the actual counting rate (Fig. 34–1). In this figure the broken line indicates the counting rate that would be detected if no self-absorption of beta particles occurred. The solid line is the actual counting rate (activity) of a sample as a function of thickness. The counting rate in Figure 34–1 levels off and asymptotically

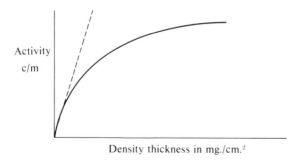

Activity
c/m

Density thickness in mg./cm.²

Figure 34–1 *Self-absorption of beta radiation.*

approaches a value at infinite thickness. Under these idealized conditions (infinite thickness) only the rays emitted by the top layer of the sample are measured.

The activity for beta-emitting substances is usually reported as corrected for infinite thickness.

Curves such as that in Figure 3–14 can be described analytically as

$$\frac{d}{A} = C + \frac{d}{A_\infty} \tag{4}$$

In equation 4, d is the density thickness of the sample in milligrams/square centimeter, A is the activity in c/m or c/s, and A_∞ is the activity at infinite density thickness.

Therefore, a plot of d/A vs. d yields a straight line, and the slope of the line is equal to $1/A_\infty$ or the reciprocal of the activity at infinite density thickness. The constant C is characteristic of the material absorbing the beta radiation.

Other methods of radioisotopic analysis are based on extrapolating the results to infinitely thin layer activity. We will be concerned with only the first case.

Experimental

The experiment will be done in two parts. First, the specific activity of barium sulfate will be established and corrected for infinite thickness. Then the quantity of an unknown sample containing nonlabeled sulfate will be analyzed by the isotope dilution technique.

Before any work is done with radioisotopes, certain precautions must be emphasized and strictly enforced by the instructor.

All the preparative work and the equipment (dishes, pipets, and so forth) should be confined to one corner of the laboratory (or in a separate laboratory) and the measurement equipment kept in a different area. In the working area, the tabletop and working trays should be covered with thick absorbent paper with a water-resistant (Kimberly-Clark Corp., Neenah, Wis., Type 201 Kimpak, Cromwell Kraft X 37). If radioactive solution is accidentally spilled, the water-resistant underlayer prevents the liquid from soaking through the paper. These absorbent papers and other wastes can be disposed of in special containers labeled as such.

In handling radioactive nuclides, rubber gloves should be worn. At the end of the preparative part of the experiment the rubber gloves should be washed before they are removed from the hands. The gloves then should be checked for radiation. If no contamination is found, the gloves can be removed and stored and then the hands are washed thoroughly.

Eating, drinking, smoking, and using cosmetics in the preparative area should be strictly forbidden. Pipetting should *never* be performed by mouth suction.

All labeled material, plus washings from pipets, and so forth, should be collected and the instructor should dispose of them according to AEC regulations.

For the study of self-absorption of a weak beta-emitting substance, ^{35}S will be used. $H_2^{35}SO_4$ may be obtained from many commercial sources, such as Tracerlab, Inc., Waltham, Mass., Nuclear-Chicago Corp., Chicago, Ill., and New England Nuclear Corp., Boston, Mass.

From a stock solution of $0.11M$ $H_2^{35}SO_4$ prepare three samples in three 30 to 40 ml. test tubes in the following proportions:

$0.11M$ H_2SO_4 (ml.)	H_2O (ml.)	$0.1M$ $BaCl_2$ (ml.)
0.2	5	1
1.0	10	2
2.0	10	3

Add the barium chloride a drop at a time to the labeled $H_2^{35}SO_4$ and water solution and heat it to boiling to aid the precipitation. Weigh three filter papers of 5 cm.² area and filter the three precipitates on these filter papers. It is necessary that the thickness of each precipitate be uniform. Wash each precipitate with a few drops of water, and dry it on the filter paper under an infrared lamp without curling the papers. Reweigh the filter papers with the precipitates.

Obtain the counting rate on each sample as follows:

1. Before connecting the scaler to a 110-volt 60-cycle outlet be sure that the *master switch* of the scaler is turned off. Also, the high voltage switch should be turned off and the high voltage adjust knob turned to the extreme counterclockwise position. Check that the Geiger-Müller tube is connected to the Geiger-Müller input jack.

2. Plug the scaler into the 110 volt line. Turn on the master switch. Allow two minutes for warm-up. The pilot lamp should indicate that the power is being supplied to the instrument.

3. Place the sample in the planchet near the mica window of the Geiger-Müller tube. Turn the count switch to the count position and turn on the high voltage switch. Increase the voltage by turning gradually the high voltage adjust knob to 50 volts above the threshold value. Count each sample for two minutes.

4. After counting each sample, turn the count switch to the stop position and push the reset switch. The scaler is now ready for operation. Turn the count switch to the on position for the new sample.

Plot the counting rate obtained vs. density thickness for the three samples. Plot the density thickness/counting rate vs. density thickness (equation 4) and determine the activity at infinite thickness and the C constant for the absorption of beta particles by barium sulfate.

For the isotope dilution technique, duplicate analysis will be performed on an unknown 0.1 to $0.2M$ potassium sulfate solution. Add 2 ml. of labeled $H_2^{35}SO_4$ $(0.11M)$ and 2 ml. of unknown sulfate solution to each of two test tubes. Add 10 ml. of distilled water and bring it to the boiling point. Slowly add 5 ml. of $0.1M$ barium chloride solution to precipitate the sulfate. Weigh out filter papers of 5 cm.² area. Filter the precipitate, wash it with distilled water, and dry it under an infrared lamp. Reweigh the filter paper with the precipitates.

Obtain the counting rate (activity) on both aliquots. Knowing the C constant from the self-absorption experiment, and knowing the density thickness and the activity of your isotope dilution sample, calculate the *average* activity of the diluted sample at infinite thickness by using equation 4.

Apply equation 3 to obtain the concentration of your unknown solution, using both the specific activity of the spike and that of the sample at infinite density thickness. On the basis of the agreement between the duplicate analyses, what is the precision of this technique?

Material and Equipment. $H_2{}^{35}SO_4$; potassium sulfate (unknown solution); barium chloride; absorbent filter paper with water-resistant underlayer; No. 1 Whatman filter paper 2.5 cm. in diameter; micropipet graduated to 0.01 ml.; syringe; drying planchet; infrared lamp; Geiger-Müller counter; stopwatch.

REFERENCES

1. R. T. Overman and H. M. Clark Radioisotope Techniques. McGraw-Hill Book. Co., New York, 1960.
2. G. D. Chase. Principles of Radioisotope Methodology. Burgess Publishing Co., Minneapolis, 1959.
3. A. G. Lloyd. Biochem. J., **80,** 572 (1961).
4. C. N. Hales and P. J. Randle. Biochem. J., **88,** 137 (1963).
5. E. A. Newsholme and K. Taylor. Biochim. Biophys. Acta, **158,** 11 (1968).

XI

ELECTROCHEMISTRY

35

CONDUCTANCE OF ELECTROLYTES*

Materials are usually classified as conductors, semiconductors, or insulators in terms of their behavior toward the flow of current. Conductors offer very little resistance to the passage of current, whereas insulators have great resistances. Ohm's law, which applies to conductors, states that

$$V/i = R \tag{1}$$

where V is the potential in volts, i is the current in amperes, and R is the resistance in ohms. Metals and electrolytic solutions are good conductors with low R values.

The resistance depends upon the nature of the conductor and the geometry of the conductance cell.

$$R = \rho \frac{l}{A} \tag{2}$$

where l is the distance between two electrodes, A is the area of the electrode, and ρ is a proportionality constant specific for the material, called the specific resistance.

It is more convenient to focus attention on the conductance, L, rather than on the resistance, R. The relationship is

$$R = 1/L \tag{3}$$

L, the conductance, is expressed in mhos or ohms^{-1}. In terms of conductance, equation 2 is written as

$$L = \kappa \frac{A}{l} \tag{4}$$

where κ is the specific conductance, i.e., the conductance of a tube of material 1 cm. long having a cross section of 1 cm.^{2} The units of κ are mho cm.$^{-1}$

The specific conductance of an electrolyte solution depends upon the equivalent concentration of the charged particles present and their mobilities. In order to remove at least part of the concentration dependence, a new quantity—the equivalent conductance, Λ—is introduced.

$$\Lambda = \frac{1000}{c} \kappa \tag{5}$$

where c is the equivalent concentration. Therefore, if the electrolyte were completely dissociated (that is, each charged particle behaved independently), the physical meaning of equivalent conductance would be that it is the conductance of Avogadro's number of unit positive and unit negative particles placed in a conductance cell that is large enough to accommodate 1 liter of solution and in which the electrodes have an area of 1 cm.^{2} and are 1 cm. apart.

Arrhenius explained that the equivalent conductance of the electrolytes is not independent of concentration but decreases with increasing concentration because of only partial dissociation. According to this theory, complete dissociation of the ionic species occurs only in infinitely dilute solution, and at any appreciable concentration only fractional dissociation occurs. The degree of dissociation can be calculated as

$$\alpha = \frac{\Lambda}{\Lambda_0} \tag{6}$$

where Λ_0 is the equivalent conductance in an infinitely dilute solution. Although this theory is quite useful for weak electrolytes, it has the inherent disadvantage of requiring evaluation of Λ_0. For weak electrolytes the equivalent conductance decreases quite rapidly even in dilute solutions and an accurate extrapolation to zero concentration cannot be achieved. This is in contrast to strong electrolytes, such as sodium chloride, for which a plot of Λ vs. $\sqrt{c}$ yields a straight line, and Λ_0 can be obtained accurately by extrapolation to zero concentration.

The difference in Λ_0 for pairs of salts having a common ion is approximately constant. Kohlrausch proposed that when complete dissociation exists in infinitely dilute solutions, each ionic species migrates independently. The Λ_0 can then be considered to be the sum of equivalent ionic conductances at infinite dilution, for each ionic species:

$$\Lambda_0 = \lambda_0^+ + \lambda_0^- \tag{7}$$

where λ_0^+ and λ_0^- are the equivalent conductances of the positive and negative ions, respectively, in infinitely dilute solutions. The differences between the equivalent conductances of the individual ionic species at infinite dilution, $\Delta\lambda_0$, were obtained from conductances of pairs of strong electrolytes $[\Delta\Lambda_0 = \Delta\lambda_0^+ = (\lambda_0^{K^+} - \lambda_0^{Na^+})]$ from sodium chloride–potassium chloride, sodium nitrate–potassium nitrate, and so forth; $[\Delta\Lambda_0 = \Delta\lambda_0^- = (\lambda_0^{Cl^-} - \lambda_0^{NO_3^-})]$ from sodium chloride–sodium nitrate, potassium chloride–potassium nitrate] and were found to be constant. Although only the differences in equivalent conductances can be obtained numerically ($\Delta\lambda_0$) and not the individual conductances ($\lambda_0^{Cl^-}$, and so forth), the Kohlrausch additivity law can be used to obtain Λ_0 for a weak electrolyte in the following manner:

$$\Lambda_0^{CH_3COOH} = \lambda_0^{H^+} + \lambda_0^{CH_3COO^-} = \Lambda_0^{HCl} + \Lambda_0^{CH_3COONa} - \Lambda_0^{NaCl} \tag{8}$$

The extrapolated equivalent conductances, Λ_0, of hydrochloric acid, sodium acetate, and sodium chloride can be obtained easily since they are all strong electrolytes. Therefore, Kohlrausch's law permits the calculation of the equivalent conductance of a weak electrolyte (acetic acid) in an infinitely dilute solution.

Once Λ_0 for a weak electrolyte is obtained, the degree of dissociation of acetic acid at different concentrations can be calculated by using equation 6.

For a monoprotic acid, such as acetic acid, the dissociation reaction can be written as

$$CH_3COOH \rightleftharpoons H^+ + CH_3COO^- \tag{9}$$

and the dissociation constant as

$$K_a = \frac{(H^+)(CH_3COO^-)}{(CH_3COOH)} = \frac{(c\alpha)(c\alpha)}{c(1-\alpha)} = \frac{c\alpha^2}{1-\alpha} \tag{10}$$

Dissociation constants of weak electrolytes, obtained from conductance measurements and the Arrhenius theory, are constant at different concentrations. With strong electrolytes the case is quite different. Dissociation constants calculated for hydrochloric acid for example, increase with increasing concentration. In spite of these and other difficulties, the Arrhenius theory is still applied, but only to weak electrolytes. The pK_a values ($-\log K_a$) of weak electrolytes are listed in the literature.

The behavior of strong electrolytes in conductance experiments has been successfully explained by the Debye-Hückel theory for dilute solutions (up to 0.01M). This theory treats the strong electrolytes as completely ionized particles. However, the independent motion of each ion is hindered by two effects: (a) the relaxation phenomenon of the ionic atmosphere and (b) electrophoretic mobility. A charged positive ion is, on the average, surrounded by an atmosphere of negative ions and vice versa. When the positive ion moves in one direction under the influence of the electrical field, the ionic atmosphere is distorted (Fig. 35–1). This distortion opposes the applied field and it results in a decrease in the current produced by the field.

The greater the electrolyte concentration, the greater the distortion of the ionic atmosphere; correspondingly the equivalent conductance decreases with the increase in the square root of concentration ($\sqrt{c}$).

The second effect stems from the fact that ions are highly solvated. The solvation layer, so to speak, moves with the ionic species. As the differently charged ions move in the opposite direction, their solvent layers encounter a viscous drag, which decreases the mobility of the ions. Again, this effect is more pronounced as the concentration of the electrolyte increases. The Debye-Hückel theory also predicts a square root concentration dependence for this effect.

The two effects enumerated are important only if the ionic concentration is

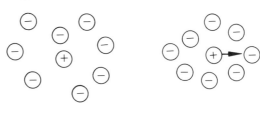

Figure 35–1 *Relaxation of ionic atmosphere. A, ionic atmosphere at rest; B, atmosphere in movement under the influence of an applied electrical field.*   A   B

appreciable. For weak electrolytes the Debye-Hückel corrections are small, and they therefore can still be satisfactorily described in Arrhenius' terms.

Moreover, at concentrations greater than 0.01M even the Debye-Hückel theory cannot explain the conductance behavior of strong electrolytes. There are a number of theories, none of them completely satisfactory for describing the behavior in more concentrated solutions. The qualitative picture is that, as the ionic concentration increases, more and more ion pairs and triplets are formed. This obviously decreases the independent movement of the individual particles in the electrical field, thus resulting in a decrease in the equivalent conductance with an increase in concentration.

These theories of Bjerrum and Fuoss differ, however, from the old Arrhenius theory. An ion pair is only a temporary arrangement due to inelastic collisions, whereas an undissociated molecule is a permanent arrangement. The ion pair and higher ionic aggregations become especially important in solvents with low dielectric constants.

To understand the conductance behavior of different electrolytes in different concentration ranges, one selects the theory that explains the phenomenon in the simplest terms. No unified theory is available at present. However, the different theories try to account for the number of particles moving independently at any given concentration.

The relationship between the conductance of an independent ion and its mobility can be established unequivocally for infinitely dilute solutions.

For an applied potential, V, the flow of current is given by equation 1. Rewriting this equation in terms of equivalent conductance for infinitely dilute solutions, we get

$$i = \Lambda_0 V = (\lambda_0{}^+ + \lambda_0{}^-)V = i^+ + i^- \tag{11}$$

where i^+ and i^- are the currents carried by the positive and negative ions, respectively.

The mobility of an ion, u, is defined as the average velocity, v, under the influence of a potential of 1 volt applied across a 1 cm. cell.

$$u = \frac{v}{V} \tag{12}$$

Since the current carried by the positive ions in one equivalent of electrolyte corresponds to the passage of N/Z^+ ions with an average velocity of v^+ carrying eZ^+ charges,

$$i^+ = eZ^+ \frac{N}{Z^+} v^+ = eNv^+ = Fv^+ \tag{13}$$

where N is Avogadro's number, e is the unit charge, and F is the Faraday constant. A similar equation can be written for i^-. Combining equations 11, 12, and 13 one obtains

$$u_0^+ = \frac{\lambda_0^+}{F} \quad \text{and} \quad u_0^- = \frac{\lambda_0^-}{F} \tag{14}$$

If one assumes that small ions are spherically symmetrical and that their frictional resistance can be given by Stokes law, the mobility can be given as

$$u = \frac{\zeta\epsilon}{6\pi\eta} \tag{15}$$

where ϵ is the dielectric constant of the solvent, η is the coefficient of viscosity of the solvent, and ζ is the zeta potential, which is related to net charge at the shear surface, Q, of a sphere with radius r.

$$\zeta = \frac{Q}{\epsilon r} \tag{16}$$

Therefore, if the ϵ and η of a solvent are known, one can predict the mobility of an ion in the given solvent for a set zeta potential from equation 15. Conversely, from the comparison of mobilities or equivalent conductances at infinitely dilute solutions of ions in different solvents, one can estimate whether the effective surface charge is the same or different in the different solvents.

Experimental

The conductivities can be conveniently measured with a Wheatstone bridge circuit (Fig. 35–2).

In this circuit we employ an audio frequency A.C. signal ($\sim$1000

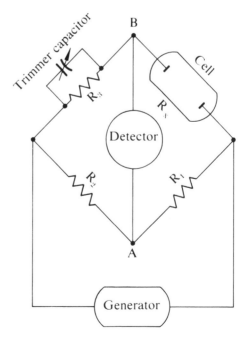

Figure 35–2 *Conductance bridge.* $\sim$ 1000 cycles/sec. A.C.

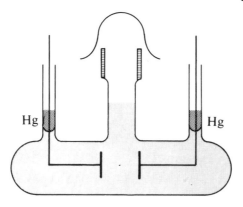

Figure 35–3 *Conductance cell.*

cycles/second). The purpose of using alternating current is to avoid polarization of the electrodes in the conductance cell, which would lead to electrode reactions. The conductance cell is a glass cell with sealed-in platinum electrodes (Fig. 35–3).

The platinum electrodes usually have a 1 cm.^{2} surface area and are set a certain distance apart, but the geometry of the cell has to be determined by calibration. The platinum electrodes extend to the side arms where they are sealed in glass. The leads of the platinum electrodes either extend beyond the glass side arm so that they can be directly incorporated into the circuit or they dip into small mercury pools. Under the latter condition the leads from the conductance bridge make contact with the mercury pools in both sides of the conductance cell. The conductance cell is filled with the electrolyte solution and thermostated before measurement.

A bridge is balanced by an oscilloscope or other detector device. The condition for balancing is that the potential at B should be the same as at A. This condition is achieved when

$$\frac{Z_1}{Z_2} = \frac{Z_x}{Z_3} \tag{17}$$

where Z is the impedance of the sections in Figure 35–2. The impedance is given by

$$Z = R + iX \tag{18}$$

where X is the reactance, R is the resistance, and $i = \sqrt{-1}$.

Rewriting equation 17 in terms of equation 18, we obtain

$$R_1 R_3 - X_1 X_3 = R_2 R_x - X_2 X_x$$
$$R_1 R_3 - R_3 X_1 = R_2 X_x - R_x X_2 \tag{19}$$

In order for equations 19 to apply to the balance of resistances

$$\frac{R_1}{R_2} = \frac{R_x}{R_3} \tag{20}$$

the phase angles have to be the same,

$$\theta_1 = \theta_2 \quad \text{and} \quad \theta_x = \theta_3 \qquad (21)$$

where θ is the phase angle between current and voltage and is given by

$$\theta_i = X_i/R_i \qquad (22)$$

The condition for balancing of resistance in Figure 35–2 is achieved by adjusting R_1 and R_2 with a slide wire and at the same time adjusting the trimmer capacitor. A more detailed account of the electronic components of the apparatus may be found in Part XIV of this book.

The same type of procedure is followed if a Wayne-Kerr universal bridge is used in the circuit. In this case, the detector is a photoelectric eye that opens completely when balance is achieved.

With the Wayne-Kerr bridge, balancing is started with the sensitivity (knob to the left) turned to the lowest value. One then selects the conductance range (mmho to $\mu\mu$mho) that causes the widest opening of the eye in the coarse detector. Then, progressively increasing the sensitivity, one balances the bridge by alternately changing the resistance and capacitance. The final balance is achieved at the greatest sensitivity. In the Wayne-Kerr bridge the conductance is read off directly.

In a Wheatstone bridge setup, R_x is calculated from the measured resistances (R_1, R_2, R_3) and from equation 20.

The solvents employed should be of low ionic conductance. In aqueous solutions deionized water should be used. This is achieved by passing the water through an ion exchange column. If this is not feasible, the distilled water should be boiled to expel all dissolved carbon dioxide and capped while it is still hot. In any case, the conductance of the water should be less than 5×10^{-6} mhos (5×10 μmhos). If this cannot be achieved, *the conductance of the solvent must be subtracted from the conductance of the solution.*

In order to determine the conductance cell constant, prepare (very accurately) a potassium chloride solution (one-tenth of one equivalent/liter) in deionized water. You need only about 100 ml. of solution. Thermostat the solution at 25° C. and measure its conductance. The specific conductance of 0.1N potassium chloride solution is 0.012886 mho cm.$^{-1}$ at 25° C. Using this, calculate the conductance cell constant (A/l) from equation 4. *Use this cell constant for future calculations.*

Prepare four solutions of silver nitrate in deionized water. The highest concentration should be 0.1N silver nitrate. Similarly prepare four solutions of silver nitrate in 25 percent (w/w) dimethyl sulfoxide (DMSO) in water. (Alternatively, dimethylformamide [DMF]-water, dioxane-water, acetone-water, and so forth can be used.) The highest concentration in the organic solvent–water mixture should be 0.05N.

Also prepare four solutions of various concentrations of acetic acid in deionized water and four solutions of acetic acid in DMSO (or DMF dioxane, acetone, and so forth). Measure their conductances. Calculate and tabulate the equivalent conductance of all 16 solutions. Plot the Λ vs. $\sqrt{c}$ for the solutions.

Compare the behavior of silver nitrate and acetic acid in the different solvents. Can you account for the change in the conductance of silver nitrate in the different solvents solely on the basis of different coefficients of viscosity and dielectric constants? Use your data for the coefficients of viscosity, η, and dielectric constants, ϵ, of the solvents obtained in Experiments 20 and 15, respectively. The equivalent conductance of the silver ion in water at 25° C. is given as $\lambda^0_{Ag^+} = 63.5$ and that of the nitrate ion is $\lambda^0_{NO_3^-} = 70.6$ in (mhos/centimeter)/(equivalents/liter) in the *Handbook of Chemistry and Physics*.

Calculate the degree of dissociation, α, for acetic acid in water, using λ_0 values from the literature (*Handbook of Chemistry and Physics*).

Estimate the λ_0 values for acetic acid in DMSO from that obtained for water and by using equations 14, 15, and 16.

Calculate the α's for acetic acid in DMSO (or DMF, acetone, dioxane, and so forth).

Calculate the equilibrium constant of acetic acid in water and in DMSO.

What conclusions can you draw regarding the effect of the solvent on the conductance?

Material and Equipment. Potassium chloride; silver nitrate; acetic acid; deionized water; dimethyl sulfoxide DMSO (or dimethyl formamide [DMF], dioxane, acetone, and so forth); conductance cell; conductance bridge (Wheatstone bridge circuit or Wayne-Kerr universal bridge); thermostat with heater, stirrer, and thermoregulator.

REFERENCES

1. G. M. Barrow. Physical Chemistry. McGraw-Hill Book Co., Inc., New York, 1966, Chapter 21.
2. T. Shedlovsky. Conductometry. In Weissberger, A. (ed.). Physical Methods of Organic Chemistry. 3rd Edition, Vol. I, Part IV, Interscience Publishers, Inc., New York, 1960, Chapter XLV.

36

TRANSFERENCE NUMBER*

When two electrodes dipping into an electrolyte solution are connected to a source of direct current (e.g., a battery) the ionic species begin to migrate toward the electrodes of opposite charge. The ions, because of their migration, carry electricity, which closes the circuit. However, the different ionic species do not necessarily participate equally in the transfer of electricity. The doubly charged ion carries twice as much electricity as the monovalent ion. But different ionic species having the same charge may have different migrational velocities, which cause differences in concentration at the two electrodes.

In order to illustrate this, let us imagine that the usual electrolytic apparatus is divided by imaginary walls into three compartments: a cathode, an anode, and a middle compartment.

According to Faraday's law, one gram equivalent weight of anion and one gram equivalent weight of cation are discharged at their respective electrodes for every 96,500 coulombs (1 faraday) of electricity supplied. Thus, although equivalents of anion and cation disappear from the anode and cathode compartments by electrolysis, the unequal migration of ions causes unequal concentrations in these compartments.

The results of equal and unequal migrational velocities are illustrated in Table 36–1. In case *a*, ten equivalents of anions and of cations are evenly distributed in the three compartments. If we pass 10 faradays of electricity through the solution and the ionic velocities are equal, five equivalents will move from each compartment, pass through the middle compartment, and arrive at the electrode compartment of opposite charge, as shown in case *b*. However, during the passage of 10 faradays of electricity, ten equivalents of cations were discharged at the cathode and ten equivalents of anion at the anode. The final results are illustrated in case *c*. The middle compartment has its original concentration. In both the anode and the cathode compartment the concentration is reduced but, because of the equal ionic velocities, the concentrations in these two compartments are equal.

TABLE 36–1

	Anode	Middle Compartment	Cathode
a	10+ 10−	10+ 10−	10+ 10−
b	←10− 5+ 5+→ 15− ←5−	5+→ 10+ 5+→ ←5− 10− ←5−	15+ 10+→ 5−
c	5+ 5−	10+ 10−	5+ 5−
d	10+ 10−	10+ 10−	10− 10−
e	←10− 7+ 3+→ 17− ←7−	3−→ 10+ 3−→ ←7− 10− ←7−	13+ 10+→ 3−
f	7+ 7−	10+ 10−	3+ 3−

In case d, we start out similarly with ten equivalents of cations and anions distributed evenly. Again we pass through 10 faradays of electricity. However, in this case the anions migrate much faster than the cations ($7/3 = 2.33$ times faster). The result of the uneven migration during the passage of 10 faradays is given in case e. The result of discharge of ten equivalents of cations from the cathode compartment and ten equivalents of anions from the anode compartment is illustrated in case f. Again the concentration in the middle compartment is unchanged. The concentrations in the anode and cathode compartments are diminished and they are unequal.

The transference number is defined as the number of equivalents of anion or cation transferred during the passage of 1 faraday of electricity.

$$t_{Na^+} = \frac{\text{No. of equivalents Na}^+}{\text{No. of faradays electricity}} \tag{1}$$

This is an operative definition and, thus, the numerical value can be obtained from experimentation. For strong electrolytes in dilute solutions, the transference number is also equal to the fraction of the total electric current carried in the solution by the ionic species. The latter is also the definition of the *transport number*. Again, the constituent transference number and the electric transport number are equal only for strong electrolytes in dilute solution, for which the Debye-Hückel-Onsager theory is applicable.

Since our experiment is designed for strong electrolytes, we use the transference number obtained experimentally and apply the theoretical derivations obtainable from the definition of the electric transport number. Thus,

$$t_i = \frac{u_i m_i |Z_i|}{\sum_i u_i m_i |Z_i|} = \frac{\lambda_i m_i |Z_i|}{\sum_i \lambda_i m_i |Z_i|} = \frac{\lambda_i c_i}{\Lambda c} \tag{2}$$

where t_i is the transference number of ionic species i, u_i is its mobility in cm.

sec.$^{-1}$/volt cm.$^{-1}$, λ_i is its equivalent conductance in cm.^{2} ohm^{-1} gram equivalent^{-1}, m_i is its molarity, c_i is its normality and $|Z_i|$ is the absolute value of the ionic charge. The Λ and c are the equivalent conductance and the normality of the electrolyte as a whole.

The relationships between these terms in equation 2 are as follows:

$$c_i = m_i |Z_i| \tag{3}$$

$$\lambda_i = u_i F \tag{4}$$

$$\Lambda c = \sum \lambda_i c_i \tag{5}$$

where F is the Faraday constant (96,500 coulombs gram equivalent^{-1}). It is obvious from the definition of the transport number as a fraction of current carried that

$$\sum t_i = 1 \tag{6}$$

Therefore, if only one uni-univalent electrolyte is present,

$$\lambda_+/(\lambda_+ + \lambda_-) = t_+ = 1 - t_- \tag{7}$$

Transference numbers as defined here are called Hittorf transference numbers. They are calculated on the assumption that the solvent (water) remains stationary during the electrolysis. "True transference numbers" are the result of correcting the Hittorf numbers for the extent of solvent migration. In dilute solution the two are very nearly the same, but in concentrated solutions they may differ.

Transference numbers change with concentration, the greatest change being observed with electrolytes that form unstable complex ions. Smaller variations result from interionic attraction since the mobility of anions and cations decreases unequally as the ionic strength is increased.

For strong electrolytes the Debye-Hückel-Onsager theory predicts that

$$t_\pm - t_\pm^\circ = [(Z_+ + |Z_-|)t_\pm - |Z_\pm|][\mu^{1/2}\sigma/\Lambda_0] \tag{8}$$

where t and Z are, respectively, the transference number and the charge of the subscripted constituent ($+$ for cation and $-$ for anion) and the superscript $^\circ$ denotes transference number at infinite dilution. The Λ_0 is the equivalent conductance in infinitely dilute solution and obtainable from Kohlrausch's law:

$$\Lambda_0 = \lambda_0^+ + \lambda_0^- \tag{9}$$

The σ is a theoretical constant.

$$\sigma = 41.2/\eta(\epsilon T)^{1/2} \tag{10}$$

where η is the viscosity coefficient, ϵ is the dielectric constant of the solvent, and T is the absolute temperature.

The ionic strength, μ, in equation 8 is defined as

$$\mu = \tfrac{1}{2} \sum_i m_i Z_i^2 \tag{11}$$

where m_i is the molarity of species i.

Equation 8 reduces for a 1:1 strong electrolyte to the form

$$t_\pm - t_\pm^\circ = (2t_\pm^\circ - 1)\sigma\sqrt{m}/\Lambda_0 \tag{12}$$

Equation 12 predicts that a straight line will be obtained if $t_\pm$ is plotted against $\sqrt{m}$. Whether the slope of such a plot will be positive or negative depends on the relative magnitude of $t_\pm^\circ$, σ, and Λ_0.

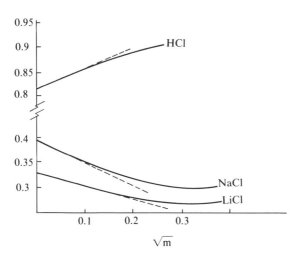

Figure 36–1 *Transference numbers of cations (H^+, Na^+, Li^+) vs. $\sqrt{m}$ for different 1:1 electrolytes.*

In essence, the plots of transference numbers of cations (for example) vs. the square root of molarity are not straight lines but curves, as in Figure 36–1. However, the curves approach the limiting slope predicted by equation 12 for most aqueous solutions at very low concentrations. The few nonaqueous data reported in the literature indicate that equation 12 may be used to a certain extent if the dielectric constant of the solvent is not too low.

In the present experiment, the transference number of silver ion will be measured by employing a solution of silver nitrate and a silver anode and a silver cathode. As electricity passes through the circuit, silver ions migrate into the cathode region and are deposited on the electrode, while nitrate ions migrate out of the cathode compartment. At the anode compartment, nitrate ions enter while silver ions leave, and simultaneously metallic silver leaves the anode to become silver ions. The anode compartment loses silver ion by migration but gains some by solution of the electrode. The final concentration in equivalents, c, for the anode compartment is

$$c_{\text{final}} = c_{\text{initial}} + c_{\text{electrodes}} - c_{\text{migration}} \qquad (13)$$

and for the cathode compartment is

$$c_{\text{final}} = c_{\text{initial}} - c_{\text{electrodes}} + c_{\text{migration}} \qquad (14)$$

Experimentally, the number of equivalents of electricity supplied to the electrodes is measured by a coulometer. The number of equivalents of silver ion present in the compartments initially and the number present finally are determined analytically. The number of equivalents that migrate can be calculated from equations 13 and 14. The transference number is then calculated by using equation 1.

Experimental

The experimental apparatus is illustrated in Figure 36–2. The glass apparatus consists of three compartments, each having a stopcock in the

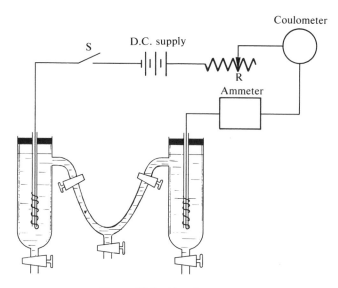

Figure 36–2 *Hittorf apparatus.*

bottom to permit drainage of the liquid from the compartments. The total volume of the unit is about 100 to 150 ml. The three compartments are separated by stopcocks or clamps. If clamps are used, the middle compartment is connected to the electrode compartments by Tygon tubing and, when the compartments are to be separated at the end of the experiment to prevent diffusion from one compartment to another, pinch or screw clamps may be applied to the tubing.

The electrodes are of heavy silver wire, which is cemented into 6-mm. glass tubing with sealing wax or other cement. Five to ten turns of the wire are wound around the glass tubing as shown to increase the electrode area. The wire should be coated with cement from the point where it leaves the tube to the first turn. A D.C. source of 30 to 45 volts is necessary so that a current of 10 to 20 milliamperes is obtained. If a 110-volt D.C. source is available, a variable resistance R(1000 ohm) is included in the circuit; otherwise, a smaller one may be used for the 45-volt current source. To measure the current approximately a milliammeter (0 to 100) is included. This need not be an accurate meter. A more detailed account of the electronic components of the apparatus may be found in Part XIV of this book.

To determine the number of faradays that pass through the solution a silver coulometer should be employed. (Other coulometers, such as copper or gas, may be used, but they are less accurate.) The silver coulometer, providing an accuracy of ± 0.1 percent, is illustrated in Figure 36–3.

A platinum crucible is used as a cathode. Its weight is obtained after drying it in a 150° C. oven. The platinum crucible is filled with 20 percent aqueous silver nitrate. A silver wire mesh serves as an anode. This is suspended in a porous ceramic cup that is permeable to the electrolyte and prevents any anode slime or broken off silver wire from falling into the platinum crucible. During the transference experiment silver whiskers will

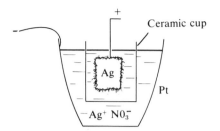

Figure 36–3 *Silver coulometer.*

be deposited in the cathode (platinum crucible). After the experiment is completed the porous ceramic cup with the anode is removed, the silver nitrate solution is slowly decanted, and the platinum crucible is washed gently with distilled water. *Care must be taken not to remove the deposited silver whiskers by the washing with the stream of water.* The crucible is dried at 150° C. and its weight is obtained.

From the difference between the initial and the final weight the weight of the deposited silver is obtained. The number of faradays is calculated. (One faraday deposits 107.880 grams of silver.)

For the first experiment, 250 ml. of 0.05N silver nitrate in water is prepared. About 75 ml. of this solution is reserved for the analysis of the initial silver ion content. Fill the Hittorf apparatus with silver nitrate solution. The arms connecting the electrode compartments to the middle compartments should be open. Be sure that no air bubbles are trapped, which will prevent the closing of the circuit. Put the electrodes in place. Make all necessary electrical connections.

Adjust the rheostat so that its entire resistance is included in the circuit. Close the switch (*S* in Figure 36–2) and adjust the resistance so that a current of 15 to 20 milliamperes is obtained. The current is allowed to flow for 100 to 120 minutes or until 0.001 to 0.003 faraday has passed through the circuit. Calculate the time necessary from the milliammeter reading. While the experiment is proceeding, prepare the solutions necessary for the analysis: 500 ml. of 0.05N potassium thiocyanate, 100 ml. of 6N nitric acid, and 25 ml. of 40 percent ferric alum indicator.

At the end of the experiment, the switch is opened and the stopcocks or clamps separating the compartments are closed. The solutions from the three compartments are drawn off into clean preweighed Erlenmeyer flasks and stoppered. The weight of the solution from each compartment is obtained by weighing the Erlenmeyer flask to its nearest 0.1 gram. Determine the grams of silver deposited in the coulometer and hence the number of faradays passed.

The apparatus is washed and cleaned and another run is prepared with 0.025N silver nitrate in 25 percent (w/w) DMSO (or DMF, acetone, dioxane, and so forth) in water solution. A new silver coulometer is set up, the necessary electrical connections are made, and the second experiment is run as before until about 2 to 3 millifaradays have passed. At the end the solutions from the three compartments are drained off, weighed, and analyzed as before.

During the second experiment, analysis of the four aqueous solutions (the initial, and the solutions from the three compartments) is performed by the Volhard method:

Pipet 25-ml. aliquots of each sample into clean beakers. Add 1 ml. of ferric alum indicator solution, acidify with 5 ml. of 6N nitric acid, and titrate with a standard 0.05N thiocyanate solution. The first color change to orange-red takes place before the equivalence point is reached; continue titrating until a permanent brownish tinge is obtained upon vigorous shaking.

Similarly, the analysis of the silver ion in the DMSO-water solvent is also obtained by the Volhard technique.

From the titration data and from the weight of the samples, calculate the number of equivalents of silver present in the initial solutions and the final solutions from each compartment. The concentration of the solution in the middle compartment should be the same as the initial solution within an experimental error of ± 1 per cent. If this is not achieved, mixing between the compartments has taken place and the experiment must be repeated with less current and for a shorter time.

Obtain the transport number of the silver ion by using equations 1, 13, and 14.

Obtain from equation 2 the λ_{Ag^+} and compare it with values in the literature. The Λ values necessary to solve equation 2 have been obtained in Experiment 35 on conductances of 0.05N silver nitrate in water and in 25 percent (w/w) DMSO in water (DMF in water, acetone in water, dioxane in water, and so forth).

Knowing the η and ϵ values for water and for the DMSO-water mixture from Experiments 15 and 20, calculate σ from equation 10. With the aid of σ and Λ_0 obtained from the previous conductance experiment, employ equation 12 to determine whether silver nitrate in 25 percent DMSO-water obeys the Debye-Hückel-Onsager theory.

Material and Equipment. Silver nitrate; silver wire; 6-mm. glass tubing; dimethyl sulfoxide (DMSO) (or dimethylformamide [DMF], acetone, dioxane, and so forth); potassium thiocyanate; nitric acid; ferric alum indicator; six Erlenmeyer flasks; 25-ml. pipet; Hittorf apparatus; platinum crucible; silver wire mesh; porous ceramic cup; D.C. power supply to provide 20 to 25 milliampere current; milliammeter.

REFERENCES

1. M. Spiro. In Weissberger, A. (ed.). Physical Methods of Organic Chemistry. 3rd Edition, Part IV, Interscience Publishers, Inc., 1960, Chapter 46.
2. I. M. Kolthoff and E. B. Sandell. Textbook of Quantitative Inorganic Analysis. 3rd Edition, The Macmillan Co., New York, 1952, p. 455.

37

POLAROGRAPHY: DETERMINATION OF THE STABILITY CONSTANT OF A COMPLEX METAL ION

Polarography is a technique for measuring current-voltage relationships in a chemical reaction in which the anodic and cathodic processes are essentially separated and only the cathode determines the current-voltage relation. This is accomplished by using a large layer of mercury as the anode and small mercury droplets as a cathode coming from a mercury reservoir with a certain hydrostatic head and forming at the end of a capillary tube at a rate of one drop in three to five seconds. The schematic diagram of the polarograph is given in Figure 37–1.

The dropping mercury electrode (DME) acts as a microelectrode because its current density is much greater than that of the large mercury pool in the anode. This means that the dropping mercury electrode is readily polarized and it determines the current-voltage relationship.

An electrode is considered to be *polarized* when it maintains a certain potential with little or no flow of current. For example, a platinum electrode dipping into a solution of zinc ions when short circuited with a calomel reference electrode will take up the potential of the reference electrode with no flow of current. However, when a sufficiently high emf is applied, the zinc starts to be deposited on the platinum electrode. The emf applied exceeds the decomposition potential of zinc and after some zinc has been plated out, the platinum electrode becomes depolarized. The potential of a *depolarized* electrode follows the Nernst equation.

$$E = E^0 + \frac{0.05915}{2} \log [Zn^{++}] \qquad (1)$$

In equation 1 the activity of the deposited zinc does not appear since it is taken

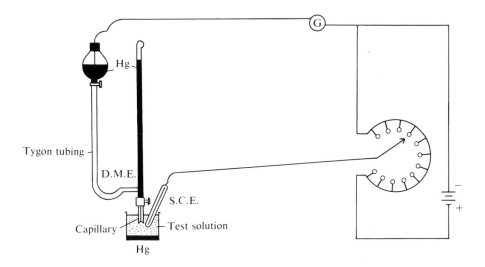

Figure 37–1   *Dropping mercury electrode apparatus for polarography.*

as unity. The number in front of the log term represents the RT/nF term at 25° C. As is evident from equation 1, the potential of the electrode depends upon the activity (concentration) of the ion. While the electrode remains depolarized, the reversibility of the reaction is maintained; the passage of current does not change the value of the potential from its reversible value. The voltage-current relationship with a depolarized electrode is given by Ohm's law. Therefore, a linear relationship is observed, as indicated in Figure 37–2.

In each case of solutions with different concentrations of zinc ion no current will flow until the decomposition potential of the solution is reached.

With a microelectrode the situation is different, especially when the solution is not stirred during the reduction of zinc ion. Under these conditions the zinc ion reduced at the microelectrode is replaced from the bulk by diffusion. The rate at which the fresh zinc ion is supplied depends on only the diffusion process. As the depletion of zinc ion at the microelectrode sets up a concentration gradient, $C_0 - C$, between the microelectrode, C_0, and the bulk, C, the rate of diffusion to the surface

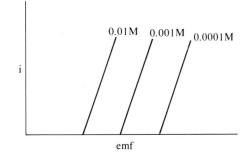

Figure 37–2   *Current-voltage diagrams in the reduction of zinc ion with a depolarized electrode. (The concentration of the zinc ion in the solutions is indicated by the molarities.)*

increases according to Fick's first law of diffusion.

$$J = -D \, \nabla C \qquad (2)$$

In equation 2, J is the flux density (number of molecules flowing through a square centimeter of area per second), ∇C is the concentration gradient, and D is the coefficient of diffusion.

Adjusting equation 2 to our situation,

$$\frac{\partial s}{\partial t} = \frac{A D}{\delta} (C - C_0) \qquad (3)$$

where $\partial s / \partial t$ is the rate of diffusion, A is the area of the microelectrode, $(C - C_0)$ is the difference in concentration between bulk and surface, and δ is the thickness of a hypothetical diffusion layer around the microelectrode.

When equilibrium is finally established, the rate of reduction of zinc ion is equal to the rate of diffusion (a diffusion-governed process). If i is the current and n is the electrons transferred at the cathode (2 in the case of Zn^{++}), the rate of discharge, $\partial r / \partial t$, is

$$\frac{\partial r}{\partial t} = \frac{iA}{nF} \qquad (4)$$

At equilibrium, equations 3 and 4 are equal. Hence,

$$i = \frac{DnF}{\delta} (C - C_0) \qquad (5)$$

In the case of a microelectrode, the current-voltage curve will look like Figure 37–3.

The current will not change with increasing applied potential until the decomposition potential is reached. At this point a linear increase occurs, as illustrated in Figure 37–2. However, as the applied potential is increased, the concentration overpotential increases rapidly, and the current reaches a limiting value when $C_0 \ll C$. Under such conditions, C_0 is negligible and the limiting current is proportional to the bulk concentration, C (equation 5). Therefore, the limiting current plateau can be used for quantitative analysis of the bulk concentration of the electrolyte.

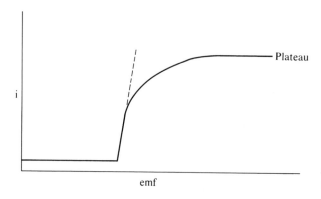

Figure 37–3   *Current-voltage curve for reduction of Zn++ with a microelectrode.*

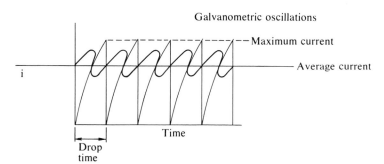

Figure 37-4 *Schematic diagram of current-time curves with a dropping mercury electrode (DME).*

With the dropping mercury electrode as the microelectrode, the current-voltage curves are not smooth curves. As the droplets of mercury form and are detached from the capillary, there is a periodic variation in the current, as illustrated in Figure 37–4.

Since the applied potential is varied linearly with time in most polarographic experiments, such oscillations will also be recorded on the current-voltage curve (Fig. 37–5).

Since the limiting current plateau is used for the quantitative determination of the oxidant, the oscillations are averaged out to a smooth curve in the curves for both the supporting electrolyte and the oxidant.

We have seen that at the limiting plateau the current is proportional to the concentration of the oxidant (equation 5). This is the case if the only contribution to the limiting current is from the diffusion current. However, the migration of ions also contributes to the observed current. For example, if the oxidant is a cation, zinc, its migration current will add to the cathodic diffusion current, but if the zinc is in the complex ion form $[Zn^{++} \cdot 4\,en^-]^=$, it will deduct from the cathodic diffusion current.

In order to minimize the contributions of the migration current of the supporting electrolyte, potassium chloride is added to the solution. If the supporting electrolyte

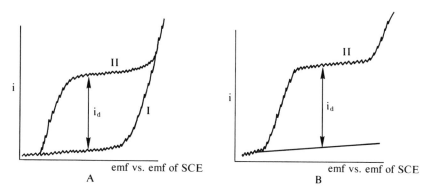

Figure 37-5 *Current-voltage curve. A, curve I is the curve of the supporting electrolyte and curve II is that of the oxidant Zn^{++} solution. B, the same polarogram but the extrapolation technique was used in determining i_d.*

is in large excess (for example, 99 percent of the total ions), approximately 99 percent of the cathode migration current will be transported by the potassium ions. Thus, the contribution of the other cations to the migration current will be reduced.

The limiting diffusion current, i_d, of the oxidant is obtained when the current of the supporting electrolyte is subtracted from that of the oxidant, as illustrated in Figure 37–5A. Another technique to use is the extrapolation of the i_r branch and the determination of the i_d from the difference (Fig. 37–5B). In this case no polarogram of the supporting electrolyte is needed.

The limiting diffusion current thus obtained depends on a number of parameters besides the concentration. Equation 6, derived by Ilkovic, gives the theoretical relationship:

$$i_d = 607nD^{1/2}Cm^{2/3}t^{1/6} \tag{6}$$

In this equation i_d is the average diffusion current in microamperes during the lifetime of the mercury drop, D is the diffusion constant of the oxidant in cm.^{2} sec.$^{-1}$, n is the number of electrons transferred in the reduction process, C is the bulk concentration of the oxidant in millimoles/liter, m is the mass of mercury flowing through the capillary per second (units in milligrams/second), and t is the drop time in seconds (i.e., the time interval between successive drops).

Since the parameters n and D are specific for the system under investigation, the Ilkovic equation may be rearranged:

$$I_d = 607n\,D^{1/2} = \frac{i_d}{Cm^{2/3}t^{1/6}} \tag{7}$$

I_d is called the diffusion current constant. The instrumental parameters are m and t.

Parameter m is influenced by the hydrostatic head of the mercury reservoir. Although the size of the droplets is determined by the capillary, an increase in the hydrostatic head will increase the number of drops per second; thus it will increase m. The drop time, t, varies with the applied emf, following the electrocapillary curve of mercury (Fig. 37–6).

The drop time also increases as the potential is increased, reaching a maximum at -0.52 volt and decreasing thereafter. The drop time for the Ilkovic equation is usually measured at the half-wave potential (see later).

Thus, if m, t, and D are known, the concentration of the oxidant can be calculated

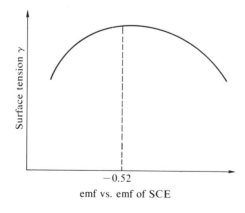

−0.52

emf vs. emf of SCE

Figure 37–6 *Electrocapillary curve for mercury.*

from the diffusion current. Or, if the concentration is known and m and t are determined, the diffusion constant of the oxidant can be obtained by using the Ilkovic equation.

There is, however, one little understood surface effect in the dropping mercury electrode polarograms that causes the current to increase to a maximum before dropping back to the level of the limiting diffusion current plateau. This is usually eliminated by using surface active materials, such as gelatin or fuchsin.

Besides the i_d parameter, the half-wave potential—the potential applied when the current is half of the limiting current—is of primary importance in chemical studies.

$$E_{1/2} = E \quad \text{when} \quad i = \tfrac{1}{2}i_d \qquad (8)$$

The significance of the half-wave potential is evident considering the reaction

$$\text{Oxidant} + ne^- \rightleftharpoons \text{reductant} \qquad (9)$$

The reversible potential at the electrode-solution interface is given by the Nernst equation.

$$E = E^0 + \frac{0.0591}{n} \log \frac{[\text{ox}]_i}{[\text{red}]_i} \qquad (10)$$

The subscript in equation 10 indicates that the concentration refers to that at the mercury-solution interface.

The observed current depends on the concentration gradient, according to equation 5.

$$i = K([\text{ox}] - [\text{ox}]_i)D_{\text{ox}}^{1/2} \qquad (11)$$

The n, m, and t parameters are included in K, the capillary constant. At the diffusion current plateau, the concentration of the oxidant at the interface, $[\text{ox}]_i$, is reduced to negligible quantities. Hence,

$$i_d = K[\text{ox}]D_{\text{ox}}^{1/2} \qquad (12)$$

and

$$[\text{ox}]_i = \frac{i_d - i}{K D_{\text{ox}}^{1/2}} \qquad (13)$$

Most metals form amalgam with the dropping mercury electrode. The concentration of the reductant is proportional to the observed current.

$$i = K[\text{red}]_i D_{\text{red}}^{1/2} \qquad (14)$$

Substituting equations 13 and 14 into equation 10

$$E = E^0 + \frac{0.0591}{n} \log \frac{i_d - i}{i} + \frac{0.0591}{n} \log \left(\frac{D_{\text{red}}}{D_{\text{ox}}}\right)^{1/2} \qquad (15)$$

The second term of equation 15 is zero at the half-wave potential (equation 8) and, therefore,

$$E_{1/2} = E^0 + \frac{0.0591}{n} \log \left(\frac{D_{\text{red}}}{D_{\text{ox}}}\right)^{1/2} \qquad (16)$$

The polarographic half-wave potential is directly related to the standard potential of the oxidation-reduction system, but it has the additional term containing the diffusion constants. Thus, the half-wave potential is a characteristic of the oxidation-reduction system under consideration.

The oxidant does not have to be a simple cation; it may be a complex formed according to the stoichiometric relationship

$$ox + pL \rightleftharpoons ox\, L_p \tag{17}$$

$$K_{comp} = \frac{[ox\, L_p]}{[ox](L)^p} \tag{18}$$

where L is the complexing ligand, p is the number of moles of ligand in the complex, and K_{comp} is the (equilibrium) stability constant of the complex.

For this situation, equation 10 takes the form

$$E = E^0 + \frac{0.0591}{n} \log \frac{[ox\, L_p]}{K_{comp}(L)^p[red]} \tag{19}$$

and the half-wave potential is

$$E_{1/2} = E^0 + \frac{0.0591}{n} \log \left(\frac{D_{red}}{D_{ox}}\right)^{1/2} - \frac{0.0591}{n} \log K_{comp}(L)^p \tag{20}$$

Thus, from equations 16 and 20

$$E_{1/2}(\text{simple ion}) - E_{1/2}(\text{complex}) = \frac{0.0591}{n} \log K_{comp}[L]^p \tag{21}$$

In practice, the half-wave potentials of the complex ion are obtained for a number of solutions in which the ligand concentration (L) is varied (but it is always in excess of the metal ion concentration).

Since the last term in equation 20 can be expressed as

$$-\frac{0.0591}{n} \log K_{comp}(L)^p = -\frac{0.0591}{n} \log K_{comp} - \frac{0.0591 p}{n} \log (L) \tag{22}$$

a plot of $E_{1/2}$ against $\log (L)$ will give a straight line and the slope of the line will be $-0.0591 p/n$. Thus, the stoichiometric complexing number, p, is obtained.

With the knowledge of p and the establishment of $E_{1/2}$ for the simple ion, one can calculate the stability constant, K_{comp}, from equation 21.

Experimental

The complexing parameters of the copper-glycinate system (see Keefer) will be studied with the polarographic technique. The reduction of the complex ion to amalgam can be represented by the following equation:

$$Cu\, L_p^{(n-pb)} + ne^- + Hg \rightleftharpoons Cu(Hg) + L^{-b}$$

where L^{-b} is the glycinate ion and Cu(Hg) is the amalgam formed.

TABLE 37–I.   Composition of Solution for Polarographic Study

SOLUTION	POTASSIUM GLYCINATE (*Moles/liter*)	CUPRIC NITRATE (*Moles/liter*)	POTASSIUM NITRATE (*Moles/liter*)	METHYL RED Percent
1	0.00	5×10^{-4}	5×10^{-1}	0.003
2	0.01		4.8×10^{-1}	
3	0.03		4.4×10^{-1}	
4	0.05		4.0×10^{-1}	
5	0.06		3.8×10^{-1}	
6	0.08		3.4×10^{-1}	

In order to obtain the parameters p and K_{comp}, five solutions with different ligand concentrations have to be made. The five solutions should contain 0.01, 0.03, 0.05, 0.06, and 0.08M glycine. Thus, the total composition of six solutions to be measured should be those given in Table 37–1.

The potassium glycinate solution should be made by combining the desired amount of glycine with *less than equivalent* potassium hydroxide. For example: For 0.05 mole of glycine, 0.04 mole of potassium hydroxide is sufficient.

The methyl red is used as maximum suppressor.

The supporting electrolyte, potassium nitrate, is sufficient to make solutions of constant ionic strength ($\mu \sim 1.0$).

The solutions should be thermostated in a 25° C. bath.

The schematic diagram of a recording polarographic apparatus is given in Figure 37–7. In order for the recorded potential to be measured against a standard potential, a calomel electrode of low resistance is used. This standard calomel electrode, SCE, can be homemade on a large surface of sintered glass. Then it is dipped into the test solution. The dropping mercury

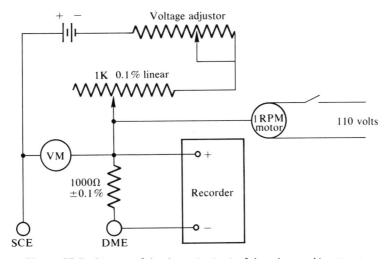

Figure 37–7 *Diagram of the electronic circuit of the polarographic apparatus.*

electrode, DME, is also dipped into the test solution. The cup containing the test solution is thermostated at $25 \pm 0.1°$ C.

The motor drives the slide wire and, thereby, supplies an applied emf that varies linearly with time. Thus the applied emf is recorded on the ordinate of the chart paper. If the voltage scanned is 1.50 volts in ten minutes, a recorder chart speed of 3 inches/minute will scan 12 blocks on the chart paper per minute and each block will represent 12.5 mv. The recorder is 10 mv. (full scale); hence, the full scale on the abscissa will correspond to 10 microamperes. A more detailed account of the electronic components of the apparatus may be found in Part XIV of this book.

1. For the first procedure take solution 1 with only cupric nitrate and supporting electrolyte, and pour about 75 ml. into the thermostated electrode vessel, which should have a layer of mercury at the bottom. Both the dropping mercury electrode and the standard calomel electrode should dip into the solution.

Set the mercury head in the dropping mercury electrode so that there is a drop approximately every four seconds. Allow the timing motor to scan to one of its limits. Stop the motor.

Turn on the voltage switch and adjust it so that the meter registers 1.50 volts. Turn the voltage switch to the other position, 0 volts. Connect the negative electrode terminal to the dropping mercury electrode and the positive electrode terminal to the standard calomel electrode. Connect the recorder terminals to the recorder. (Red wires are positive and black wires are negative terminals.)

Set the recorder for 10 microamperes full scale and the speed switch to 3 inches/minute. Set the time switch to minutes and turn on the timing motor.

The recorder will plot the polarogram, scanning the 0 to 1.5 volt range in ten minutes.

At the end, stop the motor and so forth.

2. Insert the sintered glass bubbler into the solution and allow dry nitrogen gas from a tank to bubble through it for ten minutes to remove the dissolved oxygen from the first run.

After ten minutes, stop the flow of nitrogen and repeat the procedures in section 1 to record the polarogram of cupric nitrate in the absence of oxygen.

3. Bubble nitrogen through solution 2 for ten minutes. After that, record its polarogram. Repeat the procedure with all the other solutions.

Determine the diffusion current, i_d, from the polarograms for each solution, using the extrapolation techniques illustrated in Figure 37–5B.

Obtain the $E_{1/2}$ values for the oxygen, the copper, and the complexes with different ligand concentrations.

Assuming that the free ligand concentration (L) at equilibrium is essentially the same as the initial glycinate concentration, plot the $E_{1/2}$ values obtained for solutions 2 through 6 against log (L).

Calculate the p parameter from the slope of the plot.

Employing this p value, obtain the stability constant of the complex, K_{comp}, by using equation 21.

What kind of complex is formed by copper and glycinate?

On the basis of the stability constant, predict whether the same complex will exist when the initial ligand concentration exceeds 0.1M or whether you would expect a complex with a greater number of ligands.

Estimate the error introduced by the approximation that the equilibrium ligand concentration is the same as the initial glycinate concentration. How would you correct this error? What additional measurements would you make?

On the basis of comparison of the polarograms of cupric nitrate in the presence and absence of dissolved oxygen, appraise whether deoxygenation with nitrogen was a necessary step.

Material and Equipment. Cupric nitrate; glycine; potassium hydroxide; potassium nitrate; methyl red indicator; mercury; tank of compressed nitrogen gas; recording polarograph; thermostat with heater, stirrer, and thermoregulator; sintered glass filter to disperse nitrogen gas.

REFERENCES

1. H. H. Willard, L. L. Merritt and J. A. Dean. Instrumental Methods of Analysis. 4th Edition, D. Van Nostrand Co., Inc., Princeton, N.J., 1966, Chapter 25.
2. L. Meikes. Polarographic Techniques. Interscience Publishers, Inc., New York, 1955.
3. T. M. Kolthoff and J. J. Lingane. Polarography. 2nd Edition, Interscience Publishers. Inc., New York, 1952.
4. H. Schmidt and M. von Stackelberg. Modern Polarographic Methods. Academic Press, New York, 1963.
5. C. N. Reilley and D. T. Sawyer. Experiments for Instrumental Methods. McGraw-Hill Book Co., New York, 1961.
6. R. M. Keefer. J. Am. Chem. Soc., **68**, 2329 (1946).

38

DISSOCIATION CONSTANT DETERMINATION BY POTENTIOMETRIC TITRATIONS*

The dissociation constant of a monoprotic acid, such as acetic acid, is derived from the equilibrium constant of the reaction. For the solvents water and dimethyl sulfoxide (DMSO) the reactions are

$$\overset{(A)}{CH_3COOH} + H_2O \leftrightarrows \overset{(A^-)}{CH_3COO^-} + \overset{(H^+)}{H_3O^+} \tag{1}$$

$$\overset{(A)}{CH_3COOH} + \overset{CH_3}{\underset{CH_3}{\diagdown}} SO \leftrightarrows \overset{(A^-)}{CH_3COO^-} + \overset{(H^+)}{(CH_3)_2SOH^+} \tag{2}$$

Thus, the equilibrium constants are

$$K_1 = \frac{a_{A^-} \cdot a_{H^+}}{a_A \cdot a_{H_2O}} \tag{3}$$

$$K_2 = \frac{a_{A^-} \cdot a_{H^+}}{a_A \cdot a_{DMSO}} \tag{4}$$

where a is the activity of a species present in the mixture. In the abbreviated form, A refers to the undissociated acid and A^- is its conjugated base; H^+ represents the solvated hydrogen ion and therefore it represents different species in equations 3 and 4.

348

The dissociation constant is

$$K_a = \frac{a_{A^-} \cdot a_{H^+}}{a_A} \tag{5}$$

Thus, for reaction 1, $K_a = K_1 \cdot a_{H_2O}$, whereas for reaction 2, $K_a = K_2 \cdot a_{DMSO}$. In dilute solutions the activity of the solvent is approximately unity. Therefore, $K_a = K$.

There are a variety of experimental methods that can be used to determine the dissociation constant of an acid. The most commonly used is potentiometric titration, which consists of titrating the acid with a strong base and simultaneously measuring the electromotive force, emf, by use of a potentiometer.

The emf is a potential difference under the influence of which a *reversible* reaction occurs at two electrodes. The key word in this statement is reversible because according to the first law of thermodynamics, a reversible transfer of electricity is the maximum electrical work obtainable from such a cell reaction. Thus, it is related to the free energy of the reaction through the well known Nernst equation.

$$\Delta G = -nFE \tag{6}$$

where ΔG is the free energy of the reaction when n equivalents of product are formed during the transfer of nF coulombs of charge under the potential difference of E. The F is 96,500 coulombs or 1 faraday, corresponding to the transfer of one equivalent of charge. The E is given in volts, nFE is in joules and, therefore, if ΔG is desired, expressed in calories, the result must be divided by 4.1840 (joules per calorie).

The simplest process that illustrates the working of emf is in a galvanic cell. For example, a zinc electrode dipping into a zinc sulfate solution of concentration C_1 is connected through a wire to a copper electrode dipping into a copper sulfate solution of concentration C_2. A potassium chloride salt bridge completes the circuit.

The electrode reactions in this galvanic cell are

$$Zn \rightarrow Zn^{++} + 2e^-$$
$$Cu^{++} \rightarrow Cu - 2e^-$$

The potential difference under the influence of which this reaction occurs is the emf and this is measured by a potentiometer. (There is an additional potential difference in this particular galvanic cell, which is at the boundaries of the two solutions and is known as the liquid junction potential. It occurs because the rates of diffusion of zinc and copper ions are not the same, and this sets up a potential gradient.) In the

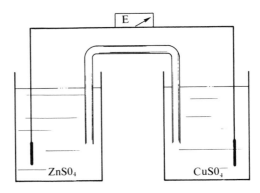

Figure 38–1 *Galvanic cell.*

case in which $C_1 \neq C_2$ an additional potential gradient results from the rate of diffusion of the sulfate ions across the boundary, which is due to the concentration gradient. Hence, the measured potential difference includes both the emf and the liquid junction potentials. The magnitude of the latter is reduced by the use of a potassium chloride bridge. This is usually an agar gel containing a high concentration of potassium chloride. Since the concentration of potassium chloride is high compared to C_1 and C_2, the liquid junction potentials are determined by the rates of diffusion of potassium and chloride ions into the adjoining solutions. Because the mobilities of potassium and chloride ions are about the same, the liquid junction potential is minimized but not completely eliminated.

Returning to the original reactions,

$$\Delta G = \Delta G^0 + RT \ln K_a \tag{7}$$

and through the Nernst equation (6)

$$E = E^0 - \frac{RT}{nF} \ln K_a \tag{8}$$

In these equations ΔG^0 and E^0 refer to the standard conditions i.e., the free energy change and the emf that would be observed if all reactants and products were at unit activity throughout the reaction.

The dissociation constant, K_a, is given in terms of activities. If the activity of all species except the hydrogen ion is given in terms of concentrations and activity coefficients, γ,

$$a = C\gamma \tag{9}$$

then

$$K_a = \frac{a_{H^+} C_{A^-} \gamma_{A^-}}{C_A \gamma_A} \tag{10}$$

Taking logarithms of both sides of the equation and including the activity coefficients into a new dissociation constant, K', we obtain

$$pK' = -\log K_a(\gamma_A/\gamma_{A^-}) = pH - \log \frac{C_{A^-}}{C_A} \tag{11}$$

We shall discuss later the conditions under which the activity coefficients can be incorporated into the new constant K'. For the present, the operative aspects of equation 11 will be followed.

When an acid is titrated with a strong base either the potential or the pH is measured as a function of the quantity of titrant base, q. The result is given in Figure 38–2. The scales of the ordinates, both E and pH, are made equidistant by properly selecting the electrode for the pH meter so that E is proportional to pH and $\log [C_{A^-}/C_A]$ is a constant.

The end point of such a titration is at maximum slope, i.e., dpH/dq = maximum. When an inverse derivative (dq/dpH) is taken as the variable of pH, a maximum of the curve will appear at the point at which

$$pK' = pH \tag{12}$$

This is illustrated in Figure 38-3.

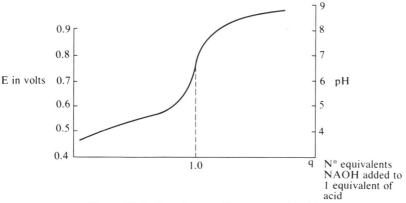

E in volts

Figure 38–2 *Titration curve for a monoprotic acid.*

From the end point of the titration the number of equivalents of acid in the solution can be calculated. Furthermore, if the titration curve of the solvent is subtracted from the titration curve of the solution, the resulting "difference titration curve" can be used to calculate the pK' value by using equation 11. For example: Let us assume that a 0.1N monoprotic acid is titrated with a strong base. The difference titration curve shows that when 0.02 equivalent of base was added, the pH of the solution was 4.5. The base was used to convert $A \rightarrow A^- + H^+$; therefore the concentration of A^- was 0.02, whereas the concentration of A was 0.08. Hence,

$$\log \frac{C_{A^=}}{C_A} = -0.60206 \text{ and } pK' = 4.5 + 0.6 = 5.1$$

Furthermore, when the titration data are presented in the form of a buffer capacity curve, as in Figure 38–3, a maximum is obtained.

This maximum indicates that a large amount of base can be added at this pH without appreciably altering the pH of the solution. It is the buffer capacity of the solution and it has great biological significance. For example, blood has maximum buffer capacity at pH 7. This means that slight changes in the production of metabolic acids or bases do not affect the pH of the solution, thus protecting the viability of cells, which are sensitive to changes in pH. The maximum in the buffer capacity curve then can be used to obtain the pK' value of an acid.

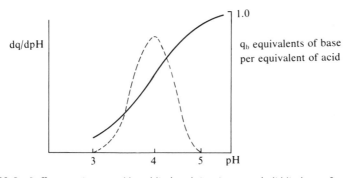

dq/dpH

Figure 38–3 *Buffer capacity curve (dotted line) and titration curve (solid line) as a function of pH.*

In the foregoing discussion we made a number of approximations that should be clarified.

1. First, we assumed that pH $= -\log a_{H^+}$.
2. Second, the buffer capacity $\Gamma = dq_b/d\text{pH}$ is also the contribution of two terms (see reference 3)

$$\Gamma_{\text{total}} = \Gamma_{\text{solvent}} + \Gamma_{\text{acid}} \tag{13}$$

Therefore, if the pK' of an acid is to be obtained from titration curves, the titration curve of the solvent must also be determined and subtracted from the total titration curve.

3. We also assumed that the total volume of the solution does not change appreciably during titration. This condition can be fulfilled if a relatively strong base or acid (1N or more) is used and a microburet is employed for the titration. Otherwise, there will be an additional term in equation 13 that contributes to the total buffer capacity (see reference 3) but this is usually small.

4. As expressed in equation 11, the $K' \neq K_a$, since K' contains a ratio of activity coefficients. In order for K' to be a constant, this ratio must be constant. However, activity coefficients are dependent upon concentration. In dilute solution with low ionic strength (less than 0.1 for univalent ions) the ratio is sufficiently constant for the approximation to be valid. Moreover, in this range the activity coefficients can be calculated according to the Debye-Hückel theory and, therefore, the pK' value can be corrected to approximate the true pK_a value. According to the Debye-Hückel theory, the activity coefficient of an ion, γ_i, with charge Z_i is

$$-\ln \gamma_i = \frac{AZ_i^2\mu^{1/2}}{1 + Ba_i\mu^{1/2}} \tag{14}$$

In this equation, there are two constants, A and B, that depend upon the dielectric constant of the solvent and the experimental temperature:

$$A = \frac{e^3}{(\epsilon kT)^{3/2}}\left(\frac{2\pi N}{1000}\right)^{1/2} = \frac{1.8245 \times 10^6}{(\epsilon T)^{3/2}} \tag{15}$$

and

$$B = \frac{(8\pi Ne^2)^{1/2}}{(1000\epsilon kT)^{1/2}} = \frac{50.290 \times 10^{-8}}{(\epsilon T)^{1/2}} \tag{16}$$

where e is the protonic charge, k is the Boltzmann constant, N is Avogadro's number, ϵ is the dielectric constant, and T is the absolute temperature. The ionic strength, μ, in equation 14 is defined as

$$\mu = \tfrac{1}{2}\sum C_i Z_i^2 \tag{17}$$

where C_i is the molar concentration and Z_i the respective charge of ionic species i.

The last term that must be defined in equation 14 is a_i. This is the effective ionic diameter of species i, including the solvation layer. It is only estimated and Kielland[4] lists values for a large number of ions in water. The inaccuracy of the estimation of a_i makes equation 14 not very reliable. It is especially hard to estimate accurately this value when the extent of solvation of an ion varies from solvent to solvent. Usually the effective ionic radii range between 3 and 8 Å.

Because of this inaccuracy, the shorter form of the expression for the activity coefficient is often used, especially at low ionic strength, i.e., less than 0.05.

$$-\ln \gamma_i = AZ_i^2\mu^{1/2} \tag{18}$$

This is called the Debye-Hückel limiting law. Even with the correction of the activity coefficients calculated from the Debye-Hückel law, the corrected pK' values are not equal to pK_a because of the inaccuracies in the first and fourth assumptions. However, at low ionic strength they are within a few hundred units of pK_a.

In this discussion, we have described the theoretical basis on which pK_a values for an acid can be obtained from buffer capacity or titration curves. Two more aspects must be discussed before giving the experimental conditions. The first aspect is the potentiometric technique and the second aspect is the measurement of pH in different solvents as related to the potentiometric technique.

As was mentioned before, the dissociation constant and the thermodynamic functions related to it refer to a reversible process. The emf, or potential difference, must be measured in such a way that the process is not changed from a reversible to an irreversible one. Therefore, a voltmeter cannot be used because it draws a small current. In the potentiometric measurement the potential of the cell is opposed by a variable potential, which is adjusted so that no current will flow through a galvanometer. A simple setup is given in Figure 38–4.

A slow discharge battery is used as a source of the variable potential in series with resistance R, A, and B. It is important that no appreciable change in the voltage of the battery occur during the measurement.

In Figure 38–4 a standard Weston cell having 1.0184 volts at 25° C is used to

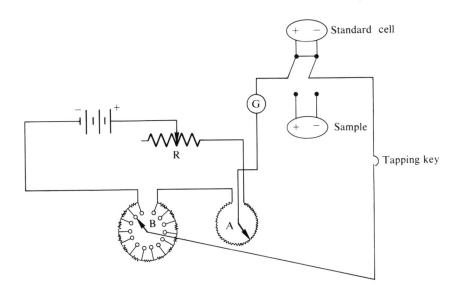

Figure 38–4   *Circuit diagram for a potentiometer.*

adjust the resistance. The slide wire, A, and the resistance box, B, are calibrated in volts ($E = IR$, hence the requirement that a steady current, I, flow through the circuit). This condition is met by adjusting A and B to register the total potential of the standard cell and then putting the throw switch to the standard cell. The variable resistance, R, is adjusted so that the galvanometer, G, registers zero when the tapping key is depressed. Once this setting of R is obtained, the throw switch is reversed to include the sample cell in the circuit. Now the slide wire, A, and the resistance box, B, are adjusted so that no current flows through the galvanometer. The potential of the cell (in volts) is then read off from A and B.

The cell in which titration is done has two electrodes. The most commonly used in aqueous solutions is the glass electrode in connection with the calomel electrode. Glass electrodes have very high resistance, 10^8 ohms and more, and therefore, in a circuit as in Figure 38–4 no deflection of the galvanometer would be observed even though the potentiometer is not balanced. (Regular galvanometers have a sensitivity of about 0.04 μamperes/millimeter.)

In commercial pH meters (which are potentiometers used with glass electrodes) an electron tube amplifier is used instead of the galvanometer as a null-point indicator. A more detailed account of the electronic components of the apparatus may be found in Part XIV of this book.

The cell having a glass electrode and a calomel reference electrode can be described schematically as

$$\text{Ag/AgCl, 0.1N HCl/glass/sample//KCl sat. Hg}_2\text{Cl}_2\text{/Hg}$$

The mechanism of the glass electrode is such that reversible transfer of hydrogen ions takes place through a thin glass membrane between the hydrochloric acid solution and the sample to be measured. No other ion should be transferred through the glass membrane. However, at high pH (low hydrogen ion concentration of the sample) some small cations, such as lithium and sodium, may pass through the membrane and, therefore, glass electrodes are not used with these cations present above pH 12. In such a high pH range, special "alkali" glass electrodes should be used.

For pH measurements in nonaqueous media, a glass electrode can be used in connection with the silver–silver chloride electrode under two conditions. One condition is that sodium chloride or hydrochloric acid should dissociate completely in the solvent and the other is that the silver chloride should not dissociate. The silver–silver chloride electrode should operate reversibly. These conditions are met in DMSO and other organic solvents, such as methanol, ethanol, dioxane, or acetonitrile.

Since the calomel electrode (or silver–silver chloride electrode in the nonaqueous system) is used as the reference electrode, the potential of the glass electrode is related to the pH, according to equation 8.

$$E = \text{constant} + \frac{RT}{2.303} \text{pH} \tag{19}$$

The constant in equation 19 is evaluated by using standard solutions, the pH of which is well known, and by obtaining the potential with the working glass electrode. This is the most commonly used technique in aqueous solutions for which such standards are readily available. In nonaqueous solutions, the glass electrode has to be calibrated on a standard material, the dissociation constant of which is well known from other types of measurements, such as conductance and spectrophotometric experiments.

Experimental

The dissociation constants of acetic acid will be measured in water and DMSO (or dimethylformamide [DMF], acetone, dioxane, and so forth). Potassium chloride will be the neutral salt used to keep the ionic strength relatively constant during the titration.

Prepare 200 ml. each of 0.05M potassium chloride solution in water and 0.001M potassium chloride solution in DMSO. Dissolve approximately 0.01 equivalent of glacial acetic acid in 100 ml. each of the aqueous and DMSO solutions.

Titration should be performed on 50 ml. aliquots of aqueous solutions of both the solvent and the acetic acid, using a commercial pH meter calibrated in both millivolts and pH units (such as Beckmann Model G or H, or Radiometer). A commercial glass electrode and calomel electrode can be used, preferably isolated from the potentiometer (pH meter) by a Faraday cage.

First the potentiometer is standardized by using buffer solutions at pH 4.0 and pH 7.0. Standard buffer solutions may be obtained commercially or prepared in the laboratory from recrystallized reagents. (For formulation of pH standards, see the Handbook of Chemistry and Physics and other references.) A magnetic stirrer is employed and a microburet is dipped into the solution. A 5N potassium hydroxide solution is used as a titrant. The normality of the potassium hydroxide is determined by titration against standard oxalic acid or phthalic acid solution.

The titration is performed by adding small increments (0.01 to 0.02 ml.) of titrants to the solution and, after a few seconds of stirring, measuring the resulting pH. (The magnetic stirrer should be turned off during the pH measurements.) Thus, *duplicate* titration curves are obtained on both the solvent (0.05N aqueous potassium chloride) and the acetic acid solution.

For the titration curves of DMSO the potentiometer used in aqueous titration is employed. One of the electrodes is a glass electrode (for example, Beckmann No. 290, containing an inner silver–silver chloride electrode in an aqueous sodium chloride solution) and the other is an isolated silver–saturated silver chloride–saturated sodium chloride electrode in DMSO. The latter reference electrode is prepared by dipping a silver plated platinum wire into a saturated solution of both silver chloride and sodium chloride in DMSO. The silver electrode can be prepared by fusing the platinum wire into the bottom of the cell and plating it from a 0.05 argentocyanide solution

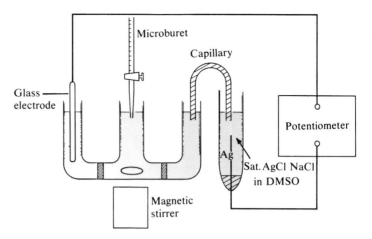

Figure 38–5 *The electrolytic cell for titration in DMSO.*

in the conventional way. After washing the plated silver wire with distilled water and then absolute ethanol, and drying it, add DMSO solution saturated with silver chloride and sodium chloride. (Since the solubility of silver chloride in DMSO increases in the presence of sodium chloride by the formation of soluble complexes, the student should be careful to ensure a full saturation.)

The assembled electrolytic cell has three compartments, as shown in Figure 38–5. The three compartments are separated by fritted glass disks. The microburet dips into the middle compartment, and a small magnetic stirrer is also present there. The glass electrode dips directly into one of the other compartments, while the reference electrode is connected to the third compartment through a small capillary tube filled with DMSO saturated with silver chloride and sodium chloride. The whole electrolytic cell is filled with test solution.

Before the titration curve is obtained, the potentiometer must be standardized similarly to that done for the aqueous solutions. However, no commercial pH standards are available for the solvent DMSO (or for other organic solvents, for that matter). There are certain values in the literature that can be used as standards (Kolthoff, Chantooni, and Bhowmik). For example: A mixture of 4.6×10^{-3}M salicyclic acid and 4.6×10^{-3}M tetraethylammonium salt of salicyclic acid in DMSO should give a pH = $pa_H = 6.8$; a mixture of 1.3×10^{-3}M 3.5 dinitrophenol and 1.3×10^{-3}M tetraethylammonium salt of 3,5-dinitrophenol in DMSO has a pH = $pa_H = 10.5$.

Other pH values may be obtained for different concentrations of a large number of acids by using pK values reported in the literature. For example, values can be obtained for DMF and methanol solvents from the article by Clare et al. and for acetonitrile from the papers of Kolthoff et al. (references 8 and 9).

Sometimes it is necessary to use a Weston standard cell as a bucking potential in series with the unknown emf to bring the observed reading on scale. These two standards can be used to adjust the potentiometer to register directly in pH units also in DMSO. After the potentiometer has been standardized, the solvent (DMSO with 0.001N potassium chloride) and the solution (0.1N acetic acid in DMSO) should be titrated with 1N tetraethylammonium hydroxide in DMSO from a Gilmont microburet. After the addition of an increment (0.2 ml.), sufficient mixing should be allowed to have equilibrium in all three compartments of the electrolytic cell. The pH readings should be taken until a steady value is reached. Duplicate titration curves are thus obtained on the solvents and on the solutions.

For acetic acid in water and in DMSO, subtract the titration curve of the solvent from that of the solution and plot the difference titration curve.

Plot the buffer capacity curves obtained from the titration curves.

Calculate the pK' value from several points on the titration curve by using equation 11. Average these values.

Compare the average values to the pK' value obtained from the buffer capacity curves.

Correct the pK' value by using activity coefficients obtained from equation 18 and thus calculate the dissociation constants of acetic acid in water and in DMSO. The Born relationship, taking into account only the difference between the dielectric constants of water and DMSO, would predict a $\Delta pK'$ value of $\sim$2 for acetic acid [$\Delta pK = pK'_{H_2O} - pK'_{DMSO}$].

Does your result indicate that the Born prediction is correct? If not, what other effects may influence the difference in pK' values for the two solvents?

Material and Equipment. Potassium chloride; acetic acid; dimethyl sulfoxide (DMSO) (or dimethylformamide [DMF], acetone, dioxane, and so forth); standard buffer solutions for calibration; silver chloride; sodium chloride; platinum wire; silver cyanide; salicylic acid; tetraethylammonium salicylate; 3,5-dinitrophenol and tetraethylammonium salt of 3,5-dinitrophenol; tetraethylammonium hydroxide; commercial pH meter (such as Beckmann Model *G* or *H* or Radiometer) with glass and calomel electrodes; Faraday cage; magnetic stirrer; electrolytic cell with three compartments separated by fritted glass disks; microburet calibrated to 0.005 ml.

REFERENCES

1. I. M. Kolthoff and H. A. Laitinen. pH and Electro Titrations. 2nd Edition, Wiley & Sons, Inc., New York, 1952.
2. C. Tanford and S. Wawzonek. In Weissberger, A. (ed.). Physical Methods of Organic Chemistry. 3rd Edition, Vol. I, Part IV, Interscience Publishers, Inc., New York, 1960, Chapter 44.

3. F. A. Bettelheim. Ann. New York Acad. Sci., **106,** 247 (1963).
4. J. Kielland. J. Am. Chem. Soc., **59,** 1675 (1937).
5. I. M. Kolthoff and B. B. Reddy. Inorganic Chem., **1,** 189 (1962).
6. I. M. Kolthoff, M. K. Chantooni, Jr., and S. Bhowmik. J. Am. Chem. Soc., **90,** 23 (1968).
7. B. W. Clare, D. Cook, E. C. F. Ko, Y. C. Mac and A. J. Parker. J. Am. Chem. Soc., **88,** 1911 (1966).
8. I. M. Kolthoff, S. Bruckenstein and M. K. Chantooni. J. Am. Chem. Soc., **83,** 3927 (1961).
9. I. M. Kolthoff and M. K. Chantooni. J. Am. Chem. Soc., **85,** 2195 (1963).

XII

SURFACE PHENOMENA

39

INSOLUBLE MONOLAYERS AT MOBILE INTERFACES

The surface of a liquid or the interface between two immiscible liquids is a region of discontinuity in which great variations occur in the intensity and shape of the electrical field as one proceeds from the surface. This region is usually extremely thin, and preferential orientation of molecules occurs in it. (Although in the present experiment we shall deal with layers at mobile interfaces, similar considerations can be applied to solid interfaces. However, a complicating factor enters there, namely, the inhomogeneity of solid surfaces.)

Films may be formed by condensation from a vapor phase or by spreading of a liquid at a liquid-air interface. Similarly, films may be formed at interfaces between two immiscible liquids either by adsorption from solution or by spreading.

Such films may be of three types: (a) a monolayer, which is a layer of monomolecular thickness; (b) duplex films, which are so thick that their energy (of interfaces) is independent of the energy of the surface; (c) multilayers, which are thicker than monolayers but thinner than duplex films.

In the following discussion, attention is focused on insoluble films of monomolecular thickness. These monolayers have been studied for the past 80 years, and they have yielded a wealth of information on molecular structure, thermodynamics and kinetics of reactions at interfaces, and thermodynamics and kinetics of transport phenomena through interfaces. In connection with the latter, many biological membrane phenomena are studied with the aid of monomolecular models. Furthermore, the use of monolayers is widespread in reducing evaporation from water reservoirs in arid areas of the world. These are just a few of the many ramifications of the study of monolayers at mobile interfaces.

When a monolayer is spread on the surface of a liquid there is a lowering of surface tension, which is called surface pressure, π.

$$\pi = \gamma_{water} - \gamma_{film} \qquad (1)$$

where the γ's are the surface tensions. This surface pressure can be obtained from simple measurements.

If a clean sheet of platinum or mica is suspended in a pure liquid, its position may be balanced through the use of an analytical balance. When a monolayer is spread on the surface and the surface tension is thus decreased, the immersed mica sheet will rise until the decrease in the downward pull, $-\Delta G_d$, is balanced by the decrease in the buoyancy, $-\Delta G_u$, of the immersed part of the mica sheet.

$$\Delta G_d = \Delta G_u \tag{2}$$

Hence,

$$2p\,\Delta\gamma = gptw\,\Delta h \tag{3}$$

where p is the perimeter of the immersed mica sheet of w width and t thickness.

$$p = 2(w + t) \tag{4}$$

The downward pull on the sheet is $p\gamma_{\text{water}}$ in water and the decrease in the downward pull is $2(\Delta\gamma)p$. The g in equation 3 is the gravity constant, and Δh is the difference in the immersion heights caused by the decrease in surface tension. If the heights are measured by a cathetometer and the dimensions of the mica sheet are known,

$$\Delta\gamma = \pi = \tfrac{1}{2}gtw\,\Delta h \tag{5}$$

The surface pressure can be calculated from equation 5.

Now the surface pressure is a function of the area covered by the monomolecular layer. When the area covered by the monolayer is decreased by compression under isothermal conditions, the surface pressure increases. The behavior of the insoluble monolayer film can be represented in a π-σ diagram, where π is the surface pressure and σ is the area occupied by one molecule (Fig. 39–1). The σ can be calculated if one knows the area occupied by the monolayer, A, and the number of molecules

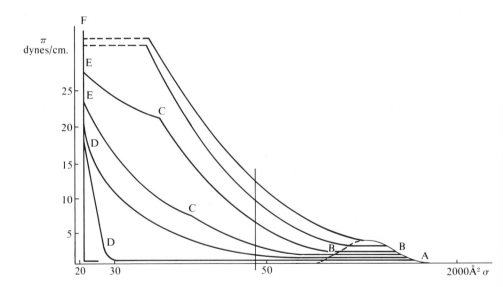

Figure 39–1   *Isotherms of surface pressure vs. area per molecule for a monomolecular layer.*

in the monolayer, N. The value of N can be obtained from the knowledge of the amount of material spread, m, and the molecular weight, M.

$$N = \frac{m}{M} \times 6.02 \times 10^{23} \tag{6}$$

and

$$\sigma = \frac{A}{N} \tag{7}$$

The π-σ plots are the analogues of the pressure-volume diagrams of bulk (three-dimensional) phases.

With very large molecular areas and with very low surface pressures (A–B, Fig. 39–1) the monolayer behaves like a perfect gas, $\pi\sigma = kT$. If the molecular areas are smaller, an equation of state of an imperfect gas, such as the van der Waals equation, can describe the π-σ behavior of the monolayer.

At a certain point, condensation of vapor occurs, which appears as a horizontal portion of the π-σ diagram. At this point the gaseous monolayer is in equilibrium with a condensed or liquid phase. However, this liquid phase of the monolayer has relatively *high compressibility*, and it is termed an expanded liquid phase (C–B).

On further compression with a second order transition, the expanded liquid phase enters an intermediate liquid phase, which has very high compressibility. The intermediate liquid phase domain is indicated by C–D in Figure 39–1.

On further compression it becomes a true condensed liquid phase, which has low compressibility. This is indicated by the domain of D–E in Figure 39–1.

At the greatest compressions before the monomolecular layer breaks, there is a domain with very low compressibility—an almost linear portion of the π-σ curve. This is the domain of E–F in Figure 39–1, and it can be classified as a super liquid or solid phase.

Harkins (reference 1) found that in certain cases more than one solid phase exists, just as in the three-dimensional case more than one stable crystal form of a compound exists.

As one can see from the π-σ diagram in Figure 39–1, not all those phases may exist at any temperature. For example, at a sufficiently low temperature only gas, condensed liquid, and solid phases are exhibited by a monolayer at different stages of compression.

Whether the transition from one phase to another is a first or a higher order transition can be evaluated by searching for discontinuity in certain properties. For example, the surface free energy is represented by equation 8.

$$dG_s = dE - S\,dT - T\,dS + P\,dV + V\,dP - \gamma\,d\sigma - \sigma\,d\gamma \tag{8}$$

Substituting

$$dE = T\,dS - P\,dV + \gamma\,d\sigma \tag{9}$$

into equation 8, we obtain

$$dG_s = -S\,dT + V\,dP - \sigma\,d\gamma \tag{10}$$

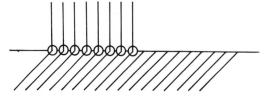

Figure 39–2 *Orientation of molecules in a monolayer.*

Therefore,

$$\left(\frac{\partial G}{\partial \gamma}\right)_{T,P} = -\sigma \tag{11}$$

since, according to equation 1, $\delta\gamma = -\delta\pi$ the criteria for a first order transition are

$$\left(\frac{\partial G}{\partial \pi}\right)_{P,T} = \sigma \quad \text{or} \quad \left(\frac{\partial G}{\partial T}\right)_{P,\sigma} = -S = q/T \tag{12}$$

This means that if there is a discontinuity in the surface area at constant pressure and temperature, or if there is a discontinuity in entropy, hence in heat absorbed, at constant pressure and area, a first order transition is at hand.

A second order transition is characterized by

$$\left(\frac{\partial^2 G}{\partial \pi^2}\right)_{P,T} = \left(\frac{\partial \sigma}{\partial \pi}\right)_{P,T} = -\sigma k \tag{13}$$

where k is the compressibility. Hence, if there is a discontinuity in the compressibility but not one in the area, a second order transition is at hand.

Regardless of the types of transitions that have occurred at the high surface pressures and small molecular areas, the monomolecular layer reaches a solid type of condensed phase.

In this phase the molecules are packed in crystallike, two-dimensional lattices with special orientations. For example, when oleic acid (which has a long hydrophobic tail and a hydrophilic head of COOH) forms a monolayer on the surface of water, the long axes of the molecules align perpendicular to the surface (Fig. 39–2).

Other molecules may have preferential orientation parallel to the surface. By extrapolating the linear portion of the π-σ curve to zero surface pressure (see Fig. 39–3), the area of a molecule in the closest packing can be obtained, σ_c.

Knowing the bond distances and bond angles of the molecule, one can propose

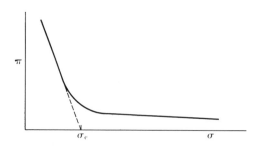

Figure 39–3 *Method for determining the area occupied by a single molecule.*

the specific orientation that would result in the molecular area suggested by the extrapolation.

Experimental

The experimental apparatus consists of an analytical balance in which one of the pans is removed and a thin mica plate (5×2 cm) is suspended through a hole in the bottom by fine wires. The mica plate is half immersed in the water in a glass trough that is about $15 \times 30 \times 3$ cm. (Fig. 39–4).

A scale extends the length of the trough and there is a movable slide, J, across the trough that can be positioned anywhere on this scale, thus limiting the area in which the monolayer is spread. The dimensions of the trough should be known accurately. Place the trough on a stand that can be leveled by leveling screws. Thoroughly clean it with cleaning solution and rinse it with distilled water.

Fill the trough with distilled water that has been equilibrated with charcoal to remove any surface-active contaminants. Before dipping the hanging mica plate into the water, rub it lightly in the vertical direction with a fine carborundum paper, rinse it with alcohol, and allow it to dry. This treatment ensures that the hanging plate will make a zero contact angle with the water. Immerse the hanging plate in water and balance it with weights in the pan of the analytical balance so that it is immersed about halfway.

Take the reading of the immersion height, h_0, with a cathetometer.

Prepare about 10 ml. of cetyl alcohol in which a small amount of spreading agent, such as hexane, is dissolved. Usually 0.5 mg. of spreading agent per milliliter of cetyl alcohol is sufficient. However, the exact concentration has to be known. Put some of this solution into a microsyringe that can deliver about 0.1 ml. with an accuracy of ± 0.00005 ml. After filling the microsyringe, expel all the air bubbles and, holding the syringe horizontally just above the center of the clean water, deliver about 0.1 ml. in individual

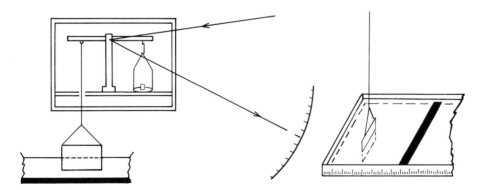

Figure 39–4 *Experimental apparatus.*

drops forced out in succession. Read the microsyringe accurately so that the volume actually delivered is known. After the formation of the monolayer, wait about five minutes to allow all the hexane to evaporate. Then move the movable slide and note its position on the scale of the trough. Observe that the spread monolayer changes the immersion height and take a reading of the new immersion height, h_1, with a cathetometer.

Calculate the $\Delta h = h_0 - h_1$ for the first position of the movable slide. Then decrease the area of the monolayer by moving the movable slide, and note its new position. Obtain the corresponding change in immersion height, Δh, with the cathetometer. Repeat this procedure about eight to ten times with successively smaller and smaller areas.

An alternate technique to the measurement of the immersion height, Δh, by using a cathetometer is the use of a reflection mirror on the analytical balance to reflect a light beam onto a circular scale.

With such a setup, first set the hanging mica plate in water and add enough weight to the pan of the analytical balance to bring the reflected light to the zero setting on the scale. When the monolayer has spread, read the deflection on the scale at different molecular areas. Then convert this deflection to surface pressure by establishing a calibration curve in which the hanging mica plate suspended in water is again balanced to zero deflection and the deflections are caused by removing small weights from the pan of the analytical balance. After obtaining a weight-deflection calibration calculate the surface pressure from

$$\pi = \frac{mx_1 g}{2yx_2} d \tag{14}$$

where m is the slope of the weight-deflection calibration curve, d is the deflection observed, y is the width of the mica plate (as in Fig. 39–5), x_1 and x_2 are the lengths of the balance arms (usually identical), and g is the gravity.

From either equation 5 or equation 14, calculate the surface pressure at each setting of the sliding scale. From the dimensions of the trough and from the known amount and concentration of the cetyl alcohol that was spread, calculate the area per molecule for each setting of the sliding scale (equations 6 and 7).

Repeat the procedure with another monolayer of cetyl alcohol, and two more monolayers of cyclic dimethyl siloxane samples, spread with cyclohexane.

The cyclic dimethyl siloxane should be provided to the students by the instructor. It can be obtained very easily by hydrolyzing dichlorodimethyl silane in water that contains sufficient sodium bicarbonate to neutralize the evolving hydrochloric acid. The hydrolytic products will yield silicone oil

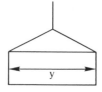

Figure 39–5 *Dimension of the hanging mica plate.*

upon heating. About 50 per cent of the silicone oil can be distilled off at
1 mm. pressure and a temperature up to 200° C. The distillate contains
cyclic polymers, largely tetramer (84 per cent), pentamer (14 per cent),
and the rest is higher cyclic polymers.

Plot the curves for surface pressure vs. area per molecule and establish
the area occupied by one molecule in the condensed alignment by the
extrapolation technique shown in Figure 39–3.

Knowing the molecular weight of cetyl alcohol, speculate on the orientation of the cetyl alcohol in the condensed monolayer. Using the assumed
orientations and known bond angles and lengths, what significant data can
you obtain regarding molecular structure? Is there any indication of hydrogen
bonding within the monolayer or to the water?

Go through the same argument with respect to dimethyl siloxane to
determine its orientation in the monolayer. Be aware, however, that the
sample provided to you may contain a mixture of cyclic polymers (84 per cent
tetramer, 14 per cent pentamer, and 2 per cent higher cyclic polymers).
What conclusions can you draw regarding orientation and molecular structure
under these conditions?

Material and Equipment.   10 ml. cetyl alcohol; *n*-hexane; dichlorodimethyl silane plus sodium bicarbonate (or cyclic polydimethyl siloxane);
cyclohexane; ethanol; carborundum paper; microsyringe ± 0.00005 ml.
accuracy; analytical balance with a hole in the bottom; thin mica plate
(5 × 2 cm.); glass trough (15 × 30 × 3 cm.) with scale and movable slide;
cathetometer (or deflection mirror and circular scale); fine wire.

REFERENCES

1. W. D. Harkins. The Physical Chemistry of Surface Films. Reinhold Publishing Corp.,
   New York, 1952.
2. A. E. Alexander and P. Johnson. Colloid Science. Oxford University Press, London,
   1949.
3. A. E. Alexander. In Weissberger, A. (ed.). Techniques of Organic Chemistry. 3rd
   Edition, Vol. I, Part I, Interscience Publishers, Inc., New York, 1959, Chapter XIII

40

WATER VAPOR
SORPTION ISOTHERMS
OF SOLIDS; HYSTERESIS

When the surface of a solid is exposed to gas in a closed system, the pressure of the gas decreases. The phenomenon is an indication that the solid sample adsorbed some of the gas on its surface. The nomenclature of this process is somewhat ambiguous. If we are certain that only the physical surface of the solid takes part in the process, we call it *ad*sorption. If the whole matrix of the solid material is penetrated by the gas molecules we call it *ab*sorption. When neither the surface nor the degree of penetration is well defined, we use the noncommittal word *sorption*. The amount of gas sorbed (determined by gravimetric measurement or by observation of the decrease in vapor pressure) is a function of the initial vapor pressure, the amount of solid present, the temperature, and the time. The time element indicates the kinetics of sorption while the time-independent elements are related to the equilibrium sorption properties. The latter usually are observed after a sufficient amount of time has elapsed and no additional changes in the pressure–gravimetric weight measurements are observable. Data obtained at a constant temperature that relate the amount of gas sorbed to the equilibrium vapor pressure are called sorption isotherms. A few shapes of sorption isotherms are given in Figure 40–1. The abscissa is the equilibrium relative vapor pressure, P/P_0, where P_0 is the vapor pressure of the pure gas at the temperature of the isotherm. The ordinate is the volume of gas sorbed per gram of solid at standard temperature and pressure.

A simple analysis of the sorption isotherm in Figure 40–1 was given by Langmuir. In the Langmuir sorption isotherm it is assumed that there is a *uniform* solid surface upon which molecules from the gas phase are adsorbed. However, the adsorption is limited to the surface of the solid and, therefore, at maximal coverage the surface of the solid is covered by a layer of the adsorbate, which has a thickness the diameter of the molecule. This is called a monolayer. The Langmuir model

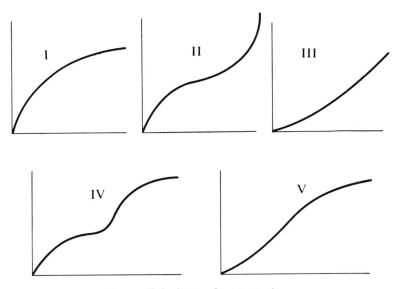

Figure 40-I *Shapes of sorption isotherms.*

visualizes the two processes: (a) the evaporation of the adsorbed gases from the surface, and (b) the condensation of the gases at the surface. The rates of the two processes are equal at equilibrium.

$$\text{Rate of evaporation} = k_1 \theta \tag{1}$$

$$\text{Rate of condensation} = k_2 P(1 - \theta) \tag{2}$$

and, at equilibrium,

$$\theta = \frac{k_2 P}{k_2 P + k_1} = \frac{bP}{bP + 1} \tag{3}$$

where k_1 and k_2 are rate constants and $b = k_2/k_1$, P is the pressure of the gas, and θ is the fraction of the surface covered by a monolayer. Since θ is a fraction, it can be expressed in terms of the monolayer

$$\theta = \frac{V}{V_m} \tag{4}$$

where V is the volume of gas adsorbed at any equilibrium point and V_m is the volume of gas in a completed monolayer. Rearranging equation 3

$$V + bVP = bPV_m \tag{5}$$

$$\frac{1}{bV_m} + \frac{P}{V_m} = P/V \tag{6}$$

Therefore, the Langmuir equation predicts that isotherms of the shape of that in Figure 40–1I, when plotted in the form of P/V vs. P, yield a straight line, the slope of which indicates the reciprocal of the monolayer. Statistical mechanical derivations of the same equation can be found in textbooks (Dole).

The Langmuir equation, being limited to the type of isotherm in Figure 40–1I, represents most of the *chemical* sorption (chemisorption) isotherms in which a chemical bond is formed between the surface and the adsorbed gas molecule. This chemical bonding obviously is limited to the monolayer. Once the monolayer is known from the Langmuir isotherm, one can calculate the surface area of the solid by assuming that there is a certain type of packing (hexagonal is the closest) and that the surface is occupied by one layer of molecules.

For type II to V sorption isotherms in Figure 40–1, the monolayer adsorption model is not sufficient. These types have been accounted for by the so-called multi-layer adsorption theory, among which the best known is Brunauer-Emmett-Teller (BET) theory (reference 3).

The BET theory proceeds in the same direction as the Langmuir theory, but it allows for the simultaneous sorption of more than one layer of adsorbate. At any equilibrium, therefore, there may be fractions of surface that are covered by more than one layer of adsorbed gas molecules. The BET theory is developed for *uniform* solid surfaces and it is assumed that at equilibrium the rate of evaporation of molecules from each layer is equal to the rate of condensation of gas molecules to form that layer. It is further assumed that the energy of sorption on the surface of the solid is unique but that the energy of sorption on any subsequent adsorbed gas molecular layer is the same as the energy of condensation of the gas. The set of equations now has the form

$$a_1 P\theta_0 = b_1\theta_1 e^{-E_1/RT}$$
$$a_2 P\theta_1 = b_2\theta_2 e^{-E_L/RT} \qquad (7)$$
$$a_3 P\theta_2 = b_3\theta_3 e^{-E_L/RT}$$

where a_1, a_2, a_3, and so forth, are the constants for the rate of condensation in the respective layers; b_1, b_2, b_3, and so forth, are the constants for the rate of evaporation from the respective layers; and θ_0, θ_1, θ_2, and so forth, are the fraction of surface covered by the respective layers of adsorbate. The P is the equilibrium vapor pressure, E_1 is the energy of adsorption, and E_L is the energy of liquification.

With the assumption that the ratios of the rate constants are the same beyond the monolayer,

$$\frac{b_2}{a_2} = \frac{b_3}{a_3} = g \qquad (8)$$

and

$$\frac{a_1}{b_1} g \sim 1 \qquad (9)$$

the summation of the set of formulas in equation 7 becomes

$$\frac{1}{V_m C} - \left(\frac{1}{V_m C} + \frac{1}{V_m}\right) P/P_0 = \frac{P}{P - P_0} \frac{1}{V} \qquad (10)$$

where V_m is the volume of the adsorbed gas in the monolayer (covering the total surface of the solid per gram of solid); P is the equilibrium pressure; V is the volume of the gas adsorbed per gram of solid at each equilibrium pressure, P; P_0 is the vapor pressure of the pure gas at the temperature of the isotherm; and

$$C = e^{E_1 - E_L/RT} \qquad (11)$$

is related to the excess energy of adsorption ($E_1 - E_L$).

In spite of the assumptions used in its derivation, the BET equation has been widely applied to physical adsorption data. The most important parameter obtainable from equation 10 is the volume of the monolayer. This is obtained from plots such as Figure 40-2.

From the intercept and the slope in Figure 40-2 the V_m is calculated. Once this is known the surface area of the solid can be calculated by assuming that there is a certain area per molecule. Nonpolar gases usually are adsorbed on the surfaces of solids when the surface area is the important parameter to be obtained (such as in the case of solid catalysts). For instance, from the BET equation the monolayer volume of nitrogen on a copper catalyst is 0.09 ml. per gram of catalyst at standard temperature and pressure. The area covered by a nitrogen molecule is 16.2 Å². Since at standard temperature and pressure 6.02×10^{23} molecules of nitrogen occupy 22.4 liters,

$$\frac{6.02 \times 10^{23} \times 0.09}{22.4 \times 10^3} = 2.41 \times 10^{18}$$

are the number of molecules in the monolayer, each occupying 16.2 Å². The total area occupied by the molecules, i.e., the surface area of the catalyst, is

$$16.2 \times 10^{-16} \text{ cm.}^2 \times 2.41 \times 10^{18} = 3.9 \times 10^3 \text{ cm.}^2/\text{gram}$$

Although the BET equation was derived for uniform surfaces (just as the Langmuir equation), in reality there is nothing that approaches uniform surface. The closest approach would be a single plane of crystal surface (such as a 111 plane), which is infinitely large compared to other planes of the crystal. Even then dislocations and faults cause the surface to be incompletely uniform. In a powder sample, one is hardly able to talk about uniformity of surface. Still, the BET equation is applicable to most physical adsorption isotherms because the nonuniformity of the surface affects mainly the C term, or energy term (equation 11), which is the least important parameter of the BET equation.

The multilayer coverage is a useful operative model up to moderate relative vapor pressures $[P/P_0 = 0.3 - 0.6]$. Up to this range most isotherms can be represented with a plot similar to the one in Figure 40-2. Above this range there is deviation

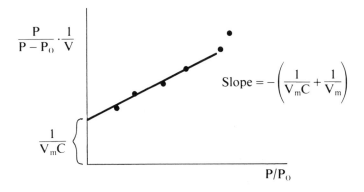

Figure 40-2 *The Brunauer-Emmett-Teller (BET) plot.*

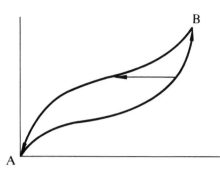

Figure 40-3 *Hysteresis of sorption isotherms.*

from linearity (the sudden upward curvature of the type II isotherm in Figure 40–1), which is explained in terms of capillary condensation. Beyond a certain multilayer coverage the crevices of the surface or the contact surfaces of the powder particles act as tiny capillaries in which preferential condensation occurs with a decrease in the vapor pressure. However, appreciable capillary condensation occurs only at relatively high vapor pressures. Therefore, the part of the isotherm obtained at low vapor pressures is customarily used for the evaluation of the surface area of the monolayer, even for adsorbents such as charcoal in which electron and optical microscopic pictures clearly demonstrate the abundance of crevices.

However, the nonuniform surfaces and especially the capillaries have another effect on the adsorption isotherms. This effect demonstrates itself in hysteresis, which is a phenomenon in which the pathway from *A* to *B* is different from the pathway from *B* to *A* (see Fig. 40–3).

In a process with hysteresis, therefore, all points between *A* and *B* are not true equilibrium points referring to a reversible process, and the width of the hysteresis loop is a measure of the irreversibility of the process. If an adsorption isotherm has proceeded through multilayer adsorption to capillary condensation, hysteresis will be apparent. The reason for hysteresis is that the same amount of gas that is adsorbed on the capillary when the vapor pressure is decreased will not be removed from the capillary at the adsorption pressure but only after the pressure has been decreased (see horizontal arrow in Fig. 40-3).

Most heterogeneous adsorbents and solids in which the physical surface is not well defined also behave in this manner. It is hard to speak about adsorption, especially in the latter case. Gases and vapors can penetrate the solid matrix fully or partially and there is a large, undefined area between strict adsorption and absorption. Because of this, vapor sorptions with partial penetration and swelling and on molecular rather than physical surfaces are referred to as *sorption* rather than adsorption.

The sample used in this experiment is a case on point. Albumin (gelatin, starch, polygalacturonic acid, Drierite, and so forth) binds water strongly on the polar sites available on the molecular (as contrasted to physical) surface. Still, the concept of a monolayer can be used under this condition. It will not have any specific relationship to the physical surface area of the solid. On the other hand, one may interpret a monolayer as being the amount of water vapor tightly bound and a multilayer as being the amount of water vapor condensed on top of the tightly bound molecules (water of hydration).

The purpose of this experiment is to study the effect of surface on the rate and equilibrium condition of water vapor sorption on serum or egg albumin (or gelatin, starch, amylose, polygalacturonic acid, pectin, Drierite, and so forth).

Experimental

A high vacuum vapor sorption apparatus (Fig. 40–4) is composed of a forepump (A) capable of producing 10^{-2} mm. mercury vacuum, traps (B) that are in a Dry ice–acetone bath to prevent water vapors entering the pump, a removable sorption chamber (C) into which the solid sample is placed, a sorbate chamber (D) for the supply of water vapor, and a manometer (E) filled with Octoil (density 0.875 grams per milliliter at 25° C.). (Numbers 1 through 8 in Figure 40–4 indicate positions of stopcocks.)

Materials needed include freshly boiled distilled water and dried sorbent samples. Any one of the following sorbents may be used. Serum (reference 5) or egg albumin (reference 6), gelatin (reference 6), starch (reference 7), amylose (reference 8), polygalacturonic acid (reference 9), and so forth, should be in two physical states: as a commercial powder and as lyophilized (freeze-dried) material. The latter should be prepared in advance by the instructional staff. A 0.5 gram sample of the powder and of the lyophilized material each should be placed in a weighing bottle and dried in a vacuum oven at 65° C. for at least one hour. Drierite should be prepared in two mesh sizes—granular, and powder about 100 mesh. Each should be oven-dried for at least one hour at 120° C. At the end of the drying, the weighing bottles are capped, weighed on an analytical balance, and stored in a desiccator until they are transferred to the sorption chamber in the high vacuum apparatus. After the transfer, the empty bottles are weighed to determine the amounts of dried material transferred.

Before the detailed procedures are given, the overall aspects of the experiment will be delineated. The water vapor sorption isotherm of the selected solid will be obtained volumetrically. This means that the amount of water vapor sorbed will be calculated from the difference between the initial and the equilibrium vapor pressure. In order to relate the difference in

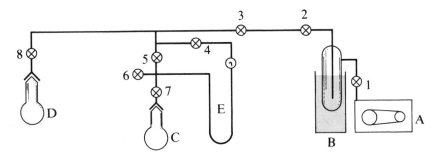

Figure 40–4 *High vacuum vapor sorption apparatus.*

pressures to the number of moles of water vapor sorbed, we assume that the water vapor behaves ideally (low pressures). However, to use the ideal gas law and its modified forms (equations 15 and 16) it is necessary to obtain the volumes of the different parts of the vacuum system.

The first step in the detailed procedure is designed to accomplish just that. The volume of the sorption chamber is measured by weighing it when it is empty and when it is filled with water. When this volume is known, the volumes of the other parts of the vacuum system are determined by letting a gas under certain vapor pressure expand from the unknown volume into the combined volume of the unknown volume plus the volume of the sorption chamber. The equilibrium pressure is measured, and the unknown volume is calculated by using Boyle's law.

After the volumes of the different parts of the vacuum system have been determined and the total volume of the system has been established, we proceed to secure the points of the sorption isotherm in the following manner:

With the sorption chamber closed, a specific amount of water vapor from the sorbate chamber is added to the system. The initial vapor pressure is then measured. The sorption chamber is opened next, and the change in the vapor pressure is monitored. The decrease in pressure is due to expansion of the water vapor into the sorption chamber and sorption of water on the solid sorbent. The decrease in pressure due to expansion can be calculated from the knowledge of the volumes, the initial pressure, and the residual equilibrium pressure in the sorption chamber. Then the decrease due to sorption is determined. The latter is the difference between the total decrease in vapor pressure and that calculated for expansion only. From the decrease in pressure due to sorption the amount of sorption is calculated.

This process is repeated many times, going always to higher equilibrium vapor pressures when the sorption branch is being followed and to lower equilibrium vapor pressures when the desorption branch of the isotherm is involved.

WARNING: Be certain that stopcocks 4 and 5 (Fig. 40–1) are open every time before atmospheric pressure is admitted to the system. This is necessary to equalize the pressure on both arms of the manometer and to avoid jumping the manometer fluid. Also be certain that stopcock 4 is closed when the system is evacuated and before water vapor pressure is applied to the system. Under these conditions the right arm of the manometer is under constant vacuum during all measurements.

The detailed procedures will now be described. While the sorbent is drying, weigh the empty sorption chamber (C) of the apparatus and then fill it up to the stopcock with distilled water of measured density. Weigh the filled chamber, and from the data calculate the volume of the chamber. Then empty and dry the sorption chamber. Place about 30 ml. of boiled distilled water in the sorbate chamber (D). Place the Dry ice–actone cooling liquid around the removable trap, and place the sorbate chamber in position on the rack.

Close all the stopcocks and start the vacuum pump. Open the following stopcocks in succession—stopcocks 4, 5, 1, 2, 3, and 8—allowing the pump to quiet down each time. After opening stopcock 8, let the water in the chamber come to a quick boil and close stopcock 8 immediately (after five to ten seconds of boiling). Continue pumping the system. Attach the sorption chamber at its place on the rack. Open stopcock 7. *Do all the stopcock operations slowly and with deliberate care, and always use both hands and wear glasses.*

Evacuate the whole system by pumping on both arms of the manometer (E). After the pump has quieted, pump for two more minutes before closing stopcocks 4, 5, and 7. Pump for one more minute and close stopcocks 2 and 3. With all stopcocks closed, wait for two or three minutes to see whether there are any leaks between stopcocks 4, 5, 6, and 7. If there is no change in the manometer, open stopcock 7 to check whether there are leaks in the sorption chamber. Probe further for leaks between stopcocks 4, 3, and 8 by opening stopcock 5. If opening all these stopcocks does not cause a change in the manometer, the system is leakproof. If a leak is discovered in any part of the system, let air into it, regrease the joints and stopcocks, and proceed to evacuate the system as before.

With only stopcocks 2, 3, 7, and 8 closed, allow water vapor to expand into the system by opening stopcock 8 until about 10 to 20 cm. of oil pressure is obtained in the manometer. Close stopcock 8 and wait until the pressure stabilizes. Record the temperature near the manometer and this initial pressure. Open stopcock 7 and allow water vapor to expand. Wait again until final pressure is reached and then record it.

Knowing the volume of the sorption chamber (from your first calculation) and the initial and final pressures at the temperature measured, use the ideal gas law to calculate the volume of the whole system (excluding the areas beyond stopcocks 2, 3, and 8).

Evacuate the whole system again, close stopcock 7, and repeat the procedure using a higher initial pressure of 20 to 40 cm. of oil. Calculate the volume of the system again. If the volume of the sorption chamber is V_A, and the total volume is V_T, the volume between stopcocks 2, 3, 6, 7, and 8 and the manometer fluid is $V_T - V_A = V_S$. Note that V_S changes as the pressure changes (i.e., as the manometric fluid level changes), but since two V_S values are calculated for two recorded manometric levels, one can calculate V_S for any manometric level.

After performing the calibration, open stopcocks 4 and 5; allow air into the system by opening stopcock 6. Remove the sorption chamber and fill it with the pre-dried solid sample, which has been weighed accurately. Close stopcock 7. Attach the sorption chamber to the rack; open stopcock 7. Pump on the whole system and the sample for 10 minutes and then begin the sorption process by closing stopcocks 4 and 7. Open stopcock 8 and let a certain amount of water vapor into the system (except to the sorption chamber). When the manometer reads about 10 to 15 cm., close stopcock 8 and allow the pressure to reach equilibrium. Take this reading, P_{i1} (initial pressure). Open stopcock 7 and allow the water vapor to be sorbed on the sample. Follow the kinetics of sorption after the opening of the stopcock by taking pressure readings every five to ten seconds. When no further pressure decrease can be observed, take the equilibrium pressure reading, P_{E1}.

Proceed to the next point on the isotherm by closing stopcock 7 and opening stopcock 8 and allow another increment (15 to 20 cm. Octoil) of water vapor into the system. Again take the initial reading, P_{i2}. Open stopcock 7 and allow the water vapor to be sorbed on the sample. Obtain the kinetics of sorption by observing the decrease in pressure with respect to time. When no further pressure change can be observed, note the equilibrium pressure reading, P_{E2}.

Repeat the procedure for the other four points on the isotherm.

Since the system is calibrated (V_A and V_S are known), the actual initial pressures (P') are calculated according to Boyle's law

$$P'_1 = P_{i1}(V_S/V_T) \tag{12}$$

$$P'_2 = (P_{i2}V_S + P_{E1}V_A)/V_T \tag{13}$$

$$P'_3 = (P_{i3}V_S + P_{E2}V_A)/V_T \tag{14}$$

and so forth

The amount of water vapor sorbed by the sample at each equilibrium pressure is then calculated in terms of moles of water

$$\Delta n = \frac{1}{RT}(P'_1 V_{T1} - P_{E1}V_{TE1}) \tag{15}$$

where P'_1 is the initial pressure calculated from equations 12, 13, 14, and so forth; P_{E1} is the corresponding equilibrium pressure; V_{T1} is the total volume of the system with the manometric fluid at P'_1 position; and V_{TE1} is the total volume of the system with the manometric fluid in the P_{E1} position. Since the change in V_T with the movement of the manometer is usually small, as a first approximation, one can use $V_T = V_{TE}$.

After obtaining the sorption isotherm on the first sorbent sample, obtain three points of the *desorption* isotherm in the following manner. Read the initial equilibrium vapor pressure, P_{E1}. With stopcocks 7 and 8 closed and the pump running, open stopcocks 2 and 3 for a few seconds to permit the vapor pressure to decrease to zero in the V_S part of the system. After closing stopcock 3, allow the water vapor to diffuse into the system by opening stopcock 7 to the sorption chamber. Wait until equilibrium has been established, and then record the equilibrium vapor pressure, P_{E2}. Use equation 16 to calculate the amount desorbed.

$$\frac{P_{E1}V_S - P_{E2}V_T}{RT} = \Delta n \tag{16}$$

Repeat the procedure with the other two desorption points.
Repeat the whole procedure with the second sample.
Plot the sorption and desorption isotherms for the two samples.

Calculate the BET monolayer for each sample. For each equilibrium point, plot the kinetics of water vapor sorption for the two samples.

What kind of rate equation can you formulate for the kinetics of sorption? To what extent does the physical surface of the solid sample influence the sorption process?

Material and Equipment. 1 gram serum or egg albumin (or gelatin, starch, amylose, polygalacturonic acid, pectin, Drierite, and so forth); vacuum line; forepump; lyophilizing apparatus (not necessary if Drierite is used as a solid sorbent); vacuum oven; two weighing bottles; desiccator; high vacuum vapor sorption apparatus; cathetometer.

REFERENCES

1. I. Langmuir. J. Am. Chem. Soc., **40,** 1361 (1918).
2. M. Dole. Introduction to Statistical Thermodynamics. Prentice-Hall, Inc., Englewood Cliffs, N.J., 1954.
3. S. Brunauer, P. H. Emmett and E. Teller. J. Am. Chem. Soc., **60,** 309 (1938).
4. A. W. Adamson. Physical Chemistry of Surfaces. Interscience Publishers, Inc., New York, 1960.
5. J. M. Seehof, B. Keilin and S. W. Benson. J. Am. Chem. Soc., **75,** 2427 (1953).
6. K. S. Rao and B. Das. J. Phys. Chem., **72,** 1223 (1968).
7. M. Masuzawa and C. Sterling. J. Appl. Polymer Sci., **12,** 2023 (1968).
8. D. J. Macchia and F. A. Bettelheim. J. Polymer Sci. B, **2,** 1101 (1964).
9. F. A. Bettelheim, C. Sterling and D. H. Volman. J. Polymer Sci., **22,** 303 (1956).

41

ADSORPTION AT SOLID-LIQUID INTERFACES*

One of the theoretically most intriguing and, at the same time, industrially most important types of adsorption is that which occurs at solid-liquid interfaces when there are at least two components in the liquid phase. The adsorption on a solid surface may be preferentially specific or nonspecific, depending on the interaction between the solid and the components, as well as the interaction between the components, of the liquid phase. In studying the changes that occur during adsorption, the external variables are usually kept constant. Since most of the experiments are performed at constant temperature, we speak of adsorption isotherms. In the graphic representation of the adsorption isotherm, one usually plots the amount of component 1 adsorbed on the surface of the solid per gram (or per unit surface area) of the solid against the equilibrium composition of the solution. (The equilibrium composition is the composition of the liquid phase after it attains equilibrium in contact with the solid surface.) In a binary liquid mixture, the composition can be given in terms of mole fraction. In most cases two types of adsorption isotherms have been observed when the phenomenon is studied over the whole range of concentration. These are given in Figure 41–1.

In both types of isotherms of binary mixtures the adsorption on the solid is by necessity zero when the mole fraction is 0 or 1, because the adsorption is measured as a change between initial and equilibrium concentration of component 1 in the bulk liquid phase; obviously, there is no change when the liquid is pure component 1 or 2. That does not mean that there is no adsorption of the pure liquid on the surface of the solid. Such adsorption demonstrates itself in different packing arrangements of the molecules at the solid surface and in the bulk. However, this packing arrangement is not measured in terms of the concentration units already mentioned.

It should be clearly understood that the isotherms given in Figure 41–1 are composite isotherms because both components 1 and 2 of the binary mixture contribute to them. In physical terms this means that both components 1 and 2 are

378

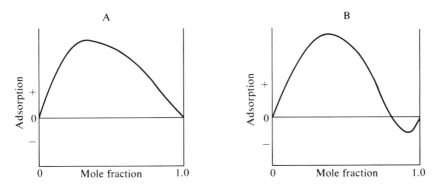

Figure 41-1 *Adsorption isotherms of binary mixtures at solid-liquid interface.*

adsorbed on the surface of the solid. The *individual* isotherms cannot be obtained experimentally unless the actual composition of the adsorbed surface layer can be analyzed. This is an almost impossible task. There are techniques by which such *individual* adsorption isotherms can be calculated using certain assumptions. We will discuss some of these techniques later. At present it is sufficient to treat the composite adsorption isotherm theoretically.

Because the composite isotherm of a binary mixture is the contribution of two individual isotherms, one can visualize the individual isotherms for the two composite types as shown in Figure 41-2.

The individual isotherms have no inflection point in cases *A* and *B* where no inflection point appears in the composite isotherm. Conversely, in cases *C*, *D*, and *E*, where there is an inflection point in the composite isotherm, inflection points must exist for the individual isotherms. Thus, the negative adsorption of component 1, as illustrated on the composite isotherm in Figure 41-1*b*, simply means that in this concentration range there is preferential adsorption of component 2 on the solid

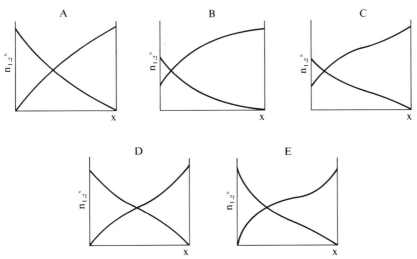

Figure 41-2 *Types of individual isotherms in Schay's classification.*

surface. The difference between the initial and equilibrium concentrations of component 1 is, therefore, a negative number.

In the literature, the composite isotherm often looks more like a Langmuir isotherm because only a limited range of concentration (dilute solutions) has been studied, and the isotherm presented is only a small, truncated part of the total isotherms as given in Figure 41–1.

In mathematical terms, the equation of the composite isotherm can be derived with minimal assumptions.

If the total number of moles in the binary mixture is initially n_0, at equilibrium the same number of molecules brought in contact with w grams of solid will be distributed so that

$$n_0 = n_1 + n_2 + wn_1^s + wn_2^s \tag{1}$$

where n_1 and n_2 are the number of moles of components 1 and 2 in the bulk and n_1^s and n_2^s are the corresponding moles of components 1 and 2 on the surface per gram of solid.

Let us designate the mole fraction of component 1 initially as x_0 and, at equilibrium, the concentration in the bulk as x.

$$x_0 = \frac{n_1 + n_1^s w}{n_0} \; ; \qquad x = \frac{n_1}{n_1 + n_2} \; ; \qquad 1 - x = \frac{n_2}{n_1 + n_2} \tag{2}$$

The amount of component 1 adsorbed is by definition

$$\Delta x = (x_0 - x) = \frac{n_1 + n_1^s w}{n_1 + n_2 + wn_1^s + wn_2^s} - \frac{n_1}{n_1 + n_2} \tag{3}$$

Finding the common denominator and simplifying the equation, we obtain

$$\Delta x = \frac{(n_2 n_1^s - n_1 n_2^s)w}{n_0(n_1 + n_2)} \tag{4}$$

Rearranging and writing the equation in terms of mole fractions,

$$\frac{n_0 \Delta x}{w} = (1 - x)n_1^s - xn_2^s \tag{5}$$

For each point of an isotherm, the initial number of moles present in the binary mixtures, n_0, is calculated. The weight of the adsorbent solid, w, is also known. The change in the mole fraction due to adsorption, Δx, is obtained by measuring the initial composition, x_0, and the equilibrium composition, x. This can be done by a number of techniques, involving refractive index, absorption coefficient, optical rotation, or titrimetric or gravimetric analysis. Thus, for each equilibrium composition, x, the adsorption $(n_0 \Delta x/w)$ is calculated, and this value is plotted as a function of x.

Equation 5 is valid over the whole range of concentration and is used extensively in the literature. It gives the limiting conditions as follows: $(n_0 \Delta x/w)$ goes to zero when $x = 0$ if $n_1{}^s = 0$, and when $(1 - x) = 0$ if $n_2{}^s = 0$. Both these limiting conditions are obvious.

However, equation 5 contains two unknowns, $n_1{}^s$ and $n_2{}^s$ and it cannot be solved unless a relationship between the two unknowns is established. The simplest relationship can be provided by the *experimental approach*.

If the solid adsorbent is suspended in a basket above a binary mixture in an evacuated and closed vessel, the amount of vapor adsorbed can be determined gravimetrically.

$$A(x) = \text{grams of vapor adsorbed/grams of solid} = n_1{}^s M_1 + n_2{}^s M_2 \qquad (6)$$

where M_1 and M_2 are the molecular weights of components 1 and 2.

The results of two adsorptions at a set concentration (mole fraction) can be combined (i.e., equations 5 and 6), and it will yield

$$\frac{n_0 \Delta x}{w} = (1 - x)n_1{}^s - x\,\frac{A - n_1{}^s M_1}{M_2} \qquad (7)$$

Upon rearrangement,

$$n_1{}^s = \frac{\dfrac{n_0 \Delta x}{w} + \dfrac{Ax}{M_2}}{(1 - x) + \dfrac{x M_1}{M_2}} \qquad (8)$$

Thus, the individual isotherms, $n_1{}^s$ vs. mole fraction and $n_2{}^s$ vs. mole fraction, can be obtained. However, one assumption is made in this derivation, the validity of which has not been proved. The difference between the two experimental techniques is that, in the case of vapor sorption of the solid, an extra interface (solid-vapor) is present which is absent in the equilibrium of solid with liquid by immersion. It is assumed that the adsorption at this interface is negligible compared to the adsorption at the liquid-solid surface. Therefore, when the previously mentioned technique is used in the evaluation of the individual isotherms, one has to prove that the assumption is correct.

The equilibrium of solid with vapor of a liquid mixture is usually a long process (sometimes a week), especially if the vapor pressure is less than 50 torr. For this reason this approach cannot be used in a laboratory experiment limited to three or four hours.

An alternate approach to establish some relationship between $n_1{}^s$ and $n_2{}^s$ would be to consider an applicable theory or model system. There are *many such* considerations in the literature, each of which has definite limitations because of approximations involved. In the following we shall consider the Langmuir monolayer theory of gas adsorption on solids for two reasons: (1) The Langmuir isotherm has a satisfactory theoretical basis, and (2) in some cases this approach yields results that are comparable to the experimental approach previously described.

The Langmuir theory as applied to a single adsorbate involves the assumption that maximal coverage of a surface is a monomolecular layer; at any concentration only a fraction of this surface is covered by the adsorbate. Let us call this fraction θ. The equilibrium established is the result of the equalization of the rates of two processes: desorption from and adsorption on the surface. The rate of desorption is proportional to the area covered.

$$\text{Rate of desorption} = k_1\theta \tag{9}$$

where k_1 is the rate constant.

The rate of adsorption is proportional to the surface area not covered by the adsorbate and to its concentration mole fraction, x.

$$\text{Rate of adsorption} = k_2 x(1 - \theta) \tag{10}$$

The two rates are equal at equilibrium:

$$k_1\theta = k_2 x(1 - \theta) \tag{11}$$

which upon rearrangement is written as

$$\theta = \frac{k_2 x}{k_1 + k_2 x} = \frac{x}{K + x} \tag{12}$$

where the equilibrium constant $K = k_1/k_2$.

The Langmuir equation is applicable to one adsorbate only; hence, the assumption must be made that each component will act independently when adsorbed on the surface. This would require ideal behavior. Keeping in mind this limitation of the theory, we can now equate the fraction of the area covered with the adsorbate, θ, to the amount of material adsorbed.

$$\theta_1 = \frac{n_1^s}{a_1}; \qquad \theta_2 = \frac{n_2^s}{a_2} \tag{13}$$

where a is the amount of material in the monolayer on 1 gram of solid. Now, using the condition that the two components act independently, equation 5 can be written

$$\frac{n_0\Delta x}{w} = (1 - x)\frac{a_1 x}{K_1 + x} - x\frac{a_2(1 - x)}{K_2 + 1 - x} = (x - x^2)\left[\frac{a_1}{K_1 + x} - \frac{a_2}{K_2 + 1 - x}\right] \tag{14}$$

where the subscripts 1 and 2 refer to components 1 and 2, both in the monolayer content and in the equilibrium constants.

Equation 14 contains four unknowns and theoretically could be solved only by selecting four points on the composite isotherm. However, the solution of a set of equations of this type is not a simple matter, because they are nonlinear algebraic equations, and their solution without a number of approximations is quite cumbersome. Furthermore, in the many cases in which such a calculation has been attempted in order to reconstruct individual isotherms from a composite isotherm, the calculated individual Langmuir type of isotherms do not correspond to those obtained experimentally by using vapor adsorption techniques described in equations 6, 7, and 8.

Now that the difficulties involved in obtaining individual isotherms are fully

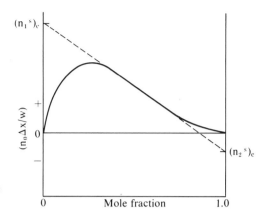

Figure 41–3 *Schay and Nagy's analysis of composite isotherm.*

understood, we can proceed to some special cases that may provide insight into the relationship of individual and composite isotherms.

Schay and Nagy point out that many composite isotherms have a linear portion (or approximately linear portion) over a limited range of concentration. When such behavior exists, the rearrangement of equation 5 will give

$$\frac{n_0 \Delta x}{w} = n_1{}^s - (n_1{}^s + n_2{}^s)x \tag{15}$$

One interpretation of the linear portion of the composite isotherm is that, over the specified range of concentration, the composition of the adsorbed layer is constant. We designate this by using the symbols $(n_1{}^s)_c$ and $(n_2{}^s)_c$, where the subscript c indicates that this value is constant. This constant composition can be calculated by the extrapolation of the linear portion of the isotherm as given in Figure 41–3.

The extrapolated line will give the value $(n_1{}^s)_c$, where $x = 0$, and $-(n_2{}^s)_c$, where $x = 1$.

We may interpret the region of constant surface composition by assuming that the surface of the solid is heterogeneous. Components are adsorbed at two kinds of sites. Site A is saturated with component 1, and site B is saturated with component 2 (Fig. 41–4). Thus, if a linear portion exists in the composite isotherm, the completed monolayer values of the individual isotherms $(n_1{}^s)_c$ and $(n_2{}^s)_c$ can be calculated.

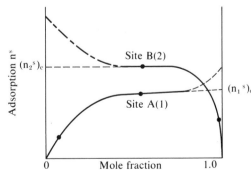

Figure 41–4 *Individual isotherms of heterogeneous adsorbent surfaces. The staggered line indicates the region of multilayer adsorption.*

Once such monolayer values are obtained, the surface area of the adsorbent solid covered by each monolayer can be estimated. This is done by making two assumptions: (a) that the shape of the molecules can be approximated by a sphere and (b) that the packing of molecules on the surface is essentially the same as in bulk—spherical close packing.

With these operative assumptions one can calculate roughly the effective area occupied by one molecule, A.

$$A = (V_{liquid}/N_{Av})^{2/3} \qquad (16)$$

where V_{liquid} is the molar volume of the liquid component and can be calculated from the density and molecular weight, and N_{Av} is Avogadro's number. If the effective area occupied by one molecule, A, is known, the surface area covered by the molecules is determined by simply multiplying it by Avogadro's number and the number of moles in the monolayer.

$$S_1 = A_1 N_{Av} (n_1^s)_c \qquad (17)$$

The total surface area is the sum of the two kinds of surfaces.

$$S_{total} = N_{Av}[A_1(n_1^s)_c + A_2(n_2^s)_c] \qquad (18)$$

Furthermore, in very dilute solutions (i.e., when $x < 0.01$) equation 5 simplifies to

$$n_1^s \sim \frac{n_0 \Delta x}{w} \qquad (19)$$

because, even though n_2^s may be appreciable, $n_2^s x$ is negligible. The same is true at the other end of the binary composition, when $x > 0.99$.

$$n_2^s \sim \frac{n_0 \Delta x}{w} \qquad (20)$$

Thus, two points on the individual isotherms can be obtained by simple inspection: one that corresponds to the saturated monolayer by using the Schay-Nagy extrapolation and another at the very dilute solution range. From these two points *approximate* individual adsorption isotherm curves can be drawn as in Figure 41–4, keeping in mind that the flat portion of the individual isotherm should extend over the range of concentration corresponding to the linear portion in the composite isotherm.

Sometimes the composite adsorption isotherm is more complex than those represented in Figure 41–1. Figure 41–5 is the type of isotherm that has been observed in a few systems of interacting miscible pairs of liquids and in many adsorption isotherms from solutions of solids (i.e., in which the binary mixture is composed of a solid solute and solvent).

Such a complex isotherm can be considered as having two parts (steps). The first part, which has linear portions in it, represents the formation of mixed monolayers

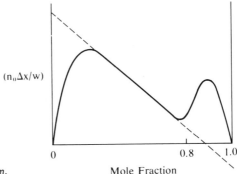

$(n_0 \Delta x/w)$

0   0.8   1.0

Figure 41–5 *Stepwise adsorption isotherm.*   Mole Fraction

on two different adsorption sites, as was discussed before. At approximately 0.8 mole fraction of component 1 a second adsorbed layer is formed, which is richer in component 1 than is the mixed monolayer. From the linear portion of the first part the constant composition of the mixed monolayer can be calculated by the Schay-Nagy extrapolation technique.

Experimental

Prepare approximately 10 grams each of 15 mixtures of dimethyl sulfoxide (DMSO) and water (or dimethylformamide [DMF]-water, DMSO-benzene, and so forth), ranging from 0.01 to 0.99 mole fraction.

Determine the refractive index of each solution and that of pure water and pure DMSO (or DMF, acetone, dioxane, and so forth) at 25° C. using, if possible, a source of monochromatic light (sodium lamp). (The use of white light allows less accuracy.) Be sure to read the refractometer accurately to the last digit obtainable (e.g., 1.4092).

Transfer exactly 5 grams of each solution into a 25-ml. Erlenmeyer flask and add to each flask exactly 1 gram of silica gel (or activated charcoal). Clamp the Erlenmeyer flasks into a shaking apparatus and lower it into a 25° C. thermostat. Shake the mixtures in the Erlenmeyer flasks for an hour, and then remove each flask from the thermostat and centrifuge its contents at medium speed for three to five minutes. Determine the refractive indices of the *supernatants* at 25° C. with the monochromatic light used with the original sample. Try to read the refractometer with the greatest accuracy.

Obtain a calibration curve by plotting the refractive indices of the original solutions against their respective mole fractions. Do this on a large (double sized) sheet of graph paper because the differences between the refractive indices of the original and equilibrium mixtures are *small* corresponding to *small* changes in concentration. In order to detect this with reasonable accuracy, draw *tangents* to the calibration curve at each point of the original composition. To draw tangents use the techniques illustrated in Figure 41–6.

For each point on the calibration curve, select a center, such as *A* in

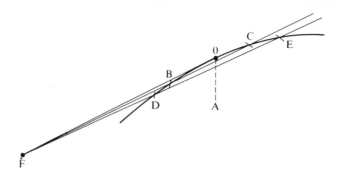

Figure 41-6 *Drawing of tangents.*

Figure 41–6, and draw two circles, one with radius r_1 intersecting the calibration curve at B and C, and another with radius r_2 intersecting the calibration curve at D and E. Connect B to C and D to E and extend the two straight lines until they intersect at F. Connect F to O, the point on the calibration curve, and the slope at O is obtained.

The Δx is obtained from the reciprocal of the slopes (tangent) at each x_0 and from the *difference* between the refractive indices $(n - n_0)$ of the equilibrium and initial compositions of the liquids.

$$\text{Reciprocal slope} = \frac{\Delta x}{\Delta n} ; \quad \Delta x = (\text{reciprocal slope})(n - n_0) \quad (21)$$

Tabulate the data in the following form:

Total No. of Moles Present	Initial Mole Fraction	Reciprocal Slope at x_0				
n_0	x_0	$\Delta x/\Delta n$	Δx	x	w	$n_0 \Delta x/w$
—	—		—	—	—	
.	.	.	.	.	.	.
.	.	.	.	.	.	.
.	.	.	.	.	.	.
.	.	.	.	.	.	.

Plot the adsorption isotherm of the liquid mixture on the solid at 25° C. ($n_0 \Delta x/w$ vs. x).

What is the nature of this composite isotherm? Is it a simple composite or a composite isotherm with more than one step in it?

Does it exhibit approximately linear behavior over a limited range of concentration? *If it does,* calculate the $(n_1^s)_c$ and $(n_2^s)_c$ by using the Schay-Nagy extrapolation (Fig. 41–3). If $(n_1^s)_c$ and $(n_2^s)_c$ can be calculated, also obtain n_1^s and n_2^s from the parts of the composite isotherm representing dilute solution by using equations 19 and 20.

With the aid of these points, try to reconstruct *roughly* the individual isotherms.

Assuming that $(n_1^s)_c$ and $(n_2^s)_c$ each represent a monolayer on the silica gel surface (mixed monolayer on two different adsorption sites), calculate the surface area of silica gel (or activated charcoal) by using equations 16, 17, and 18.

What explanation can you advance for the different adsorption behavior at the different sites?

Material and Equipment. 20 grams of silica gel (or activated charcoal); dimethyl sulfoxide (DMSO) (or dimethylformamide [DMF], acetone, dioxane, and so forth); 20 test tubes used also as centrifuge tubes; 20 Erlenmeyers of 25-ml. capacity; five medicine droppers, lens paper; refractometer; sodium lamp; circulating pump; thermostat bath with heater, stirrer, and thermoregulator; shaking apparatus; centrifuge.

REFERENCES

1. J. J. Kipling. Adsorption from Solutions of Non-Electrolytes, Academic Press, New York, 1965.
2. G. Schay and L. G. Nagy. J. Chim. Phys., **58**, 149 (1961).
3. J. J. Kipling. Quart. Rev., **5**, 60 (1951).

42

ZONE (GEL) ELECTROPHEROSIS OF PROTEINS

A solution of macromolecules having dimensions greater than 100 Å can be classified as a colloidal system. In such a system the physical chemical phenomena occurring at the interfaces of the macromolecules (surfaces) are of utmost importance because the surface area of such particles is large compared to their bulk. Two surface effects stabilize colloidal systems: the solvation layer and the electric double layer. Without these two effects the macromolecules or particles would undergo inelastic collisions because of Brownian motion, which would result in aggregation and eventual separation of phases (precipitation).

The solvation layer is composed of the solvent molecules that are tightly attached to the colloidal particles and that move with the particles; therefore, the macromolecule and its solvation layer act as a kinetic unit. One can then consider this solvation layer to be a protective layer that prevents the colloidal particles from touching each other.

The second stabilizing effect is the electric double layer. The surface of colloidal particles is always charged. These charges may originate from the chemical structure of the macromolecule. For example, in proteins a number of acidic (—COOH) and basic (—NH$_2$) groups are present. These groups are ionized and, therefore, protein molecules have a number of charges on the surface, depending on pH and ionic strength.

A neutral particle may acquire charges by preferential adsorption of ionic species from the medium. As a matter of fact, the charges on the surface may be the result of both presence of intrinsic ionizable groups and adsorption.

Accounting for all the charges on the surface, we are particularly interested in the so-called net surface charges, which represent the difference between the number of positive and negative charges on the surface. If the net surface charge is zero (that

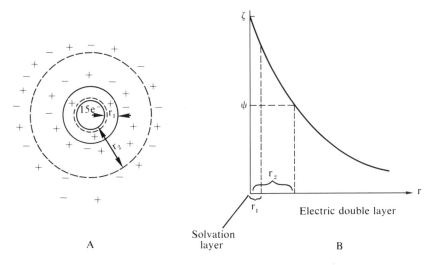

Figure 42-1 *Nature of the electric double layer: A, distribution of charge; B, potential energy diagram.*

is, the number of positive charges equals the number of negative charges), we speak of the isoelectric point of the colloidal particle. At its isoelectric point the colloid will not migrate in an electric field. In any other situation net positive or net negative charges on the surface will cause the macromolecule to move toward the cathode or anode, respectively, when an electric field is applied.

However, in the system as a whole, electroneutrality is maintained. The net immobile surface charges of a colloidal particle are balanced by a cloud of mobile counterions. The distribution of these counterions is such that there is a potential gradient that diminishes as one proceeds from the surface (Fig. 42–1).

The ζ potential is the work necessary to bring a charge to the surface of the particle from infinite distance. This potential decays linearly within the solvation layer and logarithmically beyond it. If the colloidal particle is spherical, one can imagine a sphere around it with radius r_2, at the boundary of which the potential is half the potential at the surface of the particle. This imaginary layer surrounding the particle is called the electric double layer, which is defined as a boundary beyond which *ad infinitum* there is a net total positive charge and within which there is a net total negative charge (Fig. 42–1A). These two domains, divided by the imaginary boundary of the electric double layer, balance each other electrically. Such an electric double layer stabilizes the colloidal system because the repulsive forces of the electric double layer around each colloidal particle prevent inelastic collisions.

The two parameters of the electric double layer (namely, the zeta potential and the thickness of the double layer) are influenced by the ionic strength of the medium (Fig. 42–2). Although the number of charges on the surface is the same at each ionic strength, both the zeta potential and the thickness of the electric double layer decrease with increasing ionic strength. This is the salting out effect frequently used in isolating specific proteins from a mixture. With increasing concentration of salt the electric double layer progressively diminishes and is finally removed, first from

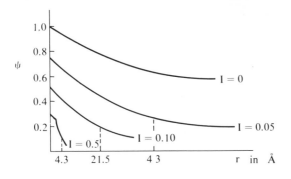

Figure 42–2 *The effect of ionic strength on the electric double layer.*

proteins that have small surface charge density; these proteins then aggregate and precipitate from solution.

Another way to separate proteins or any colloidal particles with different surface charge densities is to allow them to migrate in an electric field. Electrophoresis (i.e., the migration of colloidal particles in an electric field) can be performed in the absence or presence of a stabilizing medium. If no stabilizing medium is present, the phenomenon is called free boundary or moving boundary electrophoresis. If a stabilizing medium is present, it is called zone electrophoresis.

The free boundary method is applicable when the isoelectric point of a colloid particle and the mobility are to be determined. However, the disadvantages are that the moving boundary is not stable, incomplete separation of different migrating ions occurs, and it requires large samples and relatively expensive equipment. With zone electrophoresis, in which the colloidal particle migrates through a stabilizing medium, such as filter paper, starch gel, or polyacrylamide gel, the mobility cannot be determined directly. However, this can be used for clear separation of the components, using small samples and inexpensive equipment.

When a colloidal particle having a net charge, Q, is in an electric field of E field strength or potential gradient, the force, F, producing the migration is

$$F = QE \qquad (1)$$

Since the field strength is

$$E = \frac{V}{d} \qquad (2)$$

where V is the potential in volts and d is the distance between the electrodes, equation 1 also has the form

$$F = \frac{V}{d} Q \qquad (3)$$

This force is counteracted by the friction the particle encounters during its migration. The balance of forces results in a constant velocity of migration. If the particle is spherical and is suspended in a medium with η viscosity coefficient, the frictional force is given by Stokes' law

$$\frac{V}{d} Q = 6\pi\eta r v \qquad (4)$$

where r is the radius of the particle and v is the migrational velocity. The mobility of a particle is defined as the velocity in unit potential gradient and from equation 4

$$u = \frac{v}{E} = \frac{vd}{V} = \frac{Q}{6\pi\eta r} \qquad (5)$$

Thus the mobility has units of (cm./sec.)/(volt/cm.) or cm.^{2} sec.$^{-1}$ volt^{-1}.

For a charged sphere the relationship between the net charge, Q, and the zeta potential, ζ, is

$$\zeta = \frac{Q}{\epsilon r} \qquad (6)$$

where ϵ is the dielectric constant and r is the radius of the sphere.

For a spherical particle, the Hückel equation gives the following relationship

$$u = \frac{\zeta\epsilon}{6\pi\eta} \qquad (7)$$

and the zeta potential can be determined from mobilities.

These considerations are for spherical particles only, to which Stokes' law is applicable. For rod shaped particles, equations similar to equation 7 have been developed. If a rodlike particle is moving with its long axis parallel to the direction of migration, the Smoluchowski equation is applicable

$$u = \frac{\zeta\epsilon}{4\pi\eta} \qquad (8)$$

and if the long axis is perpendicular to the direction of migration

$$u = \frac{\zeta\epsilon}{8\pi\eta} \qquad (9)$$

We have already seen from Figure 42–2 that the ionic strength affects the zeta potential and, therefore, the mobilities. The mobility is inversely proportional to the square root of ionic strength.

$$u \sim \frac{1}{(I)^{1/2}} \qquad (10)$$

and

$$I = \tfrac{1}{2} \sum c_i z_i^2 \qquad (11)$$

where c_i is the concentration of ionic species i, and z_i is its charge.

Therefore, if ζ potentials are determined from mobilities, the ionic strength of the medium always has to be specified or, even better, the zeta potential should be determined at different ionic strengths and extrapolated to zero ionic strength.

As was indicated before, mobilities can be determined directly by moving boundary electrophoresis, which involves observing the migration of a boundary at different times under unit field strength. When the migration occurs in a stabilizing medium, additional effects are present.

For one, the migrating species may have preferential adsorptive capacity toward the supporting medium: filter paper, agar gel, and so forth. This would result in separation of the different ionic species more on the basis of selective adsorption than on the basis of mobilities. Ideally, the supporting medium should not adsorb any of the migrating species. Thus, one should select a medium in which this effect is minimized. One way to tell whether adsorption plays a major role in zone electrophoresis is to look for tailing. If the colloidal species are stained after electrophoresis and comet-like tailing effect follows most of the boundaries, adsorption is a major effect, and a different supporting medium should be used.

The second effect may be the heat produced during the flow of current. This may be demonstrated by an increase in the temperature of the medium, which can be detected by an increase in the current. If temperature changes are allowed, the mobility is greatly affected (i.e., mobility increases with increasing temperature).

$$\log u = A/T + B \tag{12}$$

where A and B are constants. The second effect of heat is the evaporation of the solvent. Both can be avoided by sufficient thermostating the supporting medium.

The third effect is peculiar to a specific supporting medium. If the surfaces of the filter paper or other supporting medium have fixed charges on them, they cannot migrate in an electrical field; thermodynamic compensation occurs, with the effect that the solvent molecules surrounding the gel migrate. This is called electroendosmosis. Electroendosmotic mobility is related to the ζ potential of the supporting medium through equation 8. This may be relatively high with glass beads or filter paper and almost nil in polyacrylamide gels. The electroendosmotic effect may be calculated by observing the mobility of a neutral particle, such as dextran, during electrophoresis; then this has to be added to the observed mobility of the colloidal particle.

$$u_{correct} = u_{colloid} + u_{dextran} \tag{13}$$

The supporting medium influences the mobility in two other ways. In each case the mobility is retarded. One reason for this is that the pathway in a gel, for example, is not a straight pathway as in free boundary electrophoresis (i.e., the channels may meander). This is called the tortuosity factor. The second effect is that the channels in a gel are not uniform—certain parts are constrictive. This may result in a delaying sieving effect, thus decreasing the mobility.

All these effects can be calculated by using pure colloids, the mobilities of which are known from moving boundary electrophoresis, and observing their migration in different supporting media with varying channel size (concentration of gel), ionic strength, and so forth.

Even though under such standardizing conditions a tortuosity factor, sieving factor, or other factor may be found for a particular supporting medium at constant temperature and set ionic strength, these factors may not be used unequivocally in reference to other colloids, the parameters, such as size and ζ potential, of which are not known. The sieving factor for a small globular protein, such as albumin, may be quite different from the sieving factor for a large random-coil linear polymer, which will encounter many more constrictive obstructions.

The best one can do is to run a standard material (serum albumin) in the same zone electrophoresis experiment as the unknown colloidal species and determine the relative mobilities of the components of the colloidal mixture (i.e., relative to albumin).

Experimental

The apparatus for horizontal zone electrophoresis is shown in Figure 42–3.

It is essentially composed of the following: two silver–silver chloride electrodes, A, dipping into concentrated potassium chloride solution, B; filter paper or agar-agar bridge, C, connecting compartments B and D; polyacrylamide gel, F, held between two supporting Plexiglas plates, G, through which thermostated water may be circulated; and a buffer-soaked filter paper bridge, H, connecting compartment D and the polyacrylamide gel.

The polyacrylamide gel slab is $6 \times 40 \times 250$ mm. in which three samples may be run simultaneously. The polyacrylamide gel is prepared as follows:

Prepare 1 liter of tris-borate buffer by dissolving 11.1 grams of tris buffer, 0.96 gram of ethylendiamine tetraacetate (EDTA), and 0.365 gram of boric acid in distilled water. Bring the volume to 1 liter. The pH is 9.2.

Weigh 7.5 grams of Cyanogum 41 (A.E.C. Apparatus Corp., Philadelphia, Pa.), which contains acrylamide and bisacrylamide in specific proportion.

Dissolve the Cyanogum 41 in 150 ml. of tris-borate buffer. Add 0.3 ml. of dimethylaminopropionitrile and 300 mg. of ammonium persulfate initiator and mix it thoroughly. (Make certain that the ammonium persulfate is fresh and check the C.P. purity; otherwise the gelling may take hours.) Turn the two Plexiglas support plates into vertical position. Place a sponge strip at the bottom to support the polyacrylamide gel slab. Pour the polymer solution between the Plexiglas plates and adjust the set screws so that the plates are separated uniformly by 6 mm. Apply slight vacuum at the top of the plates to remove any air bubbles trapped by pouring the polymer solution. Allow tap water to circulate in the Plexiglas support plates to remove the heat of polymerization. Turn the plates into horizontal position after

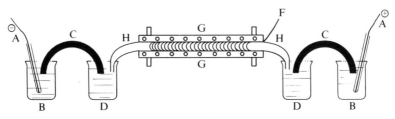

Figure 42–3 *Apparatus for horizontal zone electrophoresis.*

about 20 minutes of polymerization and allow gel formation to proceed for an additional 20 minutes.

Set up the electrophoretic apparatus as indicated in Figure 42–3. Fill the two compartments, *D*, with tris-borate buffer, and after soaking filter paper in this buffer, establish a bridge, *H*, between compartments *D* and the polyacrylamide gel, *F*. One may use either one or two filter papers as a bridge. One filter paper will provide a smaller current, hence a slower movement and cleaner separation of colloidal particles. However, because of time limitations for this laboratory experiment, it is recommended that two filter paper bridges be used—one touching the top and the other the lower surface of the gel on both sides of the horizontal apparatus. This way, 20-milliampere current can be conveniently drawn with 300-volt potential. Also fill compartments *B* with potassium chloride solution and allow the system to reach equilibrium by passing D.C. current at 300 volts for 20 minutes.

Three samples will be run: blood serum, bovine serum albumin, and dextran. The blood serum will be used as such; approximately 2 ml. each of a 2 percent solution of serum albumin in tris-borate buffer and a 2 percent solution of dextran in tris buffer will be prepared.

After equilibration of the polyacrylamide gel, remove the top Plexiglas support. Cut three wedges transversely in the gel with a razor blade about two-thirds the distance from the end of the plate in which direction the migration of the negatively charged particles is expected to occur. Cut Whatman No. 3 MM filter paper strips (wicks) the sizes of the wedges in the gel. Soak these strips of filter paper in the respective samples and place them in the wedges. Take note that the dextran sample runs along one edge of the gel because it will require special staining. Close the wedges (slots) by pushing the polyacrylamide gel, and replace the top Plexiglas support. Use a circulating pump to run thermostated water (25° C.) through the Plexiglas supports. Reapply the 300-volt potential and *note the zero time.* Allow the electrophoresis to run for $2\frac{1}{2}$ hours. Watch for and note any appreciable change in the current registered during the electrophoresis.

While the electrophoresis is in process, prepare the following staining and destaining solutions: *for proteins,* dissolve 1 gram of Amido Black 10B in 100 ml. of 7 percent acetic acid (the destaining compound); *for dextran,* prepare 0.5 percent potassium permanganate in 1N sodium hydroxide.

In the meantime, determine the dielectric constant and the viscosity coefficient of the tris-borate buffer as was done in Experiments 15 and 20. (If this is not feasible, as a rough approximation you can assume the ϵ and η values for water.)

To end the electrophoresis shut off the current and note the time. Cut out one-third of the gel where the dextran sample was expected to run. Remove the wick and place this part of the gel in a tray containing permanganate solution. Allow two to five minutes for the stain to penetrate; decant the stain and allow the slab to dry. When it dries, the position of the dextran will show up as a yellowish white line in a pink background.

After the wicks have been removed, dip the remaining two-thirds of the gel, where the protein samples were expected to run, into a tray containing the Amido Black stain for two minutes. Decant the stain and rinse the gel three or four times with 7 percent acetic acid solution for 30 minutes.

The protein fraction should show up as black bands in a yellowish white background.

If the position of the protein bands is not clear, leave the sample overnight in 7 percent acetic acid solution.

From the voltage, the time of the electrophoresis, and the distance the bands traveled, calculate the observed mobility for each band.

Correct the observed mobilities for electroendosmotic effect, if any (equation 13).

Calculate the relative mobilities of each band of the blood serum sample by assuming that the absolute mobility of the bovine serum albumin in the tris-borate buffer is -7.0×10^{-5} cm.^{2} volt^{-1} sec.$^{-1}$.

Use equation 7 to calculate the approximate value for the ζ potential for each compound in the blood serum by assuming that they are spherical.

Discuss which components of the blood serum are the most stable colloids on the basis of the data you obtained.

Material and Equipment. 10 grams of acrylamide (Cyanogum 41); 15 grams of tris buffer; 1 gram of ethylenediamine tetraacetate (EDTA); 0.5 gram of boric acid; 1 ml. of dimethylaminopropionitrile; 0.5 gram of ammonium persulfate; 1 ml. of blood serum; 20 mg. of bovine serum albumin; 20 mg. of dextran; 1 gram of Amido Black 10B stain; acetic acid; 1 gram of potassium permanganate; sodium hydroxide; potassium chloride; agar; electrophoretic apparatus with two silver–silver chloride electrodes; D.C. power supply and two supporting Plexiglas plates; thermostat bath with heater, stirrer, and thermoregulator; circulating pump; sponge strip; Whatman No. 3 MM filter paper.

REFERENCES

1. G. Zweig and J. R. Whitaker. Paper Chromatography and Electrophoresis. Vol. I, Academic Press, New York, 1967.
2. M. Lederer. An Introduction to Paper Electrophoresis and Related Methods. Elsevier Publishing Co., Amsterdam, N.Y., 1955.

XIII

MACROMOLECULES

43

OSMOTIC PRESSURE

Among the colligative properties, osmotic pressure is of primary importance in the calculation of molecular weight if the number of dissolved particles in a unit volume is low. For ideal solutions the van't Hoff equation relates osmotic pressure, π, to molecular weight, M.

$$\pi = \frac{RTc}{M} \tag{1}$$

where c is the concentration, T is the absolute temperature, and R is the gas constant.

Osmotic pressure is especially useful for evaluation of the molecular weights of macromolecules, in which case the solubilities in terms of molarities are very small. However, polymer solutions are not ideal solutions and the relationship of osmotic pressure to concentration generally can be expressed as

$$\pi = \frac{RT}{M}c + B^*c^2 + C^*c^3 + \cdots \tag{2}$$

in which B^* and C^* are called the second and third virial coefficients. Osmotic pressures of moderately dilute solutions can be expressed using only the first two terms on the right-hand side of equation 2. The second virial coefficient contains a parameter reflecting the interaction between solvent and solute. The greater the value of B^*, the better the solvent for a particular solute. Negative B^* values correspond to poor solvents. Rewriting equation 2,

$$\pi/c = \frac{RT}{M} + B^*c + \cdots \tag{3}$$

A plot of π/c vs. c has $\dfrac{RT}{M}$ as the intercept of π/c at $c \to 0$ and B^* as the slope of the curve (Fig. 43–1). Since osmotic pressure is a colligative property, care must

399

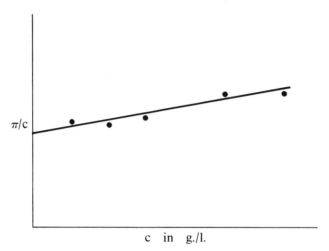

Figure 43–1 *Plot of π/c vs. concentration.*

be taken in the determination of the molecular weights of polyelectrolytes, such as proteins, not to average the number of polymeric molecules and their counterions.

To avoid the presence of counterions as independent particles, the ionization of the protein is suppressed with a buffer at the isoelectric point (i.e., the point at which the number of positive charges of the protein equals the number of negative charges).

Experimental

A stock solution of phosphate buffer is prepared by dissolving monopotassium phosphate in water to make a 0.1M solution and adjusting it to pH 4.6 with a 0.1N potassium hydroxide solution. Five different solutions are prepared having concentrations between 7 and 30 grams per liter of serum albumin in phosphate buffer. The osmotic pressure measurements are started with the solution of highest concentration.

A Fuoss-Meade osmometer consisting of metal blocks into which Visking cellulose semipermeable membranes are placed on two sides is the solution chamber (Fig. 43–2). The semipermeable membranes are held rigidly in place by Teflon rings and the outer part of the metal block. Two capillary tubes are inserted at the top to complete the assembly.

Any residual liquid is first removed from the osmometer with a long syringe, and it is washed with 2 ml. of the solution to be analyzed. After it is washed, the osmometer is emptied again and finally filled with the solution. Care must be taken that air bubbles are not trapped in the solution chamber. This can be achieved by proper positioning of the osmometer

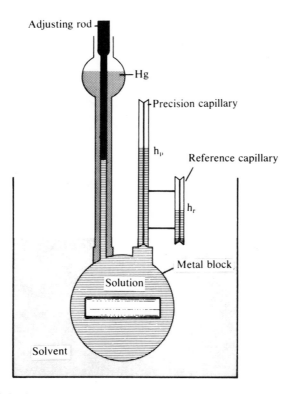

Figure 43–2 *Osmometer. Osmotic pressure (π) is the difference between the height of the menisci of the precision capillary and the reference capillary ($h_p - h_r$).*

during filling so that the air bubbles rise in the capillary and subsequently are expelled from there. A metal rod is inserted into the sampling arm to help set the initial level in the capillary.

The osmometer is now placed in a thermostated solvent bath containing enough liquid so that the lower end of the reference capillary on the side is about 0.5 to 1.0 cm. below the surface of the liquid. The solvent bath is placed into the thermostat. A small amount of mercury is poured into the cup above the sampling arm to enclose hermetically the adjusting rod. The adjusting rod is pushed down to displace a couple of drops of liquid in the precision capillary and then raised again to adjust the initial height in the precision capillary about 2 or 3 cm. above the expected osmotic pressure heights. By use of a cathetometer, the osmotic pressure is read (at two-minute intervals for about 20 minutes) in terms of the difference in heights of the menisci of the reference and precision capillaries.

The level of the precision capillary is set about 1 cm. below that of the last readings by lowering and then raising the rod. Readings are again taken at two-minute intervals for 10 to 15 minutes.

Plot the two curves of capillary height vs. time. The two curves tend to converge at the equilibrium osmotic pressure (Fig. 43–3).

Repeat the entire procedure for the other four solutions.

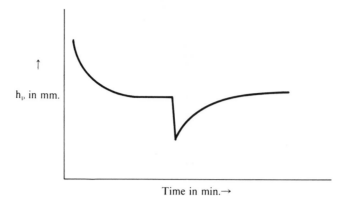

Figure 43–3   *Osmotic pressure equilibration curve. Equilibrium position is approached from both high and low initial osmotic pressure readings.*

Calculate the molecular weight of serum albumin and the second virial coefficient. Comment on the latter in view of the fact that you used an isoelectric medium for the solvent. Be certain that you have converted the observed difference in heights of the menisci into atmospheres by using the density of the solution. In dilute solutions it usually is sufficient to take the density of the buffer solution as the density of the polymer solution. What is the error introduced by this approximation? How does it affect the molecular weight?

Material and Equipment. Monopotassium phosphate; potassium hydroxide; 5 grams of serum albumin; Visking cellulose semipermeable membranes; Fuoss-Meade type of osmometer with filling syringe and adjusting rod; thermostated bath with heater, stirrer, and thermoregulator; cathetometer; mercury.

REFERENCES

1. P. J. Flory. Principles of Polymer Chemistry. Cornell University Press, Ithaca, N.Y., 1953, pp. 269–282.
2. C. Tanford. Physical Chemistry of Macromolecules. Wiley & Sons, Inc., New York, 1961, pp. 210–238.

44

MOLECULAR WEIGHT DISTRIBUTION BY GEL FILTRATION

Most synthetic polymers, as well as isolated biopolymers, with the exception of proteins, are collections of molecules with the same primary structure but different molecular weights.

One of the most characteristic features of macromolecules is that the physical properties depend more on the size of the molecule than on its composition. For example, there is greater difference between paraffin and polyethylene film than between polyethylene film and polystyrene film. In the former case, the primary structure is the same but the polyethylene chain may be 100 times longer than the paraffin molecule. In the latter case, the primary structures are quite different, but the molecular sizes may be of the same magnitude.

Therefore, it is of primary importance that the size of the molecule and the distribution of these various sized molecules should be known in each sample so that the physical properties may be controlled by blending.

There are a number of ways by which a polymer sample may be fractionated (selective precipitation by temperature and solvent gradients, and so forth) but in recent years a chromatographic variant, gel filtration or gel permeation chromatography, has come into use. This chromatographic technique employs cross-linked polymers in the form of insoluble but highly swollen particles (gel). The most frequently used gels are polystyrene cross-linked with divinylbenzene (Styragel, from Dow Chemical Co., Midland, Mich.), cross-linked dextran (Sephadex, from Pharmacia Fine Chemical, Uppsala, Sweden), and a copolymer of acrylamide and methylenebisacryl amide (Bio-Gel, from Bio-Rad Laboratories, Richmond, Calif.). Others include cross-linked cellulose, starch, agarose, polyvinyl alcohol, polymethacrylate, vulcanized rubber, and porous glass powder.

The cross-linked polymeric gels act as molecular sieves and they are available in different bead sizes. The different degrees of cross-linking create different pore sizes in these gels. A highly cross-linked gel, for instance, may exclude large molecules and allow only small molecules to penetrate its pores. These large molecules travel with the solvent only in the interstices of the beads. The separation and, therefore, the gel filtration process may be visualized as molecular sieving.

The polymer solution under study is introduced into a column of cross-linked polymer gel, and it is eluted with the solvent. Molecules that are too large to penetrate the pores are eluted first because they travel the shortest distance (interstices). Somewhat smaller molecules may penetrate some pores and be excluded from others, and they are eluted next. The smallest molecules penetrate all pores, and, therefore, their elution path is the longest; hence, they are eluted last.

Laurent and Killander showed that a model of the cross-linked gel, made up of a three-dimensional network of randomly distributed rigid rods, can explain the gel filtration behavior of many proteins (globular particles). The same theory can also be applied to random coils.

In order to understand this theory the student must first become familiar with the terminology used in gel filtration.

The experiment is performed on a column 100 cm. long and 2 cm. in diameter, containing swollen gel beads. The polymer sample to be fractionated is introduced into the column, and it is eluted with the solvent. Fractions of eluents are collected, for example, in 25-ml. batches, and each is analyzed for its content by an appropriate technique, such as ultraviolet absorption, differential refractive index, or some colorimetric analysis of a chemical reaction.

The result is then plotted as concentration of polymer in the different fractions vs. eluent volume (Fig. 44–1).

In a diagram such as Figure 44–1, reference is made to three different elution volumes: the actual elution volume of species i, V_i; the total volume, V_t; and the void volume, V_0. The void volume is the volume of the interstices, that is, the elution volume that carries with it the polymers that are too large to penetrate the pores of the gel. Together with V_t it is the specific property of a column of swollen gel beads. In essence, it is obtained by chromatographic analysis of an aqueous solution of a very large polymer of known molecular weight (for example, dextran, for which $M = 3 \times 10^6$).

The elution volume in which these large molecules appear is termed the void

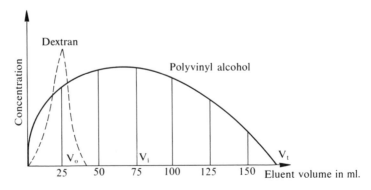

Figure 44–1   *Chromatography of a polyvinyl alcohol on a Bio-Gel P-10 column.*

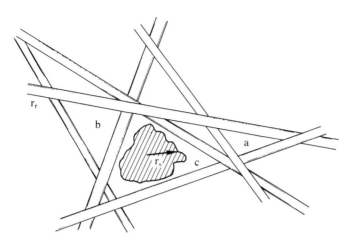

Figure 44-2 *Model of gel filtration.*

volume, V_0. The total volume is the elution volume necessary to carry the smallest molecule that penetrates all pores of the gel beads. On hydrophilic columns this is usually determined by the elution of tritiated water. The activity of the eluents is measured by a liquid scintillation counter. The volume necessary to elute the entire tritiated water sample is the total volume, V_t.

Each fraction of a polymer obtained with different elution volumes, V_i, is characterized by a partition coefficient, K, between the gel phase and the liquid phase.

$$K = \frac{V_i - V_0}{V_t - V_0} \tag{1}$$

The Laurent and Killander model is presented in Figure 44-2. In this model we have uniform, rigid rods of radius r_r, randomly distributed in a three-dimensional network. The concentration of the gel may be expressed in centimeters length of rod per cubic centimeter, L. If the polymer molecule is a sphere (such as a globular protein), it can be represented with molecular radius r_s, as in Figure 44-2. Some of the pores, such as a, exclude this molecule while others, b and c, allow its passage through the gel. With this model the partition coefficient can be written as

$$K = \exp\left[-L(r_s + r_r)^2\right] \tag{2}$$

The validity of this model has been shown in many experiments on globular particles. Moreover, it can be applied to random coils or even to rod shaped polymer molecules, except that the actual molecular parameters must be restated in terms of the radius of an equivalent hydrodynamic sphere. This hydrodynamic sphere is the average volume element swept out by a rotating tumbling coil or rod.

Thus, if the radius, r_r, and the length, L, of the rigid, uniform rods forming the gel network are known, using experimental K values, the radius of the equivalent hydrodynamic sphere of the eluted polymer, r_s, can be calculated. Conversely, the dimensions of the gel network, r_r and L, can be obtained by gel filtration analysis of two different globular particles of known sizes, r_s, and obtaining the corresponding

K values by using equation 1 and by solving simultaneously equations of the type of equation 2.

A more pragmatic approach without molecular models is to obtain the relationship between elution volume and molecular weight. One of the simplest techniques is to isolate the fractions obtained in gel filtration and to determine the viscosity of each fraction. The viscosity of a polymer solution is dependent upon the concentration and the molecular parameters. We define the ratio of the viscosity coefficient of a solution to that of a solvent as the relative viscosity number, η_{rel}.

$$\eta_{rel} = \frac{\eta_{solution}}{\eta_{solvent}} \tag{3}$$

The viscosity numbers are related to the time of efflux from a viscometer and to the specifications of the viscometer (radius and length of capillary, volume of liquid). If we use the same viscometer the relative viscosity may be determined from the efflux times, t, and from the densities, ρ.

$$\eta_{rel} = \frac{t_{soln}\rho_{soln}}{t_{solvent}\rho_{solvent}} \tag{4}$$

For most dilute polymeric solutions $\rho_{soln}/\rho_{solvent} \sim 1$. The specific viscosity coefficient, η_{sp}, that is, the contribution of the polymer molecules to the viscosity of solution is given by

$$\eta_{sp} = \frac{\eta_{soln} - \eta_{solvent}}{\eta_{solvent}} = \eta_{rel} - 1 \tag{5}$$

This is concentration dependent and may be expressed as

$$\eta_{sp}/c = [\eta] + k'[\eta]^2 c \tag{6}$$

In equation 6 the $[\eta]$ is the intrinsic viscosity that would be the contribution of a single polymer molecule to the viscosity of the solution at infinite dilution. The $[\eta]$ is obtained by plotting η_{sp}/c against concentration, c (Fig. 44-3) and extrapolating to zero concentration.

From the slope of the line, k' can be obtained (see equation 6). This is the Huggins constant and it is a characteristic of solvent-polymer interaction at a set temperature.

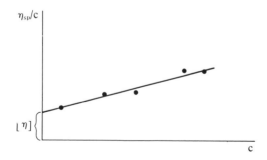

Figure 44-3 *Plot of* η_{sp}/c *vs. concentration.*

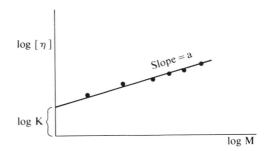

Figure 44–4 *Staudinger plot.*

The intrinsic viscosity is the property related to molecular parameters through the Mark-Houwink-Staudinger equation.

$$[\eta] = KM^a \tag{7}$$

In this equation, M is the molecular weight (average molecular weight in case of polydispersity); a is a shape factor, which is 0.5 for spheres, 0.75 for random coils, and 1 or greater for rigid rods; and K is a function of the polymer-solvent interaction parameters as well as of temperature.

The constants in equation 7 are obtained by measuring the molecular weights of different fractions of a polymer by light scattering, ultracentrifugation, or osmometry and by determining the intrinsic viscosity of each fraction from plots, such as that given in Figure 44–3. When these are obtained $\log [\eta]$ is plotted against $\log M$ (Fig. 44–4). The intercept of such a plot is $\log K$ and the slope is the constant a.

Conversely, once these constants, K and a, are known for a polymer-solvent system at a set temperature, intrinsic viscosity values evaluated from viscosity measurements can be used to determine the molecular weight.

In this experiment polygalacturonic acid will be fractionated by gel filtration. The intrinsic viscosity of each fraction will be determined from viscosity measurements. From the intrinsic viscosity values and from the known constants of the Mark-Houwink-Staudinger equation for this system the molecular weight of each fraction will be obtained. The average molecular weight of the unfractionated sample will also be obtained.

Since these molecular weights are average molecular weights, their value will depend on the type of measurements (osmometry, and so forth) through which the Mark-Houwink-Staudinger constants were obtained.

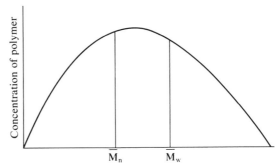

Figure 44–5 *Molecular weight distribution and number and weight average molecular weights.*

Osmometry gives number average molecular weight, $\overline{M_n}$, whereas light scattering gives weight average molecular weights, $\overline{M_w}$, defined by equations 8 and 9.

$$\overline{M_n} = \sum_i n_i M_i / \sum_i n_i \qquad (8)$$

$$\overline{M_w} = \sum_i g_i M_i / \sum_i g_i = \sum_i n_i M_i^2 / \sum_i n_i M_i \qquad (9)$$

where n_i is the number of moles, M_i is the molecular weight, and g_i is the weight in grams of species i.

Their relation in a molecular weight distribution curve is given in Figure 44-5.

Finally, the average molecular weight obtained on the unfractionated samples will be compared to the molecular weight distribution.

Experimental

This experiment, if carried out as described, may take two laboratory periods. If necessary, the second part may be done in one laboratory period together with a short experiment, such as viscosity of gases, polarizability, or adsorption at solid-liquid interfaces.

Sephadex G-50 cross-linked dextran gel will be used for the gel filtration. Prepare a 500-ml. stock solution of 0.155M sodium chloride. Weigh 25 grams of dry Sephadex G-50 and add it slowly, with constant stirring, to 100 ml. of 0.155M sodium chloride solution.

The chromatographic column is a Pyrex tube about 30 to 40 cm. long and 3 cm. inside diameter. The lower part of the column is a sintered glass filter of medium porosity (see Fig. 44–6). Attach Tygon tubing to the lower part of the column and insert a screw clamp or stopcock to control the flow.

Pour the slurry of Sephadex G-50 in 0.155M of sodium chloride into the chromatographic tube and allow the gel to settle. Wash it with additional 0.155M sodium chloride solution but take care not to allow the column to run dry.

Prepare 25 ml. of 1.6 percent polygalacturonic acid solution in 0.155M sodium chloride. The polygalacturonic acid will not dissolve, but a steady suspension should be prepared by using a magnetic stirrer.

Figure 44–6 *Chromatographic column.*

Set the electrodes of a pH meter into the suspension and determine the pH of the suspension. If the material suspended is really polygalacturonic acid, a pH between 3 and 4 will be obtained. Slowly add 0.5M sodium hydroxide. At the beginning the pH will not change, but the suspended particles will go into solution. Be sure the rate of stirring is constant so that no local alkali pockets form in the solution. After most of the suspended particles have been dissolved, the pH will change rapidly. Adjust the solution to a final pH of 6.0 with 0.1N sodium hydroxide.

Using a hypodermic syringe, inject the whole solution of polygalacturonic acid thus prepared into the column of Sephadex G-50 and let the solvent flow into the column at a constant rate.

After discarding the first 10 ml. of eluent, collect ten 25-ml. fractions and number them. Be certain that the column does not run dry.

Prepare 1 liter of 0.5N hydrochloric acid in ethanol by adding concentrated hydrochloric acid to absolute ethanol.

Pour each 25-ml. fraction separately into 75 ml. ethanol solutions with constant stirring. As soon as the eluent containing polygalacturonate contacts the acidified ethanol solution a transparent gel forms. Ten solutions with gel precipitates are thereby obtained.

Filter each fraction on a Büchner filter with Whatman No. 1 filter paper, and wash it with 95 percent ethanol.

Transfer the gel of each fraction into a preweighed drying bottle by scraping it off with a spatula and washing the filter paper with 95 percent ethanol. Dry it in a vacuum oven overnight at a temperature not exceeding 50° C.

Collect the dried samples and weigh them. Calculate the amount of material in each elution volume and plot the data in a diagram similar to Figure 44-1.

Using the unfractionated polygalacturonic acid, prepare four solutions in 0.155M sodium chloride containing about 0.2, 0.15, 0.1, and 0.05 percent polymer at pH 6.

Using an Ostwald-Cannon-Fenske viscometer (Fig. 44-7) measure the efflux time of 0.155M sodium chloride solution and the four polygalacturonate solutions, using 5 ml. of liquid each time.

The efflux time is the time required for a quantity of liquid to flow through the capillary, *l*. It is timed with a stopwatch, starting when the meniscus of the liquid reaches the upper mark, *a*, and ending when it reaches the lower mark, *b*, in the viscometer.

The relative viscosities are calculated by using equation 4; the ratio of densities can be taken as 1.

The specific viscosity is calculated (equation 5) for each concentration and the data are plotted as in Figure 44-3. Obtain the values of $[\eta]$ and k' for the unfractionated sample.

Dissolve each dried fraction of polygalacturonate obtained in the gel filtration in small volumes of 0.155M sodium chloride, adjusting the pH to 6.0 with 0.01N sodium hydroxide. Thus, a solution will be obtained with known concentrations from each fraction. The concentration of each solution may vary from 0.2 to 0.02 per cent.

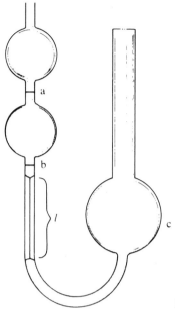

Figure 44–7 *Ostwald-Cannon-Fenske viscometer.*

Determine the efflux time in the viscometer for each solution at 25° C. Calculate the $\eta_{\rm rel}$ and $\eta_{\rm sp}$ for each solution. From equation 6 calculate the $[\eta]$ for each fraction, using the Huggins constant, k', calculated from the data on the unfractionated polygalacturonic acid sample.

From the intrinsic viscosities thus obtained, determine the corresponding molecular weights from equation 7, using the following Mark-Houwink-Staudinger constant: $K = 1.4 \times 10^{-6}$ and $a = 1.34$.

Prepare a plot of the elution volume vs. the calculated molecular weight. Is this a straight line relationship? Can gel filtration be used for molecular weight determination for a polymer solvent system?

Plot the molecular weight distribution curve as in Figure 44–5. Mark on the graph the average molecular weight of the unfractionated polygalacturonate. In your judgment is this a number or weight average molecular weight? Explain.

Material and Equipment. 25 grams of Sephadex G-50; sodium chloride; 1 gram of polygalacturonic acid; sodium hydroxide; hydrochloric acid; ethanol; Tygon tubing; Whatman No. 1 filter paper; Büchner filter; pH meter; chromatographic column with sintered glass filter and stopcock; hypodermic syringe; Ostwald-Cannon-Fenske viscometer; vacuum oven; thermostat bath with heater, stirrer, and thermoregulator; stopwatch.

REFERENCES

1. B. Gelotte and J. Porath. *In* Heftmann, E. (ed.). Chromatography. 2nd Edition, Reinhold Publishing Corp., New York, 1967, Chapter 14.
2. T. C. Laurent and J. Killander. J. Chromatog., **14,** 317 (1964).
3. H. S. Owens, H. Lotzkar, T. H. Schultz and W. D. Maclay. J. Am. Chem. Soc., **68** 1628 (1946).

45

MACROMOLECULAR PROPERTIES AS DETERMINED BY LIGHT SCATTERING

One of the most interesting features of colloidal or macromolecular solutions is the Tyndall effect: the gleaming path of scattered light that is observable at a 90 degree angle when light passes through such a solution. At other angles the scattered light has a somewhat turbid appearance. If the observation is made with a microscope at an angle of 90 degrees to the propagated light, each macromolecule larger than 10 Å in diameter appears as an independent light source and can be counted as a particle per unit volume.

The amount of light scattered by a colloidal or macromolecular solution is related to the concentration and angle of observation (experimental variables) and to the intrinsic properties of the macromolecules, such as size and shape. In order to understand how these variables influence the amount of light scattered by a solution, we shall start with the simplest system, namely, light scattered by an ideal gas.

When a particle such as a gas molecule is subjected to an electric field of strength E, an induced dipole moment, p, is produced in the particle. The magnitude of this dipole movement is proportional to the field

$$p = \alpha E \tag{1}$$

and α, the proportionality constant, is the polarizability of the particle, that is, the amount of dipole moment induced by a unit field. The electric field of a plane polarized light wave is

$$E = E_0 \cos 2\pi(\nu t - x/\lambda) \tag{2}$$

where E_0 is the maximal amplitude, ν the frequency of light, λ its wavelength, t the time, and x the location at time t along the line of propagation. Equation 2 shows the periodicity of the electric field; i.e., the same E_0 is obtained at time intervals $1/\nu$.

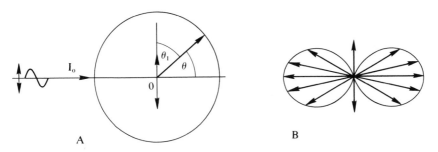

Figure 45-1 *Angular dependence of the intensity of scattered light of plane polarized incident radiation. A, the incident beam, I_0, is polarized in the plane of the drawing; the intensity of the scattered beam depends on the θ angle. The locus of the scattering particle is 0. B, the length of the arrows is proportional to the intensity of the scattered beam, i_s, at different θ values.*

The combination of equations 1 and 2 gives

$$p = \alpha E_0 \cos 2\pi(\nu t - x/\lambda) \tag{3}$$

the value of the oscillating induced dipole moment of the particle. This particle, in our first consideration, is much smaller than the wavelength of the light. The oscillating dipole itself is a source of radiation. This new radiation is the *scattered beam*. Its field strength is proportional to the second derivative of the oscillating dipole, d^2p/dt^2. Although the scattered radiation is a spherical wave, its field strength depends on the angle of observation, θ (Fig. 45–1).

Moreover, the law of conservation of energy requires that the field strength of the scattered radiation should vary with $1/r$ where r is the distance of the observer from the scattering center. Taking the second derivative of equation 3 and introducing $\sin \theta/c^2r$ as the angular and distance dependence, where c is the velocity of light, we obtain

$$E_s = \frac{4\pi^2\nu^2 E_0 \sin \theta}{c^2 r} \cos 2\pi(\nu t - x/\lambda) \tag{4}$$

where E_s is the field strength of the scattered light. However, experimentally we measure light intensity rather than field strength. The two are related as

$$I \sim E^2 \tag{5}$$

and, therefore, equation 4 is written as

$$R_\theta = \frac{i_s r^2}{I_0 \sin^2 \theta_1} = \frac{16\pi^4\alpha^2}{\lambda_0^4} \tag{6}$$

Equation 6 is the famous Rayleigh equation. It has already been shown in equation 4 that the wavelength of the scattered beam is the same as that of the incident beam. Rayleigh's equation shows that the intensity of the scattered beam depends on the inverse fourth power of the wavelength of the incident beam. In equation 6, λ_0 is the wavelength in a vacuum and was introduced instead of the expression c/ν.

The left side of the equation is the ratio of the scattered beam intensity to the incident beam intensity multiplied by the square of the distance of observation, r^2. The latter is a constant under most experimental conditions. Thus, the left side of equation 6 is referred to as the Rayleigh ratio, R_θ. The subscript θ indicates the angular dependence. The molecular parameter on which R_θ depends is the polarizability. The polarizability can be related through the dielectric constant to the refractive index, and equation 6 can be written for the intensity of scattered beam per unit volume as

$$R_\theta = \frac{4\pi^2(dn/dc)^2 Mc}{N_A \lambda_0^4} \tag{7}$$

where M is the molecular weight of the particle, N_A is Avogadro's number, c is the concentration in grams per cubic centimeter, and (dn/dc) is the refractive index gradient. It is obvious, then, from equation 7 that light scattering can be used to determine molecular weights. However, before progressing from ideal gases to dilute solutions and to macromolecular solutions, one modification has to be introduced. In the derivation of equation 7 we assumed that there is plane polarized light. Most light scattering is done with unpolarized light, which can be regarded as the superposition of the two plane polarized light beams. Thus, instead of $\sin^2\theta$ angular dependence, one has now $(\sin^2\theta_1 + \sin^2\theta_2)$ where θ_1 and θ_2 are the angles made by the line of observation with the y and z axes of the Cartesian coordinates.

$$\sin^2\theta_1 + \sin^2\theta_2 = 1 + \cos^2\theta \tag{8}$$

where θ is now defined as the angle between the incident and the scattered beam. Thus, for unpolarized light equation 7 becomes

$$\frac{i_{\theta s}}{I_0} \frac{r^2}{(1+\cos^2\theta)} = R_\theta' = \frac{2\pi^2(dn/dc)^2 Mc}{N_A \lambda_0^4} \tag{9}$$

where R_θ' is the Rayleigh ratio for unpolarized light.

Comparing Figures 45–1 and 45–2, one sees that with plane polarized light the intensity of the scattered beam is zero at a 90 degree scattering angle, but with unpolarized light the scattering intensity is a minimum but not zero at 90 degrees.

We will now progress from the ideal gas model to a dilute solution that contains only small molecules. The only modification this step necessitates is in the refractive index gradient. In deriving equations 7 and 9 we related the polarizability of a gas to the refractive index by using equation 10.

$$n^2 - 1 = 4\pi N\alpha \tag{10}$$

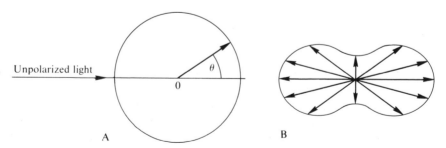

Unpolarized light

θ

0

A

B

Figure 45–2 *The angular dependence of a scattered beam of unpolarized incident radiation. A, the locus of the scattering particle is 0 and the scattering angle is θ. B, the length of the arrows is proportional to the intensity of the scattered light at different θ values.*

where N is the number of particles per cubic centimeter and n is the refractive index of the gas particle suspended in a vacuum (the refractive index is 1). In dilute solutions the medium is not a vacuum but a solvent with a refractive index of n_0. Hence, equation 10 becomes

$$n^2 - n_0^2 = 4\pi N\alpha \tag{11}$$

and equation 9 becomes

$$R'_\theta = \frac{2\pi^2 n_0^2 (dn/dc)^2 Mc}{N_A \lambda_0^4} \tag{12}$$

A further correction must be made because solutions of finite concentrations are not ideal systems. The non-ideality demonstrates itself in the dependence of the chemical potential on concentration. Introducing this non-ideal condition into equation 12 one obtains

$$R'_\theta = \frac{2\pi^2 n_0^2 (dn/dc)^2 c}{N_A \lambda_0^4 (1/M + B^*c + C^*c^2 + \cdots)} \tag{13}$$

Equation 13 differs from equation 12 in the additional terms $B^*C + C^*c^2$ where B^* and C^* are the second and third virial coefficients.

As the solutions become more and more dilute they approach ideality and with infinitely dilute solution the second and third virial coefficients become zero. Hence, equation 13 becomes equation 12 for infinitely dilute solutions.

As a final step in the progression we will note that in macromolecular solutions the size of the scattering particle is not small compared to the wavelength of the incident beam, i.e., radius of gyration, $R_G \geq \lambda_0/20$. The radius of gyration, R_G, is an average distance of a segment of the polymer chain from its center of mass and is defined by equation 14.

$$R_G = \left(\sum_i m_i r_i^2 \bigg/ \sum_i m_i \right)^{1/2} \tag{14}$$

where m_i is the mass of segment i of the macromolecule and r_i its distance from the center of mass.

If the particles are of sizes $\lambda_0/20$ and larger, one molecule provides more than one scattering center. This means that the total scattering intensity depends not only on the interference of the scattered beams coming from different particles but also on the interference of scattered beams coming from the same particle. The consequence of this is as follows: In the case of small particles we could eliminate the effect of interference by simple extrapolation to infinite dilution (ideal solution); this is not possible with macromolecules.

One can see the effect of interference coming from a single macromolecule by referring to Figure 45–3. It is obvious that at a 0 degree scattering angle there is no destructive interference between scattered beams from any macromolecular segments.

However, as the angle of observation increases, destructive interference plays an important role. Moreover, the scattering intensity is no longer symmetrical around 90 degrees as it was in Figure 45–1 and 45–2, but it has the shape given in Figure 45–4.

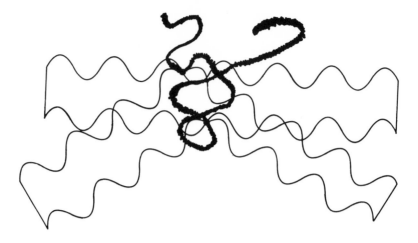

Figure 45–3 *The angular dependence of interference between two scattered beams coming from two segments of a macromolecule.*

The shape of this intensity envelope depends on the size and shape of the macromolecules. We define the particle scattering function, $P_{(\theta)}$, as

$$P_{(\theta)} = \frac{i_\theta}{i_{\theta \text{ without interference}}} \tag{15}$$

$P_{(\theta)}$ is a function of the scattering angle and the radius of gyration and it has different expressions for different shapes. For example, for a random coil

$$P_{(\theta)} = \frac{2}{x^2} \left[\exp(-x) - (1 - x) \right] \tag{16}$$

where

$$x = \frac{4\pi^2}{\lambda_0^2} R_G^4 \sin(\theta/2)$$

However, as $\theta \to 0$, $P_{(\theta)}$ becomes independent of particle shape and $P_{(\theta)} \to 1$.

$$\lim_{\theta \to 0} \frac{1}{P_{(\theta)}} = 1 + \frac{16\pi^2}{3\lambda_0^2} R_G^2 \sin^2 \theta/2 \tag{17}$$

Now equation 13 can be modified to be applicable to macromolecular solutions

$$R'_\theta = \frac{Kc}{(1/M + B^*c + C^*c^2 + \cdots) P_{(\theta)}} \tag{18}$$

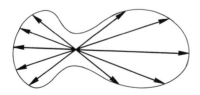

Figure 45–4 *Light scattering intensity envelope due to internal interference.*

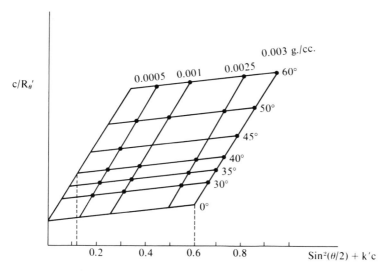

0.003 g./cc.

Figure 45-5 Zimm plot.

where K is the optical constant

$$K = 2\pi^2 n_0^2 (dn/dc)^2 / N_A \lambda_0^4 \tag{19}$$

As we indicated before, extrapolation to zero concentration eliminates the second and third virial coefficients; therefore

$$\lim_{c \to 0} \frac{Kc}{R_\theta'} = \frac{1}{M(P_{(\theta)})} = \frac{1}{M}\left(1 + \frac{16\pi^2}{3\lambda_0^2} R_G^2 \sin^2 \theta/2\right) \tag{20}$$

On the other hand, extrapolation to zero angle eliminates the radius of gyration dependence

$$\lim_{\theta \to 0} \frac{Kc}{R_\theta'} = \frac{1}{M} + B^*c + C^*c^2 + \cdots \tag{21}$$

Using double extrapolation to zero concentration and to zero angle, one can obtain three molecular parameters: molecular weight, M, radius of gyration, R_G, and the second virial coefficient, B^* (neglecting, as is valid in this case, the third virial coefficient).

This is accomplished in a Zimm plot where we plot c/R_θ' vs. $\sin^2 \theta/2 + k'c$ (Fig. 45-5). The quantity k' is an arbitrary constant to spread the plot and usually is selected so that the $k'c$ has about the same magnitude as the $\sin^2 (\theta/2)$ (usually $k' = 10$–500).

All the data are plotted and the points from the same concentration as well as the points from the same angular measurements are connected. This usually (but not always) can be achieved by drawing straight lines.

The lines are now extended in both directions. For example, the line representing the concentration 0.003 gram/cubic centimeter is extrapolated. We shall find the point corresponding to $\theta = 0$ on this line. Since $\sin^2 (\theta/2) = 0$ the abscissa at $\theta = 0$ equals $200 \times 0.003 = 0.6$ (200 being the arbitrary constant in Fig. 45–5). At the point at which the vertical line originating from 0.6 on the abscissa intersects the line representing the concentration 0.003 gram/cubic centimeter we establish the point corresponding to $\theta = 0$.

In the same way the $\theta = 0$ points are obtained on the other concentration lines. Now these $\theta = 0$ points are connected with a line and extrapolated to the y axis.

Similarly, one takes now the extrapolated θ line (e.g., $\theta = 40$ degrees). We shall find the point on this line corresponding to zero concentration. Since $k'c = 0$, $\sin^2 (20) = 0.342^2 = 0.117$. The point corresponding to zero concentration is at the intersection of the 40 degree line and a vertical line drawn from the point 0.117 on the abscissa. In the same manner, one can obtain the points on the other θ lines representing zero concentration. These points are connected and extrapolated to the y axis.

The double extrapolation, i.e., the lines corresponding to zero angle and zero concentration, should intersect at one point on the y axis (intercept).

The reciprocal of this intercept is related to the molecular weight according to equations 20 and 21. (Care must be taken to multiply the intercept by the optical constant, K).

The second virial coefficient, B^*, can be obtained from the slope of the line representing zero angle on the Zimm plot (equation 20). Again taking the slope one must remember to multiply the ordinate values by the optical constant.

The radius of gyration can be obtained from

$$R_G = \left[\frac{3\lambda_0^2/n^2}{16\pi^2} \cdot \frac{\text{initial slope of zero concentration line}}{\text{intercept}} \right]^{1/2} \tag{22}$$

where λ_0 is the wavelength in a vacuum, and n is the refractive index of the solution.

Thus, light scattering provides information on the molecular weight, the size, and the solvent-solute interaction parameter. Further information, namely the *shape of the molecule*, can also be determined from the Zimm plot by comparing the line for zero concentration to theoretical particle-scattering functions for different shapes of molecules. However, this is beyond the task of the present experiment and the student is referred instead to monographs on this subject (Stacey and Huglin).

A sample of macromolecules can be monodisperse if all the molecules have the same molecular weight. Many proteins are monodisperse.

On the other hand, when the sample contains one kind of molecule with different molecular weights, it is said to have polydispersity. Natural macromolecules, such as polysaccharides and most synthetic polymers, possess polydispersity. With such samples only average molecular weights can be obtained by a simple measurement. The molecular weight obtained in light scattering measurements is the weight average molecular weight.

$$M_w = \frac{\sum\limits_{i} g_i M_i}{\sum\limits_{i} g_i} \tag{23}$$

where g_i is the weight of fraction i possessing molecular weight M. The weight

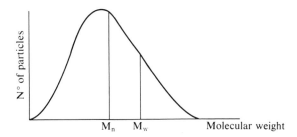

Figure 45–6   *Distribution curve.* M_n M_w Molecular weight

average molecular weight is larger than the number average molecular weight, M_n.

$$M_n = \frac{\sum\limits_i n_i M_i}{\sum\limits_i n_i} \qquad (24)$$

where n_i is the number of mols of fraction i possessing molecular weight M_i.

For a given distribution of molecular weights the relationship of M_n to M_w is given in Figure 45–6.

The light scattering measurements give weight average molecular weights because the heavier molecules contribute more to the scattering intensity than do the light molecules.

Regarding the radius of gyration in conjunction with the molecular weight, it indicates how tightly a macromolecule is packed in a given solvent.

Finally, the second virial coefficient gives an indication of the interaction between macromolecule and solvent. A large positive B value indicates a good solvent; a negative B value indicates poor solvent-solute interactions.

Experimental

The molecular parameters of glycogen will be determined by light scattering. Glycogen is a highly branched macromolecule; the condensation product of glucose, hence, the monomeric unit, is anhydroglucose. It is used as a storage material mainly in the liver, which, when needed, can quickly be hydrolyzed to glucose to provide energy.

A stock solution of glycogen should be made by dissolving about 1 gram of glycogen (accurately weighed) in 250 ml. of water. Care should be taken that all solid particles are dissolved by mechanical dispersion and stirring. An opaque solution will be obtained.

Since in light scattering measurements the critical part is that all extraneous materials (undissolved particles, dirt, smoke particles) be removed from the solution, this stock solution has to be centrifuged for about 20 minutes at 25,000 RCF $\times$ G force.

While the stock solution is being cleared by centrifugation, the light scattering measurements on the water blank can be made.

A Phoenix-Brice light scattering photometer, either 1000 *A* or 1000 *D*, may be employed. As shown in Figure 45–7, the photometer consists of: mercury lamp, *A*; a monochromator, *B*, which selects the 546 mμ mercury line; removable neutral filters, *C*, which diminish the intensity of the incident beam, if necessary; a photographic shutter, *D*, with iris control; a collimating tube, *F*, which can contain the polarizer (not used in this experiment); a diaphragm, *G*, which controls the width of the incident beam; a calibrated turntable, *H*, which turns around a stationary center holding up the scattering cell, *K*, or a calibrated reference standard, *L*; a photomultiplier, *M*, which is fixed to the turntable and thereby reads the scattered beam intensity at any angle desired; and a secondary standard, *N*, which is used only to obtain the intensity reading at 0 degrees. (This standard has a fixed position opposite the photomultiplier tube so that it is in the path of the beam only when the photomultiplier tube is at 0 degrees.)

The D.C. power supply output voltage to the photomultiplier tube is controlled by two knobs. The coarse sensitivity control knob has ten positions, which are set by the manufacturer. Each coarse setting usually is related to the previous one by a factor of 3. For instance, if the scattering intensity at a certain θ angle is recorded at 90 with a coarse setting of 10, the galvanometer reading will be 30 if the knob is switched to the setting of 9. A setting of 10 is the highest sensitivity and 1 is the lowest sensitivity. The student should establish this relationship between the settings.

There is a fine sensitivity control knob, which covers the range between the fixed sensitivity settings. This should be set by using a benzene standard and left in the same position throughout the experiment.

The galvanometer completes the instrument. It operates on 6.3 volts, and the line voltage, therefore, must be lead through a transformer. The black and red leads coming from the photomultiplier are connected to the galvanometer terminals. On top of the galvanometer there is a mechanical zero adjustment, which controls the angle of the mirror reflecting the light to the galvanometer scale. A more detailed account of the electronic components of the apparatus may be found in Part XIV of this book.

Care should be taken not to lean on the galvanometer or on the bench because a slight imbalance will cause an error in the readings.

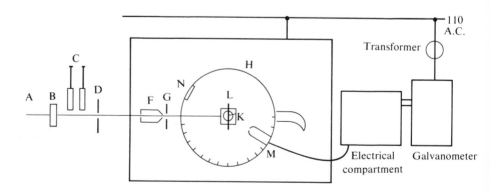

Figure 45–7 *Schematic diagram of a light scattering photometer.*

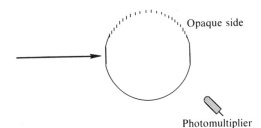

Figure 45–8 *Alignment of cell.*   Photomultiplier

The cylindrical light scattering cell should be thoroughly cleaned with cleaning solution and soap, thoroughly rinsed with distilled water, and dried. Into the clean and dustless cell, directly filter distilled water through a sintered glass filter or ultrafilter. Cover the scattering cell and place it in position so that the opaque half faces the back of the instrument. The two flat portions are aligned perpendicularly to the incoming beam (Fig. 45–8). Be sure that you do not leave fingerprints on the part of the cell that will scatter light. Using a 90 ml. capacity cell, you can handle the upper half of the cell, leaving the lower part untouched.

Be sure the power switch is off. Turn on the mercury lamp and let it warm up for five minutes. Turn on the power switch energizing the photo-multiplier power supply. Allow two minutes for warmup. Set the fine sensitivity knob to 10. Set the turntable to 30 degrees. Turn on the galvanometer switch. Take all the neutral filters out of the path of the beam. Open the shutter and set the coarse sensitivity knob to allow almost full scale deflection on the galvanometer. Close the shutter. Adjust the galvanometer to zero with the dark current knob. Open the shutter and record the galvanometer reading and the coarse setting.

Repeat the procedure with distilled water at 60, 55, 50, 45, 40, 35, and 30 degrees. Tabulate your data.

WATER

θ	Galvanometer (G)	Sensitivity (R)
90	18	1000
60	22	
55	35	
50	45	
45	50	
46	64	
35	84	

During this manipulation, also establish the relationships of the coarse sensitivity settings. For example, if at a 30-degree angle the galvanometer reading for water is 99 with a setting of 10 and the reading is 33 when the setting is switched to 9, the fixed settings are related by a factor of 3. That means, for example, that if we give a value, R, of 1000 to the setting of 10, the following table may be used.

Setting	R
10	1000.000
9	333.300
8	100.000
7	33.300
6	10.000
5	3.300
4	1.000
3	0.330
2	0.100
1	0.033

After the water blank has been measured, drain the cell and rinse it with the freshly centrifuged glycogen stock solution. Then fill the scattering cell with the centrifuged stock solution. Inspect the cell and make sure that no suspended particles are floating in the sample and that the cell is spotlessly clean and bears no fingerprints. Place the cell in position and obtain the scattering intensity of this solution again at angles of 90, 60, 55, 50, 45, 40, 35, and 30 degrees.

Perform similar measurements with four more concentrations of glycogen solutions. These concentrations should be obtained by diluting the stock solution with distilled water and again centrifuging the solutions before making the scattering measurements. If the stock solution had a concentration of 0.4 gram/100 ml., the other four solutions should have concentrations of about 0.3, 0.2, 0.15, and 0.1 gram/100 ml.

Tabulate all the primary data as shown previously.

Finally, obtain the instrument constant by measuring the scattering of triply distilled benzene in the same scattering cell at 90 degrees θ.

In order to draw the Zimm plot, the instrument response (galvanometric reading and sensitivity setting) must be converted to R_θ values.

Since R_{90} of benzene with 546 mμ of light is accurately known (16.3 $\times$ 10^{-6}), one can use this to obtain the instrument constant.

For example, if benzene has a galvanometric reading of 40 and a sensitivity of 1000 at an angle of 90 degrees,

$$\text{Instrument response} \times \text{instrument constant} = R_\theta$$

$$\frac{40}{1000} \times k = 16.3 \times 10^{-6}$$

Hence,

$$k = \frac{16.3 \times 10^{-3}}{40} = 4.075 \times 10^{-4}$$

Now the R_θ values for each solution at each angle can be obtained by subtracting the instrument response for water from the instrument response of the solution and multiplying the result by the instrument constant.

For example: For the 0.4 gram/100 ml. solution you obtained a reading at 50 degrees of $G = 40$ and $R = 10$; the instrument response is $G/R = 4.0$. For water at the same angle you obtained $G = 41$ and $R = 1000$. Hence, the instrument response was 0.041. Subtracting this,

$$4.0 - 0.041 = 3.959$$

and

$$R_{50} = 3.959 \times k = 3.959 \times 4.075 \times 10^{-4} = 16 \times 10^{-4}$$

Tabulate the R_θ values for each concentration of solution as follows:

CONCENTRATION $= 0.004$ GRAM/CUBIC CENTIMETER

θ	Sin^2 $\theta/2$	$1 + \cos^2 \theta$	R_θ	$R_\theta/(1 + \cos^2 \theta)$ $= R'_\theta$	c/R'_θ	Sin^2 $\theta/2 + 100\,c$
90	..	..	..	..	..	..
60	..	..	..	..	..	..
..	..	..	..	..	..	..
..	..	..	..	..	..	..

For any complicated calculations such as these, one can use computers; a program for the Zimm plot calculations is given in the Appendix.

After the data are calculated and tabulated, obtain a Zimm plot by plotting c/R'_θ vs. $\sin^2 \theta/2 + 100$ c.

Be cautious in connecting the points representing concentrations and angles. If it is possible, try to fit straight lines so that the extrapolation, as given in the description of Figure 45–5, may be done without difficulty. If you made an error in one of the concentrations, all the points on this concentration line will not fit the grid. Similarly, if you had a dirt spot on the scattering cell, all the points at one specific angle will not properly fit the Zimm plot grid.

Calculate the optical constant, K, (equation 19), taking the refractive index of water as 1.333 and (dn/dc) as 0.130 at 5460 Å.

Calculate the weight average molecular weight, the second virial coefficient, and the radius of gyration of glycogen, using equations 20, 21, and 22.

Using the radius of gyration as a volume parameter and the molecular weight, what is the density of the macromolecule in water? Compare this to a solid density of 1.5 grams/cubic centimeter.

Material and Equipment. 100 ml. of triply distilled benzene; 1 gram of glycogen; cylindrical light scattering cell with 90-ml. capacity; Phoenix-Brice type of light scattering apparatus; centrifuge.

REFERENCES

1. C. Tanford. Physical Chemistry of Macromolecules. Wiley & Sons, Inc., New York, 1961.
2. K. A. Stacey. Light Scattering in Physical Chemistry. Academic Press, New York, 1956.
3. M. B. Huglin. J. Appl. Polymer Sci., **9,** 3963 (1965).

46

DETERMINATION OF COHESIVE ENERGY DENSITIES OF POLYMERS BY TURBIDIMETRIC TITRATION

Many properties of polymer solutions can be described by a virial expression. For example, the osmotic pressure of a polymer solution is concentration dependent, and it can be given by the equation

$$\pi/c = (\pi/c)_0(1 + B^*c + C^*c^2 + \cdots) \tag{1}$$

where π is the osmotic pressure, c is the concentration in grams per cubic centimeter, $(\pi/c)_0$ is the osmotic pressure concentration gradient at infinite dilute solution, and B^* and C^* are the second and third virial coefficients, respectively. At infinite dilution the behavior of the solution should be close to ideality and the van't Hoff equation should be applicable.

$$(\pi/c)_0 = \frac{RT}{M} \tag{2}$$

where R is the gas constant, T is the absolute temperature, and M is the molecular weight of the polymer.

Since most dilute polymer solutions yield a straight line when π/c is plotted against concentration, the third virial coefficient is usually negligible.

The second virial coefficient, B^*, can be either positive or negative or, in some special cases, zero. Obviously, if we were dealing with an ideal solution and the van't Hoff law would be applicable, the second and third virial coefficients would be zero.

Deviation from this ideality is explained on the basis of interaction between polymer and solvent.

425

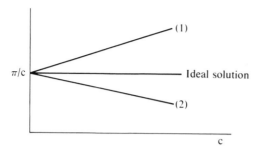

When this interaction is favorable, it is a "good" solvent and the second virial coefficient is positive. In a "bad" solvent the second virial coefficient is negative (Fig. 46–1) and the preferred interaction is between polymer segments rather than between solute and solvent.

There are many theories of the thermodynamics of polymer solutions that account for the behavior of the second virial coefficient. One of the simplest theories, but by no means the most successful, which has served as a basis for variations and modifications, is the Flory-Huggins theory (reference 1). This is based upon a statistical model in which the different polymer segments are fitted into lattices, and the solvent molecules occupy the neighboring lattice sites. In the Flory-Huggins theory equation 1 has the following form:

$$\pi/c = RT/M + RT(\bar{V}_2/V_1)(\tfrac{1}{2} - \chi_1)c + \cdots \tag{3}$$

A comparison of equations 1 and 3 makes the identity

$$B^* = M(\bar{V}_2/V_1)(\tfrac{1}{2} - \chi_1) \tag{4}$$

where $\bar{V}_2$ is the partial molar volume of the polymer, V_1 is the molar volume of the solvent, and χ_1 is an interaction parameter, a free energy term in the thermodynamic sense. As such, the interaction parameter is made up of an enthalpy and an entropy term.

$$\chi_1 = \frac{1}{z} + \frac{V_1(\delta_1 - \delta_3)^2}{RT} \tag{5}$$

where z is the coordination number, V_1 is the molar volume of the solvent, and δ_1 and δ_3 are solubility parameters.

In the following discussion we shall focus our attention on the enthalpy part of the interaction parameter, namely, the second term on the right side of equation 5.

Since we will deal with condensed phases, for all practical purposes, the enthalpy and the internal energy of the process will be the same. Hildebrand and Scott, as well as Scatchard, represent the internal energy of mixing of a binary mixture of small molecules by

$$\Delta E_M = (X_1 V_1 + X_2 V_2)(C_{11} + C_{22} - 2C_{12})\varphi_1\varphi_2 \tag{6}$$

where X_1 and X_2 are the mole fractions, φ_1 and φ_2 are the volume fractions, and V_1 and V_2 are the molar volumes of components 1 and 2, respectively. The C_{11} and C_{22} are the cohesive energy densities of pure compound 1 and 2. They represent the interaction between like molecules of compound 1 or of compound 2. They are given in terms of their energy of vaporization since the transfer from liquid to vapor phase is exactly the process that has to overcome such forces of interaction. The cohesive energy densities, then, are given by

$$C_{11} = \frac{\Delta E_{\text{vap1}}}{V_1} \qquad C_{22} = \frac{\Delta E_{\text{vap2}}}{V_2} \tag{7}$$

where ΔE_{vap1} and ΔE_{vap2} are the internal energies of vaporization of components 1 and 2, respectively. According to Scatchard's suggestion, the interaction between molecules of components 1 and 2 may be taken as the geometric mean of the interactions between molecules of component 1 and between molecules of component 2. Therefore,

$$C_{12} = (C_{11}C_{22})^{1/2} \tag{8}$$

With this assumption the energy of mixing in equation 6 can be rewritten

$$\Delta E_M = (X_1 V_1 + X_2 V_2) \left[\left(\frac{\Delta E_{\text{vap1}}}{V_1} \right)^{1/2} - \left(\frac{\Delta E_{\text{vap2}}}{V_2} \right)^{1/2} \right]^2 \varphi_1 \varphi_2 \tag{9}$$

The cohesive energy densities, i.e., the energy of vaporization of a pure compound per cubic centimeter, appear in this equation raised to the half power and since this notation occurs so frequently in theories of solutions, Hildebrand proposed to call it a solubility parameter.

$$\left(\frac{\Delta E_{\text{vap}}}{V} \right)^{1/2} = \delta \tag{10}$$

These are the same solubility parameters as in equation 5.

Although the solubility parameter as related to the energy of vaporization is an important concept and easily obtainable for small molecules, such is not the case with polymers.

The solubility parameter of the polymer, δ_3, cannot be obtained from its energy of vaporization because polymers decompose long before they reach a measurable state of vaporization.

The Flory-Huggins parameter, χ_1 (as in equation 5), obtained from osmotic pressure measurements could be used to evaluate the δ_3. However, there is very little certainty about the coordination number, z, that should be used for polymer solutions. (This can vary from 4 to 8, and so forth.)

An alternative way to evaluate solubility parameters and, consequently, cohesive energy densities of a polymer is to use ternary systems: polymer, solvent, and non-solvent, the latter being a compound with small molecular weight and with a known solubility parameter, δ_2, which does not solvate the polymer molecules.

The nonsolvent should also have approximately the same molecular size as the solvent so that the coordination number of both can be taken as being identical.

Inspecting equations 4 and 5 we see that when χ_1 becomes greater than $\frac{1}{2}$ the second virial coefficient is negative. We can go accordingly from a "good" solvent to a "bad" solvent by changing the solubility parameter, δ_1 in equation 5. This can be accomplished by adding a nonsolvent component to the mixture.

The physical significance of proceeding from a "good" solvent to a "bad" solvent is that by addition of the nonsolvent we slowly approach a phase separation, i.e., precipitation of the polymer from its solution. Just before such precipitation occurs the interaction between polymer molecules will be increasingly favorable and polymer aggregates of large colloidal sizes will form. Large colloidal particles scatter light very strongly and the solution becomes turbid. In essence, we can reach such a state (cloud point) by titrating a polymer solution with a nonsolvent.

At the cloud point the interaction parameter, χ_c, should be the same regardless of whether we use a nonsolvent of low (l) or high (h) solubility parameter (δ_l or δ_h). Therefore, for the cloud point, equation 5 can be written as

$$\chi_{cl} = \left(\frac{1}{z}\right)_{cl} + \frac{V_{ml}(\delta_3 - \delta_{ml})_{cl}^2}{RT} \tag{11}$$

$$\chi_{ch} = \left(\frac{1}{z}\right)_{ch} + \frac{V_{mh}(\delta_{mh} - \delta_3)_{ch}^2}{RT} \tag{12}$$

In these equations χ_c represents the Flory-Huggins interaction parameter at the cloud point. Subscripts l and h refer to nonsolvents with low and high solubility parameters, respectively. Therefore, χ_{cl} is the interaction parameter at the cloud point reached by the addition of a nonsolvent of low solubility parameter, δ_l. The parameter χ_{ch} is the same interaction parameter at the cloud point reached by the addition of a nonsolvent of high solubility parameter, δ_h. The V_{ml} and V_{mh} are the molar volumes of the solvent and nonsolvent mixture containing nonsolvent of low and high solubility parameters, respectively. The δ_{ml} and δ_{mh} are the corresponding solubility parameters of the mixtures.

We assume that in dilute solutions the coordination number z is the same regardless of the nature of the nonsolvent. The two interaction parameters at the cloud point are equal.

$$\chi_{cl} = \chi_{ch} \tag{13}$$

Therefore, the combined equations 11 and 12 will yield

$$(V_{ml})^{1/2}(\delta_3 - \delta_{ml})_{cl} = (V_{mh})^{1/2}(\delta_{mh} - \delta_3)_{ch} \tag{14}$$

In this equation the molar volumes of the mixtures are given by

$$V_m = V_1 V_2/(\varphi_1 V_2 + \varphi_2 V_1) \tag{15}$$

where V_1 and V_2 represent the molar volume and φ_1 and φ_2 the volume fraction of the solvent and the nonsolvent, respectively.

The solubility parameter of the mixture is obtained from

$$\delta_m = \delta_1 \varphi_1 + \delta_2 \varphi_2 \tag{16}$$

where δ_1 and δ_2 represent the solubility parameter of the solvent and the nonsolvent, respectively.

In equation 14 all the quantities can be calculated from experimental data using equations 15 and 16; hence, equation 14 can be solved for δ_3

$$\delta_3 = [\delta_{ml}(V_{ml})^{1/2} + \delta_{mh}(V_{mh})^{1/2}]/[(V_{ml})^{1/2} + (V_{mh})^{1/2}] \tag{17}$$

If the assumptions made in the foregoing derivation are correct, one would obtain a unique value of δ_3 for a polymer, regardless of the nature of the solvent and non-solvent. However, this is not always the case; therefore, a plot of δ_1, the solubility parameter of the solvent, against δ_3, the solubility parameter of the polymer, obtained from cloud point titrations is made. Usually, the deviation from a constant δ_3 value is small and the points fall on a straight line with a very large positive slope.

The knowledge of the solubility parameter, δ_3, of a polymer is important because with it one can predict the thermodynamic behavior of a polymer in a solvent that has a known solubility parameter. Equation 5 can be used for such prediction.

Experimental

Prepare 100 ml. each of 0.3 percent polystyrene solutions each in isopropylbenzene, m-xylene, toluene, benzene, chlorobenzene, and bromobenzene. Reagent grade solvents should be used. (If available, two polystyrene samples (atactic and syndiotactic) should be used. The solubility parameters of the two should be compared. If the two types are not available, any polystyrene sample in the form of powder or shaving will be sufficient. The solid polymer should be finely dispersed in order to facilitate the solvation.)

Two 25-ml. aliquots of each solution of polystyrene in the different solvents should be titrated with n-hexane to the cloud point. The standard turbidity (cloud point) is the threshold of precipitation at which a printed page becomes blurred when looking through the solution. Therefore the titration should be carried out so that the Erlenmeyer flask rests on a white printed page. Note the amount of n-hexane needed to reach the cloud point. Repeat titrations in which the agreement between the two titration values in the same solvent is not good. Note the room temperature. Solubility parameters are slightly dependent on temperature. However, the parameters given in Table 46–1 can be used for experiments performed at a temperature of 23 to 26° C. Thus, the samples do not have to be thermostated.

Using acetone, titrate two separate 10-ml. aliquots of each polystyrene solution to the cloud point.

From the titration data thus obtained and from the solubility parameters given in Table 46–1, calculate the solubility parameters of the polystyrene samples by using equations 15, 16, 17. The titration data for hexane should have the l subscript, and that for acetone should have the h subscript.

TABLE 46–I. Solubility Parameters of Solvents and Nonsolvents at 25° C

	SOLVENTS					
	ISOPROPYL-BENZENE	m-XYLENE	TOLUENE	BENZENE	CHLOROBENZENE	BROMOBENZENE
δ_1	8.53	8.80	8.90	9.15	9.50	10.0
V_1	140.2	123.5	106.9	89.4	102.3	105.5

	NONSOLVENTS	
	n-HEXANE	ACETONE
δ_2	7.29	9.81
V_2	131.6	74.0

Plot the δ_3 values obtained in different solvents against the δ_1 values.

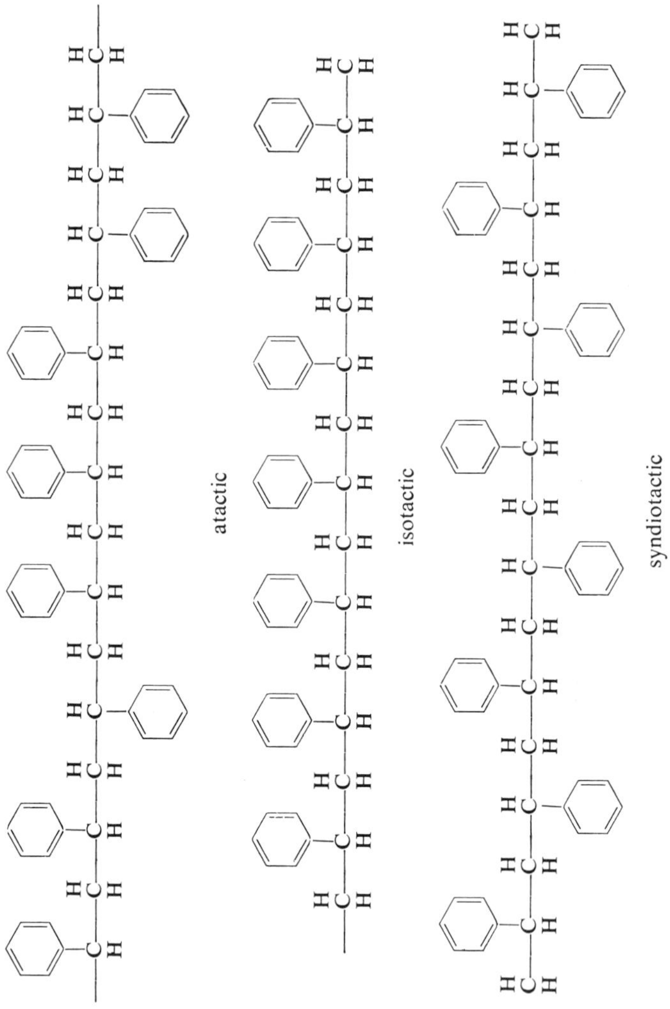

Figure 46-2 *The tacticity of polysytrene samples.*

Is δ_3 of the polystyrene a constant in the different solvents? If not, why does it increase or decrease with increasing δ_1 values?

The difference between the atactic and syndiotactic polystyrene is shown in Figure 46–2.

Explain the differences found in the δ_3 values for the two polystyrene samples on the basis of their tacticity.

Material and Equipment. 5 grams of atactic and 5 grams of syndiotactic polystyrene; 100 ml. each of isopropylbenzene, *m*-xylene, toluene, benzene, chlorobenzene, and bromobenzene; 1 liter of acetone; 500 ml. of hexane; 15 Erlenmeyer flasks of 100-ml. capacity and 15 Erlenmeyer flasks of 250-ml. capacity; two burets of 50-ml. capacity; six pipets of 25-ml. capacity and six pipets of 10-ml. capacity; thermometer.

REFERENCES

1. P. Flory. Principles of Polymer Chemistry. Cornell University Press, Ithaca, N.Y., 1953, Chapters 12 and 13.
2. J. H. Hildebrand and R. L. Scott. The Solubility of Nonelectrolytes. 3rd Edition, Reinhold Publishing Corp., New York, 1950.
3. G. Scatchard. Chem. Rev., **8**, 321 (1931).
4. K. W. Suh and D. H. Clarke. J. Polymer Sci. A1, **5**, 1671 (1967).

47

HELIX-COIL TRANSITION AS DETERMINED BY POLARIMETRY AND VISCOSITY MEASUREMENTS

Investigations of synthetic polypeptides of high molecular weights by x-ray diffraction, infrared dichroism, and similar techniques have proved that some poly-α-amino acids, such as poly-γ-benzyl-L-glutamate, are helical in the solid state. This helix very much resembles the α helix first proposed by Pauling and Corey (Fig. 47–1).

The helical structure is the result of intramolecular hydrogen bonding between carbonyl and amide groups of the peptide backbone. Every amide and carbonyl group is hydrogen bonded; the carbonyl oxygens are hydrogen bonded to the amide nitrogen, three residues away in the peptide chain (Fig. 47–2). The resulting α-helix makes a complete turn at every 3.6 residues. Each turn represents a translation of 5.4 Å along the long axis of the helix, i.e., 1.5 Å per residue.

The stability of the helical structure is such that polypeptide molecules in weakly interacting solvents preserve this structure even though the individual molecules are completely solvated.

On the other hand, in strongly interacting solvents the hydrogen bonded structure is broken up by the solvent molecules and the shape of the polypeptide chain is a random coil.

The transition between these two structures (Fig. 47–3) is of primary interest because it can relate to the all-important life processes, namely, the denaturation of proteins.

Among the many physical chemical properties, optical rotation and viscosity are the two best suited for study of transition from a helix to a coil or vice versa.

432

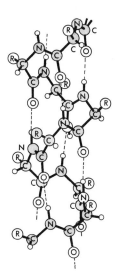

Figure 47–1 *The α-helix of a polypeptide chain.*

The reasons for the observed change in these properties during such transition will be elaborated later.

Taking poly-γ-benzyl-L-glutamate as a well known α-helical polypeptide in the solid state, helix-coil transition can be caused by varying the composition of solvent in a mixed solvent system: chloroform–dichloroacetic acid (DCA).

In such a system the helix-coil transition can be best described by following the ideas of Schellman. A mixed solvent is made of two components—a weakly interacting component, which shall be called solvent, S, and a strongly interacting component, D, which will be bound to the polypeptide chain. In the system used in this experiment, chloroform is the solvent, and dichloroacetic acid is the strongly associated component. (The symbol D is used by Schellman to indicate that such strongly associated components *de*nature many proteins, i.e., make them biologically inactive by unfolding the helical or tightly hydrogen bonded parts of proteins into random coils.)

The helix-coil transition is visualized as occurring in three steps: (1) The molecule in the helical state in the mixed solvent is transported into pure solvent, S. The free energy associated with this process is ΔG_1. (2) The molecule undergoes a helix-coil transition in pure solvent. The free energy associated with this unfolding transition process is $\Delta G_t(S)$. (3) The random coil polypeptide is transported back from pure solvent into the mixed solvent; ΔG_3 is associated with this process.

The difference between ΔG_1 and ΔG_3 is the difference between the interactions of the mixed solvent with the helical and coiled forms.

The total free energy change, ΔG_t, can be written in the general form

$$\Delta G_t = \Delta G_t(S) - \sum_i \nu_i RT \ln\left[1 + K_i(D)\right] \qquad (1)$$

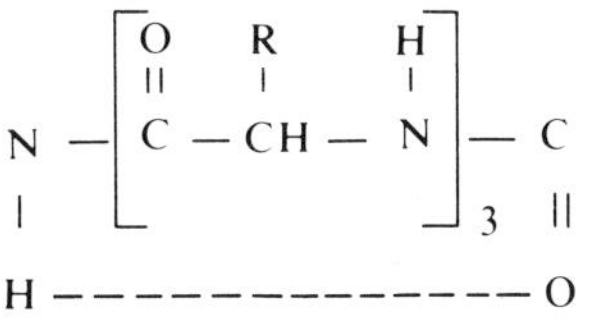

Figure 47–2 *Hydrogen bonding (– – –) in poly-γ-benzyl-L-glutamate.*

$$N - \begin{bmatrix} \overset{\displaystyle O}{\underset{\displaystyle \|}{}} & \overset{\displaystyle R}{\underset{\displaystyle |}{}} & \overset{\displaystyle H}{\underset{\displaystyle |}{}} \\ C - CH - N \end{bmatrix}_3 - C$$

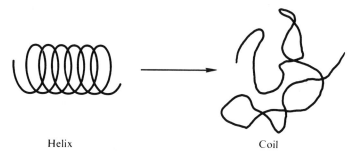

Helix   Coil

Figure 47–3 *Helix-coil transition.*

where *i* refers to the type of site brought in contact with component, *D*, by the helix-coil transition; v_i is the number of sites of type *i*; and K_i is the association constant between *D* and site *i*. The summation term on the right-hand side of equation 1 contains the ΔG_1 and ΔG_3 terms, referring to steps 1 and 3 in the scheme just described.

Equation 1 is quite complicated and one can do very little with it because the number and the type of all binding sites are not known. Moreover, Schellman's treatment is useful for dilute solutions only because all the ΔG terms are independent of concentrations; therefore, no quantitative interpretation would be forthcoming in moderately concentrated solutions even if all the binding sites were known.

Since only the prediction of the qualitative trend is being attempted, the assumption can be made that of all the sites of type *i* the only important ones are those in the peptide bond. Hence, only carbonyl and amide groups will be exposed to *D* by transition from helix to coil. Equation 1 can be rewritten as

$$\Delta G_t = \Delta G_t(S) - \bar{v}RT \ln [1 + \bar{K}(D)] \qquad (2)$$

The $\bar{v}$ is the effective number of combining sites, and for poly-γ-benzyl-L-glutamic acid it can be taken as $2n$, *n* being the number of peptide links in the polymer chain (degree of polymerization). The *K* is the effective associated or binding constant and can be taken as a reasonable average between the association constant of dichloroacetic acid and the carbonyl group and of dichloroacetic acid and the amide group. The *K* can be estimated from solubilities of model compounds in the following manner (see Bettelheim and Senatore):

The free energy of transfer of γ-benzyl-L-glutamic acid from an organic mixture or from chloroform into dichloroacetic acid can be calculated from

$$\Delta G_4 = RT \ln (N_{\text{mix}}/N_{\text{DCA}}) \qquad (3)$$

where *N* represents the solubility of the monomer amino acid derivative in the mixture and in dichloroacetic acid, respectively.

This ΔG_4 refers to the transfer of the whole glutamate molecule. In our simplified model we were looking for interactions at two sites only, namely, at the carbonyl and amide groups. The transfer of the total molecule can be represented as the sum of the parts

$$\Delta G_4 = \sum \Delta g_r \qquad (4)$$

where Δg_r represents free energy contributions from the different residual moieties of the molecule.

The Δg_r could be determined from the solubilities of model compounds. Since we are interested in the hydrogen bonding of dichloroacetic acid at the carbonyl and amide sites, these two groups may be replaced by a component that does not exhibit hydrogen bonding, such as CH_2, and the solubilities of compounds such as

$$\text{NH}_2-\overset{\displaystyle \text{H}}{\underset{\displaystyle \text{HCOH}}{\text{C}}}-\text{CH}_2-\text{CH}_2-\text{COO}-\text{C}_6\text{H}_5 \qquad \text{I}$$

$$\text{CH}_2-\overset{\displaystyle \text{H}}{\underset{\displaystyle \text{COOH}}{\text{C}}}-\text{CH}_2-\text{CH}_2-\text{COO}-\text{C}_6\text{H}_5 \qquad \text{II}$$

in chloroform or mixed solvents may be compared to those in dichloroacetic acid.

Thus, the Δg_r for the carbonyl contribution can be taken as the difference between the values for ΔG_4 of γ-benzyl-L-glutamate and of compound I for transfer between the same two systems (i.e., chloroform and dichloroacetic acid), all measurements being made at the same temperature.

Similarly, Δg_r for the amide contribution may be *half the difference* between the values of ΔG_4 for γ-benzyl-L-glutamate and compound II.

Thus, K values can be calculated from these Δg_r values and the average of K_{NH} and K_{CO} can be used in equation 2. Once this is done, the difference between the free energy of helix-coil transitions in a mixed and in a pure solvent may be obtained from equation 2.

Two models have been used frequently in explaining the helix-coil transition in mixed solvents. In the first one it is assumed that at any composition of mixed solvents a fraction of the macromolecules are helices and another fraction are random coils. This assumption makes the statistical description quite easy, although it is over-simplified. The second way to look upon a helix-coil transition is that at any composition of the mixed solvent, part of a molecule is a helix and part of the same molecule is a random coil.

Two statistical parameters govern what fraction of the molecule is helical (see Peller, and Zimm and Bragg). Parameter S is an equilibrium constant for hydrogen bonding a segment to the polypeptide chain already in helical form, and the second parameter, σ, is a factor relating to the initiation of a helix. Polypeptides of any molecular weight with low S values have random coils. Molecules with high S values tend to have helices, but this can come into play only above a certain critical length of the polypeptide chain, i.e., above the critical degree of polymerization. Below such critical degree of polymerization all molecules have random coils. Molecules of high molecular weight exist as helices with occasional random sections and disorders at the two ends of the polypeptide chain.

To detect the helix-coil transition, optical rotation is frequently used. In general, the optical rotation of a substance refers to a process in which the *change* occurs in the *direction of vibration* of linearly polarized light. (The directions of vibrations are

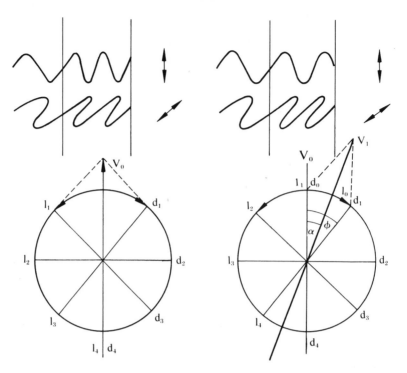

Figure 47-4 *Recombination of dextro- and levorotary components in A, an optically nonactive, and B, an optically active medium.*

perpendicular to the direction of propagation.) Linearly polarized light can be visualized as the resultant of two circular components—rotation to the right (dextrorotation) and to the left (levorotation) designated as d and l components. These components can interfere with each other. In an optically nonactive substance, the velocities of the *d* and *l* components are the same; in an optically active substance they differ. The relative velocities of electromagnetic radiation in different media are related to the refractive indices.

$$v_1/v_2 = n_2/n_1 \qquad (5)$$

An optically active substance can be considered a medium that has different refractive indices in the different directions. When a linearly polarized light passes through an optically active medium, the planar projection of the two circular components reaches a phase difference, the velocity of one component (either *d* or *l*) being greater than that of the other. The recombination of these components causes a linear vibration that is not in the same direction as the original vibration, but is rotated at a certain angle.

The terms d_1, l_1, d_2, l_2, and so forth, are the positions of the two components in the plane of recombination at the same instant of time.

The angle of rotation, α, usually refers to an optical path of 1 decimeter (10 cm.) and is expressed as

$$\alpha = 0.5\phi = (10\pi l/\lambda_0)(n_l - n_d)$$

or
$$\qquad (6)$$

$$\alpha^\circ = (1800 l/\lambda_0)(n_l - n_d)$$

where l is the length of the optical path in the medium in decimeters, λ_0 is the wave-length of the electromagnetic radiation in vacuum in centimeters, and $(n_l - n_d)$ is the difference in refractive indices.

Although the physical aspect of optical rotation is relatively easily explained as the difference in refractive indices one encounters difficulties in explaining how different molecules give rise to such anisotropic behavior, i.e., difference in refractive indices.

The usual explanation given in introductory organic chemistry (that an asymmetric center or centers are required for optical rotation) is insufficient. The Pasteur principle, which states that a molecule and its mirror image should not be superimposable, means that the molecule cannot have a center or plane of symmetry, although it may have an axis of symmetry. How this lack of molecular symmetry results in optical activity is explained on the basis of the classical theory of Born and Kuhn. According to their theory, the molecule is made up of a number of linear oscillators with fixed mutual orientation. The oscillators are coupled if their orientation is not 0 or 180 degrees, and maximal coupling is achieved if their orientation is 90 degrees. Electric coupling between two oscillators means that if a dipole is induced in one and and it vibrates with a certain frequency, this will produce a vibration in the second oscillator with the same frequency, although in a different direction. Optical rotation results when a set of coupled oscillators does not have a plane or center of symmetry.

Two other theories of optical rotation explain the phenomenon on the basis of quantum mechanics. Only a qualitative description of these theories will be given here. In the one electron theory of Eyring, optical activity is considered to be the result of the transition of a single electron from its ground state to its excited state. The electron undergoing this transition is localized in one part of the molecule, the chromophoric group. The transition can be $\pi - \pi^*$, $n - \pi^*$, and so forth. The optical rotatory power is related to matrix elements of the magnetic and electric dipole moments. The average magnetic and electric dipole moments are calculated from the perturbed wave functions and averaged over all molecular orientations.

$$\bar{m} = \int \Psi'^* m \Psi' d\tau$$
$$\bar{\mu} = \int \Psi'^* \mu \Psi' d\tau$$

(7)

The perturbed wave functions in turn are calculated from unperturbed electronic wave functions by first order perturbation techniques. When the electron (and the electronic transition) is in its local atomic field, no rotatory power exists (unperturbed electronic wave). A dissymmetric field is superimposed upon this chromophoric electron by neighboring group effects in an asymmetric molecule. These group effects may be permanent or induced dipole effects of neighboring groups, electrical field effects of ions, or overlapping of the electronic clouds of the chromophoric group with neighboring groups. Thus, the perturbed wave functions are calculated as a result of such interaction. The one electron theory of Eyring considers interactions of the electron and the neighboring groups one at a time.

The second theory, that of Kirkwood and coworkers, is a quantum mechanical formulation of the Born-Kuhn theory. The asymmetric molecule is considered as a sum of N groups, each of which possesses cylindrical symmetry. The electric and magnetic dipoles of the molecule are expressed as the sum of the contributions from the N groups. The optical activity is considered largely to be the result of an inter-action of the electric dipole of one group with that of another. In a helical molecule composed of N amide residues, which give a broad $\pi - \pi^*$ absorption band, the quasi-symmetric coupling of the individual amide chromophore excitations gives a transition with a moment directed along the axis of the helix. Antisymmetric coupling, on the other hand, produces two degenerate transition moments perpendicular to the axis of the helix.

Thus, in the α helix there will be an additional effect of the coupling of the chromo-phores over and above the intrinsic polarization due to the individual chromophores. This is demonstrated in a positive rotatory power, whereas the random coil has negative rotation.

The electric vector of the linearly polarized incident beam induces dipole moments in each of the N molecular groups. These induced dipole moments oscillate; hence they are the source of secondary radiation that is also linearly polarized but not in phase with the incident radiation. The resulting radiation propagated in the medium is the superimposition of the secondary radiation upon the incident beam. The phase difference between the primary beam and the secondary radiation leads to the rotation of the plane of polarization (see Fig. 47–4). Since the polarizability theory neglects everything but electric dipole-dipole interactions, it cannot be applied to molecules with large magnetic moments.

The optical rotation is measured using a polarimeter. Since optical rotation is dependent on both temperature and wavelength, one works with monochromatic radiation at a specified temperature. The monochromatic light is usually provided by a sodium lamp (D line). The light passes through a polarizer and enters the pol-arimeter tube as linearly polarized light (Fig. 47–5).

The polarimeter tube is usually 20 cm. long. (Its length must be ascertained for accurate calculations.) The light beam emerging from the polarimeter tube is still linearly polarized, but the plane of polarization has been changed (Fig. 47–4). To null the instrument, an analyzer or compensator is employed, the rotation of which indicates the angle of rotation.

A combination of polarizer-analyzer as a set of Nicol prisms is given in Figure 47–6.

One can see that as the second Nicol prism is rotated 90 degrees, the optical axis changes from a perpendicular (crossed) orientation to a parallel orientation.

Instead of an analyzer, a compensator is used in many polarimeters. These are

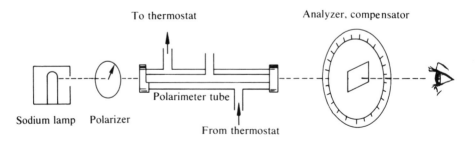

Figure 47–5 *Schematic diagram of a visual polarimeter.*

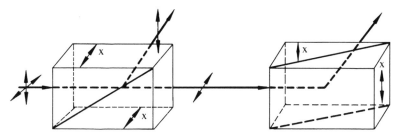

Figure 47–6 *As shown here, with crossed Nicol prisms, ⊘ ◺, no light passes through the analyzer; with parallel Nicol prisms, ⊘ ⊘, maximum light passes through them. The ↔ x shows the optical axis of the Nicol prism.*

made of quartz wedges that have different rotations, depending upon the thickness of the quartz. Therefore, by lateral positioning of the quartz wedges, the same amount of rotation of the sample is produced but in the opposite direction. Such a compensator assembly is given in Figure 47–7.

Since the human eye is more sensitive to differences in light intensities than to measurements of absolute intensities, most visual polarimeters are constructed on the half-shade field principle. Under this condition the intensities of two half-shades are matched at the zero position rather than by setting perpendicular (crossed) prisms to minimum transmittance.

The Nakamura half-shade arrangement is one of the many employed in polarimeters, and it is illustrated in Figure 47–8.

When the compensating quartz wedge is turned at an angle of 2ϵ, the fields of the two half-shades are matched, as in Figure 47–9b.

The compensator or analyzer is usually fixed in a graduated circle in which the outer circle is marked in degrees and decimal fractions of degrees, and the inner circle is rotated to compensate at zero position. A vernier is also fitted for precision work, and most commercial instruments can be read to 0.01 degree. These graduated circles are equipped with a handle for rough adjustment to a zero position and also with some device (such as a micrometer screw) for fine adjustment.

The optical rotation of the polarimeter, α, for a concentration of c, grams/100 ml. of solution, gives the specific rotation $[\alpha]_D^{25}$ with sodium D line light source at 25° C. according to the following equation:

$$[\alpha]_D^{25} = 100\alpha/lc \tag{8}$$

Movable wedges

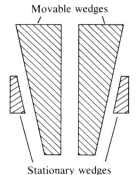

Figure 47–7 *The Schmidt-Haensch quartz wedge compensator*   Stationary wedges

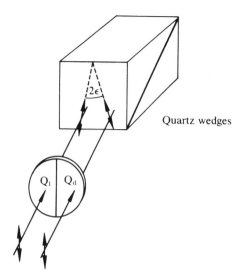

Figure 47–8 *Nakamura half-shade. The Q_l and Q_d are thin plates of quartz rotating the planes of polarization left or right by the same amount. The half-shade angle 2ϵ is twice the amount of rotation of either plate.*

where *l* is the length of the polarimeter tube. The specific rotation, therefore, is the rotation exhibited by 1 gram of an optically active substance in 1 ml. of solution having an optical path length of 10 cm.

The corresponding molecular rotation is given by equation 9.

$$[M] = M\alpha/lc \tag{9}$$

The second technique to which helix-coil transition is sensitive is viscosity measurements. According to the Standinger-Mark-Houwink equation

$$[\eta] = KM^a \tag{10}$$

The intrinsic viscosity of a polymer, $[\eta]$, is proportional to the molecular weight, M, and exponentially related to the shape of the molecule, *a*. The proportionality constant K is a function of solute-solvent interaction at a set temperature. The helix-coil transition in mixed solvents involves a change in the shape factor, *a*, the molecular weight being constant. The change in the composition of the mixed solvent affects the proportionality constant, K, and, therefore, the intrinsic viscosity is not the function of the shape factor alone. However, since the latter changes from a large number (2 to 5, depending on the length of the stiff helical rod) to 0.75 for random coil in the exponent, this will be the dominating factor and small changes in K values will be overridden by this fact.

Therefore, it is expected that as a helix-coil transition takes place a sudden decrease in the intrinsic viscosity and also in (η_{sp}/c) of the system occurs. The intrinsic

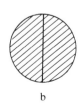

a   b   c

Figure 47–9 *Half-shade fields at rotations of: (a) less than 2ϵ, (b) an amount equal to 2ϵ, and (c) more than 2ϵ.*

viscosity is the limiting value of (η_{sp}/c), i.e., extrapolated to zero concentration. Since we are interested only in the qualitative information (at what solvent composition such transitions occur) there is no need to perform viscosity measurements at various concentrations.

Experimental

Obtain poly-γ-benzyl-L-glutamate of known degree of polymerization from commercial biochemical suppliers. A molecular weight of 1 to 3×10^5 is suitable. If the molecular weight is not known the instructor should determine it by light scattering or other techniques and provide the information. (The solubility depends on the molecular weight.)

Prepare 150 ml. of a solution of about 0.8 percent poly-γ-benzyl-L-glutamate in dichloroacetic acid and 50 ml. of 0.8 percent solution in chloroform.

CAUTION: Be very careful with dichloroacetic acid because it is extremely corrosive.

Measure the optical rotatory power of the two solutions in a polarimeter tube. Also obtain the polarimetric readings on the two solvents.

From equation 8 calculate the specific rotation $[\alpha]_D^{25}$.

In a capillary viscometer measure the efflux time of the two solvents and the two solutions using a total of 5 ml. of fluid in each case. Calculate the relative viscosity of the solution from the efflux times, t.

$$\eta_{rel} = \frac{t_{soln}}{t_{solvent}} \qquad (11)$$

In equation 11 we assumed as a first approximation that the densities of the solvent and the solutions are nearly the same. Calculate the corresponding (η_{sp}/c) where the specific viscosity $\eta_{sp} = \eta_{rel} - 1$ and c is the concentration in grams/100 ml.

From the two solutions make three mixtures, the solvent composition of which should be 65, 70, and 75 percent v/v dichloroacetic acid.

Determine the optical rotation and specific viscosity of these solutions as before using the mixed solvents as blank.

Plot both $[\alpha]_D$ and η_{sp}/c against solvent composition.

Do the two measurements indicate the helix-coil transition at the same solvent mixture composition?

Do the phenomena observed indicate an instant or a gradual transition?

What additional measurements would be necessary to obtain the $\Delta G_{unfolding}$ for the helix-coil transition?

Material and Equipment. 2 grams of poly-γ-benzyl-L-glutamate; 300 ml. of dichloroacetic acid; 100 ml. of chloroform; polarimeter; viscometer; thermostat bath with heater, stirrer, and thermoregulator; circulating pump.

REFERENCES

1. J. A. Schellman. J. Am. Chem. Soc., **62,** 1485 (1958).
2. F. A. Bettelheim and P. Senatore. J. Chim. Phys., **61,** 105 (1964).
3. L. Peller. J. Phys. Chem., **63,** 1194 (1959).
4. B. H. Zimm and J. K. Bragg. J. Chem. Phys., **31,** 526 (1959).
5. P. Urnes and P. Doty. Adv. Protein Chem., **16,** 402 (1961).
6. E. Katchalsky and I. Z. Steinberg. Ann. Rev. Phys. Chem., **12,** 433 (1961).

48

FIBER PATTERN IN X-RAY DIFFRACTION; ORIENTATION OF CRYSTALLITES IN PLASTICS

Many macromolecular compounds that in the solid state have the properties of plastics can be looked upon as two-phase systems. There are certain domains in the plastic material in which portions of adjacent macromolecules align and form a three-dimensional geometric array. These portions form the crystalline domain of plastic material.

Other portions of the macromolecules are in random entanglement with neighboring molecules and they form the amorphous domain. One macromolecule may run through more than one crystalline and one amorphous domain. The size of the individual crystallites usually is approximately 10^2 Å, but they can form superstructures by organizing themselves in certain patterns. The most frequent of the superstructures containing crystalline and amorphous domains are the spherulites commonly observable microscopically when using polarized light.

The mechanical and viscoelastic behavior of plastics is industrially the most important property. The different domains of plastics contribute differently to these properties. For this reason, the degree of crystallinity (i.e., the proportion of crystallites in the total matrix) sometimes has more influence on the mechanical properties than does the primary chemical structure of the molecules. This can be understood considering that the amorphous domains, being liquid-like structures, contribute more to the flow properties (viscoelastic behavior), and the crystalline regions provide more mechanical strength. In molecular terms this results from the large intermolecular forces operating in the crystallites because of the close packing of the chain segments.

Under stresses and strains the amorphous segments stretch, sometimes increasing the degree of crystallinity, and the crystalline regions take up certain preferential orientations. The macromolecules in the crystallites usually align in the direction of stress, causing the crystallites to be oriented. Under such conditions the mechanical properties of a piece of plastic are not the same in all directions and, therefore, the orientation of the crystallite also has an important bearing on the physical properties.

For these various reasons it is important to obtain information on the structural properties of both the amorphous and crystalline regions in order to relate these properties to mechanical behavior. The x-ray diffraction experiments yield information on the structure and orientation of crystallites only.

Monochromatic x-rays having wavelengths of the same order of magnitude as the interatomic distances can be used in the determination of the three-dimensional structure. The relationship between the diffraction angle and the interatomic distance is given by the Bragg equation

$$n\lambda = 2d \sin \theta \tag{1}$$

where n is an integer number indicating the order of reflection, λ is the monochromatic wavelength of the x-rays, d is the interatomic distance, and θ is the angle of reflection of x-rays by a family of parallel planes in the crystal (Bragg angle).

The setup for an x-ray diffraction pattern with a flat cassette camera is given in Figure 48–1. A is the source of x-rays (for example, copper target), B is a nickel foil monochromatic filter to allow only the copper $K\alpha$ radiation to pass through, C is a collimator pinhole to produce parallel x-rays, D is the sample, and E is the photographic plate. The Bragg angle can be determined from this geometry once a powder diagram is recorded on a photographic plate in the form of concentric circles, since

$$\tan 2\theta = \frac{r}{s} \tag{2}$$

where r is the radius of a concentric circle and s is the distance from the sample to the photographic film. Knowing the Bragg's angle one can evaluate the corresponding interatomic distances from equation 1.

The crystallites in an annealed polymer film are usually randomly oriented (Fig. 48–2a) and, therefore, they have a typical powder pattern. However, upon stretching, orientation occurs (Fig. 48–2b) and, instead of concentric circles, arcs are produced (Fig. 48–3). The better the orientation the smaller the arcs will be; if perfect orientation could be achieved one would get dots instead of arcs from each

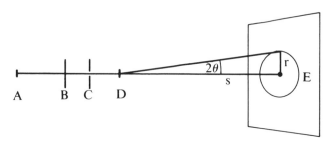

Figure 48–1 *Schematic diagram of x-ray diffraction.*

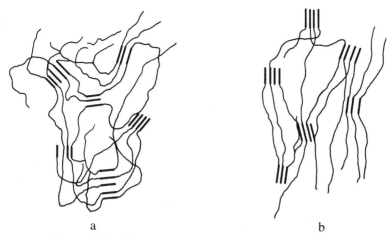

Figure 48-2 *Schematic representation of crystalline orientation.*

reflection plane, i.e., a single crystal pattern. The angular position of an arc corresponding to a certain reflection (*hkl*) indicates the average orientation of the plane in question. At the same time, the half-width of an arc (μ) gives an indication of the dispersion of the orientations about the average value. The coordinate system necessary to relate the x-ray diffraction fiber pattern to the orientation of crystallites is given in Figure 48-4. For simplicity we will restrict our attention to a special condition of the uniaxial orientation of polyethylene crystallites. These crystallites have the orthorhombic symmetry ($a \neq b \neq c$ and all angles are 90 degrees) and the macromolecules run along the *c* crystallite axis. In the special case of uniaxial orientation under observation, the *a* crystalline axis will remain basically perpendicular to the direction of stretch, Z, whereas the *b* axis will tend to align itself perpendicular and the *c* axis parallel to the direction of stretch. More and more crystallite orientation is introduced with the increase in the elongation of the film. Therefore, the degree of orientation can be followed by the evaluation of angle ε or β in Figure 48-4.

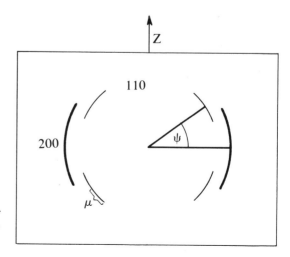

Figure 48-3 *Fiber diagram of stretched polyethylene film. Z is the direction of elongation.*

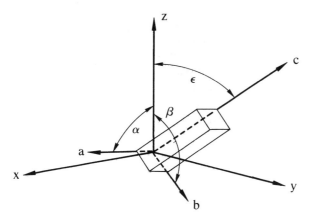

Figure 48-4 *Coordinate system of a polyethylene crystal in uniaxial orientation.*

On an x-ray diffraction fiber pattern of polyethylene films the strongest reflection is the 110 plane (Fig. 48-3). From spherical trigonometry one finds the relationship between Figures 48-3 and 48-4.

$$\cos \theta_{110} \sin \psi_{110} = \sin \Omega \cos \alpha + \cos \Omega \cos \beta \tag{3}$$

where $\Omega = $ arc tan (b/a), θ_{110} is the Bragg angle of the 110 reflection, and ψ_{110} describes its orientation as in Figure 48-3.

Since in the special case under consideration $\cos \alpha = 0$ this reduces to

$$\cos \beta = \frac{\cos \theta_{110} \sin \psi_{110}}{\cos \Omega} \tag{4}$$

Finally, in an orthorhombic unit cell

$$\cos^2 \alpha + \cos^2 \beta + \cos^2 \epsilon = 1 \tag{5}$$

Hence,

$$\cos^2 \beta = 1 - \cos^2 \epsilon \tag{6}$$

$$\cos \beta = \sin \epsilon \tag{7}$$

Therefore, from the observation of the changes of ψ with elongation one can calculate the angle of orientation of the c axis (ϵ) or of the b axis (β) as a function of elongation.

Experimental

Obtain a low density polyethylene film of about 1 mm. thickness. (About 50 to 60 per cent crystallinity in the polyethylene sample is sufficient.) Cut a strip of the film to 5 × 1 cm., and mark it with ink at 5-mm. intervals along the longer axis. Clamp the film in a jig in which it will be stretched (Fig. 48-5).

Percentage elongation is given by the following formula:

$$\text{Percentage elongation} = \frac{L - L_0}{L_0} \times 100 \tag{8}$$

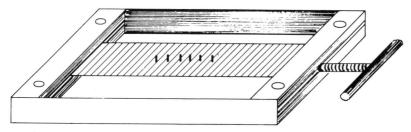

Figure 48–5 *Stretching jig.*

where L_0 is the distance between ink marks before elongation and L is the same distance after elongation.

Stretch the film to about 50 percent elongation. Check the x-ray diffraction apparatus with a fluorescent screen and focus if necessary in order to get the maximum intensity beam through the collimator. About 25 to 30 kilovolts and 15 to 20 milliamperes usually give optimum conditions.

Since exposure to x-rays is harmful, it is *imperative* that the x-ray diffraction unit be operated by or in the presence of the instructor. For that reason no detailed instructions of operating the commercial units will be given here. (The basic principles of generating x-rays were given in Experiment 22.)

Once the x-ray optics are aligned, place the jig containing the polyethylene film in a position so that the collimator will be close to the center of the polyethylene film. Fasten the sample. Fill the flat cassette camera with a no screen medical x-ray film in the darkroom. Place the camera on the tracks to provide a 5-cm. distance between the sample and the photographic plate. Start the x-ray unit and expose the camera to radiation for 1 to $1\frac{1}{2}$ hours. (If the collimator is not properly aligned and less than maximum radiation is obtained, longer exposures of 2 to 3 hours are required. Therefore, proper alignment is of primary importance.)

At the end of the exposure period turn off the x-ray unit. The photographic film is developed for five minutes, washed for 20 seconds, and fixed for ten minutes. During the developing the camera may be loaded with a new photographic film.

Wash the film after fixing and dry it overnight or in an oven.

Repeat the x-ray diffraction following two more elongations (about 100 and 150 per cent).

If a Polaroid attachment is available that works with a fluorescent screen, only two to five minutes of exposure is necessary. Under these conditions the experimenter should stretch films to eight or ten elongation positions ranging between 30 and 300 per cent and take the fiber pattern of each elongation.

The two strongest spacings appearing on the photographic films correspond to the 110 and 200 planes, respectively (see Fig. 48–3). Knowing that the unit cell is orthorhombic, calculate the a and b distances of the unit cell by using equations 1 and 2.

Measure the ψ angle as a function of elongation and calculate from equations 4 and 7 the average orientation of crystallites.

Plot ϵ and β vs. elongation. What is the relationship between molecular orientation and macroscopic extension?

Plot the half-width, μ, of the 110 reflection vs. elongation. Do the average orientation ϵ or β and the dispersion about the average orientation, μ, follow the same pattern with elongation?

Material and Equipment. Polyethylene film 1 mm. thick; stretching jig; basic x-ray diffraction unit with monochromator; collimator and flat cassette camera (or Polaroid attachment); no screen medical x-ray film (or Polaroid film); x-ray film developer and fixer.

REFERENCES

1. W. J. Moore. Physical Chemistry. 3rd Edition, Prentice-Hall, Inc., Englewood Cliffs, N.J., 1962, pp. 776–779.
2. G. M. Barrow. Physical Chemistry. 2nd Edition, McGraw-Hill Book Co., New York, 1966, pp. 811–819.
3. C. Tanford. Physical Chemistry of Macromolecules. Wiley & Sons, Inc., New York, 1961, pp. 37–73.
4. R. S. Stein and F. H. Norris. J. Polymer Sci., **21,** 381 (1956).
5. R. S. Stein. J. Polymer Sci., **31,** 327 (1958).

49

STRAIN BIREFRINGENCE
OF POLYETHYLENE

When certain polymer films, such as polyethylene or nylon, are investigated under a microscope and the film is placed between two crossed polarizers, the texture of the polymer films seems to indicate areas that are dark and other areas that are illuminated. The latter areas are said to show birefringence (another name, used principally in the British literature, is double refraction). Birefringence simply means optical anisotropy, namely, that the polarization of the system under investigation is not the same in every direction. The polarization in visible light depends only on the polarizability of the sample involved and this is related to the refractive index (see Experiment 14).

$$\frac{n^2 - 1}{n^2 + 2} = \frac{4\pi\alpha}{3}$$ (1)

The birefringence, Δ, therefore, is simply defined as the difference in refractive indices along two directions.

$$\Delta = n_1 - n_2$$ (2)

In the film already mentioned the dark areas were optically isotropic; their refractive indices were the same in different directions.

Let us assume that a macroscopic piece of film with thickness d has optical anisotropy, namely, $n_1 > n_2$ (Fig. 49–1). Light is passing through the film, which is polarized 45 degrees to the x axis of the film. It can be visualized that this plane polarized light contains two components—one along the x axis (n_1) and one along the y axis (n_2). When light enters a medium that has a greater refractive index than air, the wavelength of the light in the medium becomes appropriately shorter. The component of the polarized light along the x axis has a shorter wavelength than the component perpendicular to it ($n_1 > n_2$). Hence, the two components of the light emerging from the film are out of phase. If, therefore, the phase difference can be

449

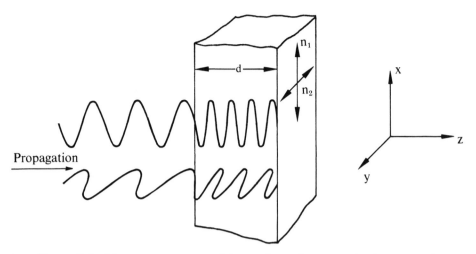

Figure 49–1 *Schematic representation of the retardation of waves in a birefringent sample.*

measured by a compensator, such as a Babinet compensator, or by the rotation of a quarter wave plate, the difference between the refractive indices in the two directions can be calculated.

The refractive index is a measure of the speed of light in the medium compared to that in a vacuum.

$$n = \frac{c}{v} \tag{3}$$

where c is the speed of light in a vacuum and v is the same in the sample.

$$c = \lambda_0 \nu \tag{4}$$

$$v = \lambda \nu \tag{5}$$

Hence,

$$n = \frac{\lambda_0}{\lambda} \tag{6}$$

where λ_0 is the wavelength of light in a vacuum (air) and λ the wavelength in the medium. The birefringence, therefore, is

$$n_1 - n_2 = \Delta = \frac{\lambda_0}{\lambda_1} - \frac{\lambda_0}{\lambda_2} \tag{7}$$

On the other hand, the number of waves in a sample of d thickness along the x axis is

$$N_1 = \frac{d}{\lambda_0} n_1 \tag{8}$$

and along the y axis

$$N_2 = \frac{d}{\lambda_0} n_2 \tag{9}$$

The retardation, N, is the difference between the number of waves passing through the sample along two directions. Hence, birefringence can be calculated from the retardation, N.

$$\Delta = n_1 - n_2 = \frac{N}{d} \lambda_0 \tag{10}$$

For slight retardation a quarter wave plate is sufficient as a compensator; for greater retardations (as much as four or five waves) a Babinet compensator is necessary.

To visualize the birefringence of a polymer network in terms of molecular structure one designs the principal polarizabilities of a bond (or its larger equivalent, the statistical segment, in a random polymer molecule) with α_{01} and α_{02} (Fig. 49–2). The former is taken as the polarizability for the electric vector parallel to the length of the bond (statistical segment) and the latter as perpendicular to it.

In a polymer molecule there are Z statistical segments (bonds) of length a. The end-to-end distance of the molecule is h. The principal polarizability along the vector, h, is α_1, and the one perpendicular to it is α_2. The optical anisotropy of the polymer molecule is represented by the difference, $\alpha_1 - \alpha_2$.

This is related to the bond or segment polarizabilities by

$$\alpha_1 - \alpha_2 = Z(\alpha_{01} - \alpha_{02})\left[\frac{3}{5}\left(\frac{h}{Za}\right)^2 + \cdots\right] \tag{11}$$

For a fully extended polymer molecule

$$h = Za \tag{12}$$

and

$$\alpha_1 - \alpha_2 = Z(\alpha_{01} - \alpha_{02}) \tag{13}$$

which is the sum of the anisotropy of the individual segments.

The more the polymer chain is coiled ($h \ll Za$), the smaller the molecular anisotropy becomes.

As the next step, the case of a single polymer molecule is extended to a molecular network. In such a network a Gaussian distribution is assumed. This means that N polymer chains, each with Z statistical segments and h end-to-end distance, are evenly distributed in the polymer matrix. It is further assumed that the components of chains change in the same ratio as the corresponding dimensions of the bulk material change when it is stretched. A Gaussian distribution of random coils in the polymer matrix would be optically isotropic and, therefore, not birefringent.

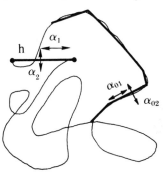

Figure 49–2 *Random coil and its statistical segment.*

Upon stretching the sample, however, optical anisotropy is introduced and with the preceding assumptions this is related to the elongation of the sample, λ, by

$$\frac{n_1^2 - 1}{n_1^2 + 2} - \frac{n_2^2 - 1}{n_2^2 + 2} = \frac{4\pi}{15} N(\alpha_{01} - \alpha_{02})\left(\lambda^2 - \frac{1}{\lambda}\right) \qquad (14)$$

where N is the number of chains per unit volume, n_1 is the refractive index in the direction of elongation, and n_2 is that perpendicular to it. The elongation, λ, is simply the normal strain.

$$\lambda = \frac{\Delta l}{l_0} \qquad (15)$$

where Δl is the change in the length of the sample upon unidirectional elongation and l_0 is the length before elongation.

Since in practice the value of $n_1 - n_2$ is small, equation 14 reduces to

$$\Delta = n_1 - n_2 = \frac{(n^2 + 2)^2}{n} \cdot \frac{2\pi N}{45} (\alpha_{01} - \alpha_{02})\left(\lambda^2 - \frac{1}{\lambda}\right) \qquad (16)$$

where n is the mean refractive index of the sample, which can be obtained using a refractometer.

$$n = (n_1 + 2n_2)/3 \qquad (17)$$

In equation 17 the assumption is stated that the two refractive indices perpendicular to the direction of elongation, n_2, are the same (unidirectional elongation–uniaxial orientation). Equation 16 would predict that from experimentally measurable values of birefringence, Δ, elongation, λ, and mean refractive index, the difference between segment polarizabilities could be calculated.

This is true only for an ideal random network of polymer chains, such as ideal rubber. However, knowing the bond of segment polarizabilities from other measurements (such as refractive indices of single crystals along the different crystal axes), one could compare calculated and experimental birefringence values to see how much an amorphous real structure deviates from ideality.

In the case of plastics the situation is more complex. One can assume that in semicrystalline polymers, such as polyethylene or nylon, the birefringence is the sum of the contribution of both crystalline and amorphous regions. In effect, three terms contribute to the total birefringence observed

$$\Delta_{\text{Total}} = X_c \Delta_c + (1 - X_c) \Delta_a + \Delta_f \qquad (18)$$

where X_c is the weight fraction of the crystalline portion of the polymer, Δ_c is the birefringence of the crystalline region, Δ_a the birefringence of the amorphous region, and Δ_f is the form birefringence. It is reasonable to believe that only the amorphous contribution will approximate the theory derived for ideal rubber. In order to determine the extent to which the behavior of the amorphous part deviates from that of ideal rubber, it is necessary to evaluate the other two terms, Δ_c and Δ_f. This can be done in the following manner:

1. Form birefringence is present when certain anisotropic particles, such as plates and needles, embedded in an isotropic medium cause birefringence that is

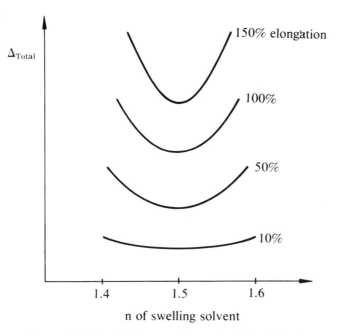

Δ_{Total}

150% elongation

100%

50%

10%

1.4   1.5   1.6

n of swelling solvent

Figure 49-3   *Form birefringence of a sample of oriented platelets.*

dependent on the shape and form of the particle. Wiener predicted the existence of such a phenomenon and one can test for it simply by swelling the polymer film at constant elongations in solvents with different refractive indices (reference 7). If the total birefringence changes with the refractive index of the swelling liquid, form birefringence is an important contributor and Wiener curves, such as those given in Figure 49–3, can be obtained. At the minima of such curves no form birefringence is present because the refractive index of the swelling medium matches the refractive index of the platelets that are perpendicular to their long axes. Therefore, birefringence measurements made in such an imbibing medium yield directly the sum of the crystalline and amorphous contributions. For low density polyethylene films this can be achieved around $n = 1.50$ by swelling the film in benzene (Bunn and de Daubeny).

2. For the contribution of the crystalline portion one can use the orientation functions of polyethylene crystallites obtained in Experiment 48. In the latter experiment x-ray diffraction of oriented polyethylene film yielded an orientation angle, ϵ, as a function of elongation. (See Figure 48–4 in Experiment 48.) The ϵ is the angle between the c axis of the crystallites (along which the polymer chains are running) and the direction of elongation. The angle ϵ can define an orientation function such as f_c.

$$f_c = \tfrac{1}{2}(3\,\overline{\cos^2 \epsilon} - 1) \tag{19}$$

The orientation function takes values from 1 (perfect orientation $\epsilon = 0$; $\cos \epsilon = 1$) to $-\tfrac{1}{2}$ ($\epsilon = 90$; $\cos \epsilon = 0$). For random orientation $\overline{\cos^2 \epsilon} = \tfrac{1}{3}$ and the orientation function is $f_c = 0$.

The crystalline portion of the birefringence for unidirectional elongation has the form

$$\Delta_c = \frac{2}{9} \frac{(n^2 + 2)}{n} (\alpha_{\parallel} - \alpha_{\perp})_c f_c \tag{20}$$

In equation 20 n is the average refractive index (1.50 for polyethylene) and $\alpha_{\parallel}$ and $\alpha_{\perp}$ are the polarizabilities of the crystallites parallel and perpendicular to the c axis. These have been calculated from the principal refractive indices of single crystals of paraffin (Holmes and Palmer). For polyethylene film equation 20 reduces to

$$\Delta_c = 0.0286(3 \overline{\cos^2 \epsilon} - 1) \tag{21}$$

Once the orientation angle as a function of elongation is known one can calculate Δ_c from equation 21. The degree of crystallinity for low density polyethylene is 56 per cent (density of 0.916). From the calculated Δ_c values and the total birefringence values obtained in benzene, Δ_a can be evaluated at different elongations.

If the amorphous material behaved as an ideal rubber, Δ_a would follow the equation derived from the Kuhn-Treloar theory (equation 16). The difference in polarizabilities $(\alpha_{01} - \alpha_{02})$ can be calculated from bond polarizabilities (Holmes and Palmer) and for polyethylene, $-CH_2(\alpha_{01} - \alpha_{02}) = 1.50 \times 10^{-25}$ cm.^{3}. The number of amorphous chains per unit volume has been estimated to be 2×10^{20} per cubic centimeter. Using these data one could plot the experimental and calculated Δ_a values (from equation 16) and compare the behavior.

Experimental

The birefringence apparatus is mounted on an optical bench. It consists of: a light source, usually a mercury lamp; a monochromator filter to obtain the 5461 Å radiation (Wratten No. 77); a polaroid with its axis oriented 45 degrees to the vertical; a sample with a vertical direction of elongation; and either a Babinet compensator for large phase differences (i.e., a thick sample having $d = 10^{-2} - 10^{-1}$ cm. and large elongations of 50, and greater) or a quarter wave plate for the 5461 Å light. (This can be used only with thin samples for which $d = 10^{-3}$ and less—elongations of less than 50 per cent in the case of polyethylene.) In either case the compensator should be set up with the slow axis of the quartz wedge or plate vertical. The apparatus also includes a Nicol prism with its polarizing axis at 45 degrees to the vertical and 90 degrees to that of the polaroid (Fig. 49–4). Differences in retardation between the sample and the blank (sample holder with pure benzene) are measured for the calculation of the birefringence.

Since the polarizer and the analyzer are set at 90 degrees for isotropic media with no birefringence, no light should reach the observer. For any birefringence in the sample the phase difference produced will allow light to pass. The calibrated Babinet compensator is adjusted to produce black

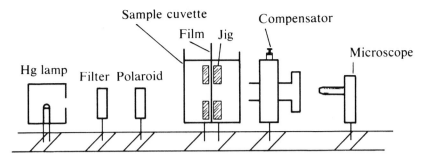

Figure 49-4 *Experimental apparatus.*

fringes and the retardation is read directly from the compensator by centering the two black fringes around the reference line. In the case of a quarter wave plate the analyzer Nicol prism has to be rotated by an angle corresponding to the retardation in order to produce a dark field. Since the combination of the quarter wave plate and the birefringence of the medium produces a difference in phase angle, $\delta/2$ (δ because of the birefringence and $\pi/4$ because of the quarter wave plate).

Therefore the birefringence will be equal to

$$\Delta = \frac{\lambda \, \delta/2}{180d} \tag{22}$$

if $\delta/2$ is obtained in angles.

Thin polyethylene films are stretched in an apparatus as described in Experiment 48, Figure 48-5. If a Babinet compensator is used, six samples

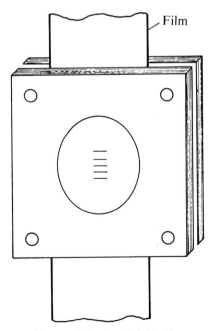

Figure 49-5 *Sample holder jig.*

should be elongated by 0, 50, 70, 100, 150, or 200 per cent (i.e., each sample elongated by a different percentage). If only quarter wave plates are available six elongations between 0 and 50 per cent should be prepared. After stretching each sample, a jig is clamped around the film to maintain its elongation (Fig. 49–5). (Jigs can be easily manufactured from aluminum plates with regular metal screws and nuts at the four corners.) Once the film is secured in the sample holder jig, it can be removed from the stretching apparatus.

The six jigs with the elongated samples in them are placed in benzene at room temperature and allowed to swell for two hours. At the end of swelling the final elongation is checked by measuring the marking. (A slight relaxation may occur because of swelling.) The thickness of the samples is determined with a micrometer.

Each jig is placed subsequently in an optical cuvette containing benzene, and the birefringence is measured. The birefringence of the optical cuvette with benzene is taken as blank. Plot the total birefringence vs. elongation.

By using equations 19, 20, and 21 and the ϵ values obtained in Experiment 48, calculate the crystalline contribution to the birefringence, Δ_c, as a function of elongation.

From the total birefringence obtained and from the calculated Δ_c values using equation 18, calculate the Δ_a values as a function of elongation. (The weight fraction of crystallites in the polyethylene is 0.56, and since the measurements have been made in benzene, $\Delta_f = 0$.)

Calculate the Δ_a values from equation 16 using $(\alpha_{01} - \alpha_{02}) = 1.50 \times 10^{-25}$ cm.^{3} and for $N = 2 \times 10^{20}$ chains per cubic centimeter. This calculated Δ_a value would represent the amorphous part of the polymer network if it behaved as an ideal rubber.

Comparing the values for real Δ_a and the Δ_a for ideal rubber as a function of elongation, what conclusions can you draw regarding the behavior of the amorphous chains in the polymer network?

Material and Equipment. Polyethylene film 1 mm. thick; benzene; stretching jig; six sample holder jigs; micrometer; sample cuvette; optical bench; mercury lamp; monochromator filter (Wratten No. 77); polarizer and analyzer set (two polaroids or Nicol prisms); compensator (Babinet compensator or quarter wave plate set in a rotating and calibrated circle); microscope.

REFERENCES

1. L. R. G. Treloar. In Stuart, H. A. (ed.). Die Physik der Hochpolymeren. Vol. III, Springer Verlag, Berlin, 1956, Chapter V.
2. W. A. Shurchliff. Polarized Light. Harvard University Press, Cambridge, Mass., 1962.

3. H. A. Scheraga and R. Signer. In Weissberger, A. (ed.). Techniques of Organic Chemistry. 3rd Edition, Vol. I, Part III, Interscience Publishers, Inc., New York, 1960, Chapter 35.
4. R. S. Stein and F. H. Norris. J. Polymer Sci., **21,** 381 (1956).
5. D. R. Holmes and R. P. Palmer. J. Polymer Sci., **31,** 345 (1958).
6. C. W. Bunn and R. de Daubeny. Trans. Faraday Soc., **50,** 1173 (1954).
7. F. A. Bettelheim and R. S. Stein. J. Polymer Sci., **27,** 567 (1958).

XIV

ELECTRONICS

DONALD T. OPALECKY
Associate Professor
Department of Chemistry, Adelphi University

Many physical chemical measurements, including electrochemical potentials, photometric data, dielectric constants, and equilibrium constants, are made with the aid of electronic devices. This section presents a few basic concepts of these measuring devices and their components.

Before one can discuss the components involved, it is important to define some of the fundamental electrical units.

Ohm. The unit of resistance to the flow of electrical current is the ohm. The international ohm is the resistance at 0° C. of a uniform column of mercury 106.300 cm. in length and weighing 14.4521 grams. The absolute ohm is now the standard in the United States. It differs only slightly from the international ohm.

Ampere. The unit of flow of electricity is the ampere, corresponding to 6.24×10^{18} electrons per second. Although the absolute ampere is now employed in the United States, the older definition is within 0.01 per cent of the currently accepted value. In other words, the ampere is the unvarying direct current that will produce 1.1180 mg. of silver per second from a silver nitrate solution under specific conditions.

Volt. The unit of electrical potential or pressure is the volt. The volt is the potential drop developed across a resistance of 1 ohm when carrying a current of 1 ampere.

Coulomb. The unit of electrical charge or quantity of electricity is the coulomb. It may be defined as the amount of charge corresponding to 1 ampere flowing for 1 second. It is also equal to the charge on 6.24×10^{18} electrons.

Farad. The unit of the capacitance or the ability to store electric charge is the farad. When a potential source is connected to two conducting electrodes separated by an insulator, charge flows and a potential difference is developed across the plates. The capacitance is the ratio of charge developed on either plate to the potential difference between the plates. If 1 coloumb is placed on the plates and 1 volt develops, this quantity is defined as the farad.

Henry. The unit of inductance or the resistance to change of electrical current in a device is the henry. If one varies the current flowing in a magnetic circuit, a

461

TABLE I. Summary of Electrical Units

Term	Symbol	Unit
Resistance	R	ohm (Ω)
Current	I	ampere (A)
Potential	E	volt (V)
Charge	Q	coulomb (Q)
Capacitance	C	farad (F)
Inductance	L	henry (H)

voltage variation will be produced in another part of the circuit. When a counter potential of 1 volt is produced by a current changing at the rate of 1 ampere per second, the inductance is said to have a value of 1 henry.

The basic relationship of the first three terms in Table 1 is given by Ohm's law.

$$E = IR \tag{1}$$

A number of methods have been developed to solve the complex arrangements of the quantities given in Ohm's law. These methods originate with Kirchhoff's conservation laws, which are:

1. The algebraic summation of current at any junction in a circuit is zero.
2. The algebraic sum of voltage drops around a closed loop is zero.

When circuit elements (resistors, capacitors, or inductors) are connected so that they are placed end to end, the arrangement is said to be in series. When circuits elements are connected across the same points, the arrangement is said to be in parallel. Figure 1 illustrates simple series and parallel resistive circuits. Complex circuits can be formed by combining both series and parallel networks.

GROUND

When one terminal of the voltage source is made common to all circuits involved, that terminal is normally used as the reference for voltage measurements. In many instruments that terminal is also connected to the metal portion or chassis of the instrument. The chassis is generally considered to be at zero potential and is called the common or chassis ground.

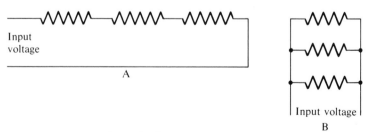

Figure I *Simple circuits: A, series; B, parallel.*

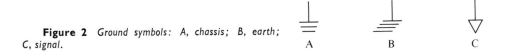

Figure 2 *Ground symbols: A, chassis; B, earth; C, signal.*

Most power lines have one terminal that is directly attached to a metal rod implanted in the earth. This type of ground is called an earth ground. In many instruments, the earth ground and common ground are interconnected.

In high quality amplifier circuits it is an accepted practice to separate the common ground from the ground used for all signals to be measured. This is called a signal or high quality ground.

The symbols for all of these grounds are shown in Figure 2.

VOLTAGE MEASUREMENTS

One of the more important measurements made in physical chemistry is that of potential difference or voltage. It may be advisable to distinguish between potential difference and voltage. The potential difference between any two points is equal to the work required to transfer a unit quantity of electricity from one point to another. The voltage is the difference in electrical pressure between two points in a circuit. If one considers any two points, the potential difference and the voltage are the same. In this chapter the two quantities will be used interchangeably.

Potential difference can be measured by employing a voltmeter (to be discussed later) or by comparison methods. In the comparison procedure, one matches the unknown voltage with one of equal magnitude. The device that generates the comparison potential is called a potentiometer.

A potentiometer is simply a voltage divider. If one considers some voltage source as E and applies it to a series arrangement of two equal value resistors, the voltage drop across each resistor will be one-half the total applied voltage, as shown in Figure 3.

$$I = \frac{E}{R + R} = \frac{E}{2R} \tag{2}$$

$$E_{AB} = I \cdot R = \frac{E}{2R} \cdot R = \frac{E}{2} \tag{3}$$

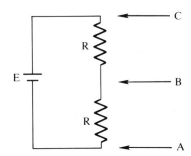

Figure 3 *Simple series circuit.*

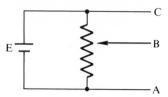

Figure 4 *Simple potentiometer.*

If each junction of a divider is made a position on a rotary switch, the division process can be extended to multiple steps. However, there are physical limits to the number of positions available on a switch, and the process cannot be extended indefinitely.

When it is desired to have continuously adjustable voltage, one selects a device constructed so that the resistive element is in contact with a movable slider, as schematically represented in Figure 4.

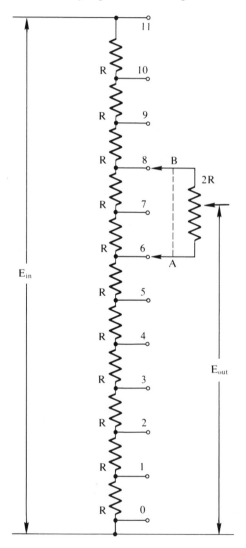

Figure 5 *Kelvin-Varley divider.*

The voltage between A and C is that of the battery, while the voltage between A and B is some fraction of that overall applied voltage. If the variable resistor (also called a potentiometer) has a linear resistive element, the fraction of rotation made will be identical to the fraction of the voltage present between A and B. This device also has mechanical limitations. The reproducibility, and eventually the resolution, is a mechanical feature of the potentiometer. For this device the maximum resolution is a function of construction; however, a good rule of thumb is about 100 parts per 360 degrees of rotation. This means about a 1 per cent resolution of the applied voltage. There are potentiometers that have either 10 (most common) or 40 total turns. This extends the resolution to 0.1 per cent for the 10-turn potentiometer or 0.025 per cent for the 40-turn potentiometer.

Highly sensitive instruments for measuring potential require a resolution in the range of microvolts when a voltage in the order of 1 volt is to be measured. Under these conditions, the resolution of a simple voltage divider or multiple turn potentiometer is not adequate. The Kelvin-Varley divider, shown in Figure 5, is one possible solution.

Consider 11 equally valued resistors (value equal to R) arranged on a switch, as indicated in Figure 5. Another resistor, attached across the two movable contacts of the switch, has a value of $2R$. The total resistance as presented to E_{in} is $10R$ ($9R$ plus $2R$ paralleled with $2R$). The lower contact of the switch, labeled A, can move from position 0 through 9, whereas the upper contact, labeled B, can move from positions 2 through 11. Since the total resistance of the divider always remains constant, independent of position, the voltage across the contacts is one-tenth of E_{in}.

Since position 0 is common to the divider, E_{in} and E_{out}, let us consider its potential as the reference or zero volts. The voltage at point 6 is then $0.6E_{in}$. The voltage at point 8 is $0.7E_{in}$ since the parallel combination in effect removes one resistor from the string of 11 resistors. If the variable potentiometer is a ten-turn device, it is able to present to the output a voltage that varies from $0.6000E_{in}$ to $0.7000E_{in}$, depending on the setting of the potentiometer. In effect the resolution is now 0.01 per cent as compared to 10 per cent for the ten-stage divider or 0.1 per cent for the turn variable potentiometer.

If one now inserts another switch arrangement between the present divider and E_{in}, with a value of each resistor equal to $5R$, the division process can be extended by a factor of ten. A number of high quality potentiometers having six or more stages of division are based on this principle.

The selection of the resistor for a particular function is dependent upon several factors. Table 2 is a list of some common resistance elements and factors affecting

TABLE 2. Common Resistance Elements and Factors in Selection

RESISTANCE ELEMENT	MAXIMUM ACCURACY	STABILITY	FREQUENCY (D.C.)	CHARACTERISTIC HIGH*	TEMPERATURE COEFFICIENT	RELATIVE COST
Carbon	Poor	Poor	Good	Good	Poor	Low
Deposited carbon	Moderate	Good	Good	Good	Moderate	Moderate
Metal film	Moderate	Good	Good	Good	Good	Moderate
Wire, wound	Excellent	Excellent	Excellent	Poor	Excellent	High

* 100 kilohertz or higher.

their selection. The choice would require consideration of all the factors listed. Most resistors used in modern circuits are of the carbon variety. As long as accuracy is not required, this is an excellent choice.

CURRENT MEASURING DEVICES

When current flows through a coil of wire a magnetic field is generated. If the coil is suspended in the field of a fixed magnet, the two magnetic fields interact, causing the suspended coil to turn. This device is called a galvanometer.

The most common moving coil meter is the D'Arsonval type, in which the coil is suspended by a fine wire to which is attached a small mirror. The wire acts as the restoring force and the mirror is used to deflect a beam of light. If the light beam is sent through several sets of mirrors, a relatively sensitive device can be made. The deflection is not proportional to the flow of current because the turning of the coil changes the orientation toward the fixed magnetic poles. However, this form of galvanometer is normally used to detect a null in the current and not as an absolute measuring device.

By changing the construction of the fixed magnet, replacing the suspension system with a spring assembly, and adding a pointer in place of the mirror, a modified D'Arsonval movement can be made for which the deflection is directly proportional to the current. Most sensitive ammeters and voltmeter movements are based on this modification.

When a more rugged movement is desired, a moving vane type of ammeter is employed. A magnetic field is generated by the current to be measured. This attracts a soft iron armature to which a pointer is attached. This form of meter movement is suitable for both A.C. and D.C. movements.

Most commercial ammeters work in the region of 0.1 to 10 milliamperes. If a greater current is to be measured, the meter is paralleled to a resistor called a shunt. The value of the shunt resistor can be determined from a simple Ohm's law calculation and a knowledge of two of the electrical characteristics of the meter (i.e., voltage sensitivity, current sensitivity, and internal resistance). Referring to Figure 6, R_m is the internal resistance of the meter, I_m is the full scale current sensitivity of the meter and I_{total} is the total current of the system.

The voltage sensitivity can be calculated from Ohm's law and is $E_m = I_m R_m$. The full scale voltage sensitivity of the meter remains constant and independent of the value of the shunt. The value of E_m also pertains to the voltage drop across the shunt resistor because it is all one circuit. The value of the shunt is thus

$$R = \frac{E_m}{I_s} \qquad (4)$$

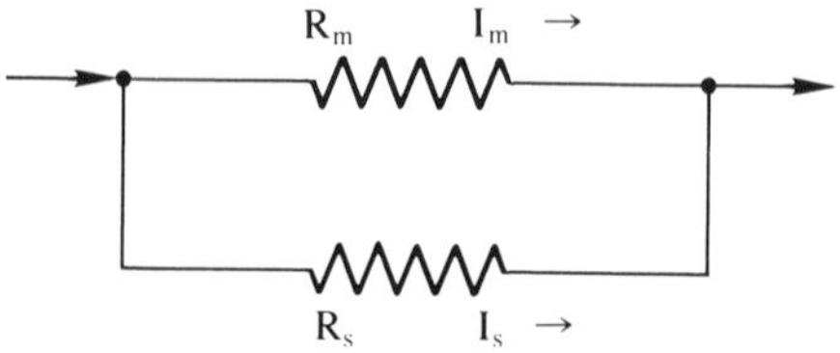

Figure 6 *Ammeter with meter shunt.*

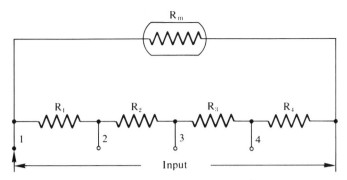

Figure 7 *Ammeter employing an Aryton shunt.*

where R_s and I_s refer to the resistance and current of the shunt, respectively. Since the total I is the sum of I_m and I_s, one arrives at

$$I_s = I_t - I_m \tag{5}$$

From equations 4 and 5

$$I_s = \frac{E_m}{E_s} = \frac{R_m I_m}{R_s} = I_t - I_m$$

so that

$$R_s = \frac{R_m \cdot I_m}{I_t - I_m} \tag{6}$$

The accuracy of most meter movements is a function of the full scale position with the accuracy being greatest at full scale. It therefore would be convenient to have a shunt mechanism by which the appropriate resistor can be selected to produce a maximum scale reading. This is generally accomplished with a circuit called an Aryton shunt. Figure 7 shows a four-position Aryton shunt in which position 1 is most sensitive and position 4 is least sensitive.

In order to establish the component values, the values of R_m and I_m must be known. A current is selected for I_{t1}, (I total at position 1), which is greater than I_m. Values for I_{t2}, I_{t3}, and I_{t4} are also selected in an increasing order. With the movable contact of the switch in position 1, the circuit is a simple meter and shunt combination. The value of R_s can be calculated from equation 6. The value R_s is simply the sum of the resistance values:

$$R_s = R_1 + R_2 + R_3 + R_4 \tag{7}$$

When the switch contact is in position 2, the potential drop across R_1 plus R_m must equal the potential drop across $R_2 + R_3 + R_4$. Under these conditions,

$$(R_1 + R_m)I_m = I_{s2}(R_2 + R_3 + R_4) \tag{8}$$

Where I_{s2} is the current in the shunt composed of R_2, R_3, and R_4, however,

$$I_{s2} = I_{t2} - I_m \tag{9}$$

so that

$$(R_1 + R_m)I_m = (I_{t2} - I_m)(R_2 + R_3 + R_4)$$

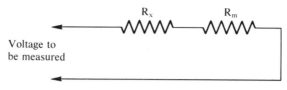

Figure 8 *Simple voltmeter.*

or

$$R_m I_m = I_{t2}(R_2 + R_3 + R_4) - I_m(R_1 + R_2 + R_3 + R_4)$$

Rearranging and substituting from equation 7 results in

$$R_1 = R_s - \frac{I_m}{I_{t2}}(R_m + R_s) \qquad (10)$$

In a like manner àn equation can be derived for each of the other positions of the switch.

$$R_1 + R_2 = R_s - \frac{I_m}{I_{t3}}(R_m + R_s) \qquad (11)$$

$$R_1 + R_2 + R_3 = R_s - \frac{I_m}{I_{t4}}(R_m + R_s) \qquad (12)$$

The value of R_4 is determined with the aid of equation 7 and the value of R_s previously established.

VOLTMETERS

A current meter is also the basis for most direct measuring voltmeters. The circuit is shown in Figure 8.

Since this is a series circuit, the current sensitivity of the meter does not change; therefore, in order to deflect the meter full scale, a voltage must be applied, which is

$$E = I_m(R_m + R_x) \qquad (13)$$

The value of R_x is selected so that the particular meter movement responds to full scale for the desired voltage. Since the voltage sensitivity of the complete circuit is less than that of the meter alone, we say that its sensitivity has been multiplied and R_x is the multiplier.

RESISTANCE MEASUREMENT

The measurement of resistance may be accomplished in a number of ways; two will be discussed here. The first of these is with an ohmmeter, a device consisting of a series combination of a current meter, a variable resistor, a battery, and the

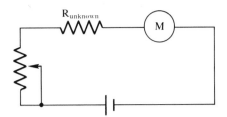

Figure 9   *Elementary ohmmeter.*

unknown resistor (Fig. 9). When the leads to the unknown resistor are left open, no current flows. When the leads are shorted ($R_{unknown} = 0$), maximum current flows. The variable resistor is adjusted so that the meter reads full scale. When an unknown resistance is now placed across the leads, some current less than full scale will flow. The value of $R_{unknown}$ can then be determined.

$$I_{\text{full scale}} = \frac{E_{battery}}{R} \tag{14}$$

$$I_{\text{deflection}} = \frac{E_{battery}}{R + R_{unknown}} \tag{15}$$

Eliminating $E_{battery}$ and solving for $R_{unknown}$,

$$R_{unknown} = \frac{R(I_{fs} - I_d)}{I_d} \tag{16}$$

The ohmmeter has two disadvantages: (1) The meter movement has limited accuracy, which is reflected in the ohmmeter, and (2) because $R_{unknown}$ is not a linear function of the deflection current (equation 16), the meter scale must be calibrated.

A much more accurate device is the Wheatstone bridge, shown schematically in Figure 10. R_s is adjusted until the null meter reads zero. This can occur only when the bridge is exactly balanced. The sensitivity of the bridge is dependent upon the sensitivity of the meter. It can be shown that

$$R_u = R_s \frac{R_2}{R_1} \tag{17}$$

If R_1 and R_2 are identical, the unknown is exactly equal to the standard. High accuracy resistors are available to place into both the fixed positions of the bridge and the variable position. Bridges accurate to 0.01 per cent are available.

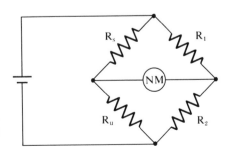

Figure 10   *Wheatstone bridge: R_s is an adjustable standard resistor; R_u is the unknown resistance; R_1 and R_2 are the fixed resistances.*

ALTERNATING VOLTAGE AND REACTIVE COMPONENTS

When a voltage with a periodic change in polarity is introduced into a circuit, two other circuit elements—the capacitor and the inductor—must be considered.

First let us consider what occurs when a voltage is applied to ideal versions of these components through a resistor. This is shown in Figure 11.

As one may observe from the diagrams, the conditions are not static but change as a function of time. When the capacitor is fully charged, the current flow is reduced to zero. For the ideal capacitor, the initial current when the switch is closed is determined by the resistor and the voltage of the battery.

The voltage across each component may be determined from the definition of each component.

$$E_L = -L \, di/dt \tag{18}$$

where E_L is the back voltage induced by change in current and L is the inductance of the coil.

$$E_C = E_{\text{battery}}(1 - e^{-t/RC}) \tag{19}$$

where E_C is the potential across the capacitor, R is the resistance in ohms, and C is the capacitance in farads.

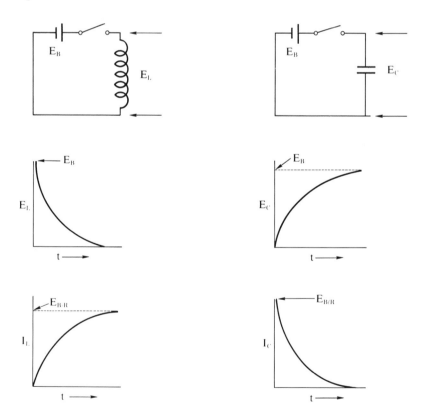

Figure II *Current and voltage wave after the closure of a switch.*

TABLE 3. Additive Laws Applicable to Series and Parallel Arrangements of Components

Element	Series	Parallel
Resistor	$R = R_1 + R_2 + \cdots$	$\dfrac{1}{R} = \dfrac{1}{R_1} + \dfrac{1}{R_2} + \cdots$
Capacitance	$\dfrac{1}{C} = \dfrac{1}{C_1} + \dfrac{1}{C_2} + \cdots$	$C = C_1 + C_2 + \cdots$
Inductance	$L = L_1 + L_2 + \cdots$	$\dfrac{1}{L} = \dfrac{1}{L_1} + \dfrac{1}{L_2} + \cdots$
X_c or X_L	$X = X_1 + X_2 + \cdots$	$\dfrac{1}{X} = \dfrac{1}{X_1} + \dfrac{1}{X_2} + \cdots$
X_c and X_L	$X = X_C - X_L$	$\dfrac{1}{X} = \dfrac{1}{X_L} - \dfrac{1}{X_c}$

When the potential difference applied to the reactive element is continuously charged in polarity, a second set of laws (equations 19, 20, and 21) may be proposed for the voltage and current relationship. These apply only when the applied voltage is a sine wave.

The A.C. resistance is called reactance and is given the symbol X. When the reactance is due to a capacitor the symbol is X_C, and when it is due to an inductor the symbol is X_L

$$X_C = \frac{1}{2\pi f C} \tag{20}$$

where f is the frequency in hertz (formerly cycles per second) and C is the capacitance in farads.

$$X_L = 2\pi f L \tag{21}$$

where L is the inductance in henries.

The overall resistance of an A.C. circuit is called impedance, which is symbolized by Z. The impedance may be evaluated with the aid of equation 21.

$$Z^2 = R^2 + (X_C - X_L)^2 \tag{22}$$

Reactive elements are handled in a manner similar to resistive elements. When the circuit is a series arrangement the voltage drop across each element is calculated. In a parallel arrangement the current through each element is determined. The basic additive laws of those components are given in Table 3.

Figure 12 shows a series arrangement of a resistor, a capacitor, and an inductor. It is generally of interest to determine the voltage across each element.

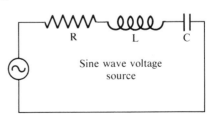

R   L   C

Sine wave voltage source

Figure 12 *Series arrangement of a resistor, capacitor, and inductor.*

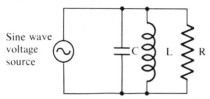

Figure 13 *Parallel arrangement of a resistor, capacitor, and inductor.*

One may calculate the total current from a knowledge of E and Z. Once these are known the potential drop across each component may easily be determined.

$$E_R = RI = R\,\frac{E}{R^2 + (X_L - X_C)^2} \tag{23}$$

$$E_c = X_C I = X_C\,\frac{E}{R^2 + (X_L - X_C)^2} \tag{24}$$

$$E_L = X_L I = X_L\,\frac{\dot{E}}{R^2 + (X_L - X_C)^2} \tag{25}$$

Since the voltages across each of these components are not in phase with each other, the algebraic sum of the voltages is greater than the applied voltage. If the vector sum of the voltages across the components is considered, its value is equal to the applied voltage.

A parallel arrangement is shown in Figure 13. Since the voltage across each

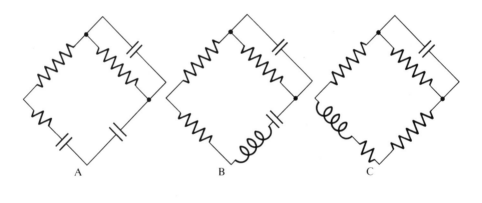

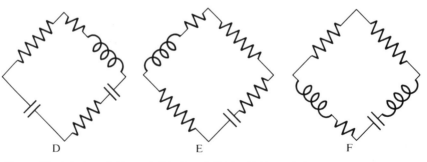

Figure 14 *Impedance bridges: A, Schering; B, Wein; C, Maxwell; D, Owen; E, Hay; F, inductance.*

element is the same, the current must be calculated.

$$I_R = \frac{E}{R}, \qquad I_L = \frac{E}{X_L}, \qquad I_C = \frac{E}{X_C} \qquad (26)$$

In any parallel combination the currents are additive and I_{total} should be $I_R + I_C + I_L$. However, I_C and I_L are 180 degrees out of phase with each other, so that I_t is equal to the vector sum of the three.

A special case exists for both series and parallel circuits when $X_C = X_L$. For the series circuit the current is determined by R alone and the current in the circuit is maximized. Then this circuit is said to be at minimum impedance. In the parallel circuit the current is minimum and the circuit is said to be at maximum impedance. The condition of $X_C = X_L$ is, of course, frequency dependent; it is called the resonance frequency of a circuit. Resonance circuits are very important in the design of high quality oscillators. Certain procedures for the determination of both inductance and capacitance are based on the resonance frequency principle.

In order to accurately measure either capacitance or inductance a bridge similar to a Wheatstone bridge may be employed. In some manner, the phase shift associated with the resistor-reactive element combination must be accounted for. To meet this requirement various bridge arrangements have been developed. A few of these are shown in Figure 14. The more common of the capacitance bridges are the Wein and the Schering bridges. The choice of bridge depends upon approximate values of the components to be measured and the availability of proper standards. Modifications of these designs constitute the basis of most modern high quality impedance measuring bridges.

A filter is another important form of reactive circuit. A low pass filter is shown in Figure 15. Low frequency signals pass with little or no attenuation, while high frequency signals are appreciably attenuated.

This may be looked at in an elementary form by stating that at high frequencies the X_L of the inductor increases, whereas at low frequencies the X_L decreases. Therefore, the potential drop across the inductor is greater at high frequencies. The capacitor, on the other hand, acts in exactly the opposite manner. At high frequencies X_C is low, whereas at low frequencies X_C is high. This in effect causes more current to flow at high frequencies than at low frequencies, which in turn causes a greater voltage drop across the inductor. This arrangement of components is the basis for most filtering circuits in power supplies.

VOLTAGE SOURCES

In the operation of all electronic devices there is a need for some source of potential for that operation. Most of these sources must be D.C. The simplest form of this

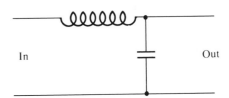

Figure 15 *Low pass filter.*

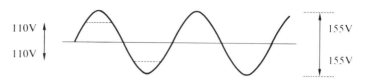

Figure 16   *Peak to peak and root-mean-square voltages for a sine wave.*

potential is a battery. The primary shortcomings are limited life and a rather poo₁
temperature coefficient.

Power companies transmit voltage as a sine wave with a frequency of 60 hertz.
The voltage at the typical laboratory receptacle has a root-mean-square value of
110 volts. In simple terms, the root-mean-square potential is equivalent to the D.C.
voltage that would give the same heating to a standard resistor. In this case approximately 310 V. peak-to-peak voltage is required to give the same heating as 110 volts
D.C. For a sine wave, the root-mean-square value is 70.7 per cent of the reference-
to-peak height value. In other words, the peak-to-peak potential is 2.828 times more
than the root-mean-square value.

In order to use A.C. voltages, they must be able to be changed from one level
to another. This may be accomplished with the aid of a transformer. In Figure 17,
three forms of transformers are shown. If the transformer shown in Figure 17A has
more turns in the output winding than in the input winding, it is called a step-up
transformer; if it has fewer turns in the output winding than in the input winding,
it is called a step-down transformer. Figure 17B shows a center tapped transformer,
whereas Figure 17C shows an autotransformer.

The output of a transformer is given by the equation

$$E_2 = E_1 \cdot \frac{T_2}{T_1}$$

where E_1 is the voltage of winding 1, E_2 is the voltage of winding 2, T_1 is the turns of
winding 1, and T_2 is the turns of winding 2.

In order to convert A.C. to D.C., one requires a device that passes current in
only one direction, such as a silicon diode. A typical rectifying circuit is shown in
Figure 18, with 18A being a half-wave and 18B a full-wave rectifier.

The diode conducts any time that the anode voltage is more positive than the
cathode. In Figure 18A this occurs once every complete cycle, making the voltage
across the resistors identical to the input, except that it occurs alternately; this can

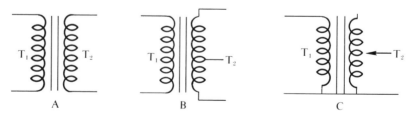

Figure 17   *Transformers: A, a step-up or step-down transformer; B, a center tapped transformer;
C, an autotransformer.*

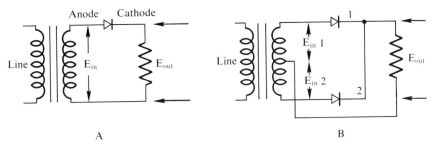

Figure 18 *Rectifiers: A, half wave; B, full wave.*

be seen in Figure 19. If E_{out} is superimposed on E_{in}, the two waves are identical except that E_{out} is missing the portion below the zero reference line.

The full-wave rectifier shown in Figure 18B functions in a like manner except that the voltage is applied alternately to two diodes. Since the reference is now the center tap of the transformer, the potentials developed across each half of the transformer will be 180 degrees out of phase. Figure 20 shows the output wave form. Since the output has a frequency twice that of the input and conducts on each half cycle, it is said to be a full-wave rectifier.

The potential developed across each resistor may be considered to be pulsating D.C. However, it differs from the input to the diode in that the wave form of the voltage never drops below the zero reference. If either of these assemblies is connected to a low pass filter as previously shown, the A.C. component will be averaged out and a D.C. potential will develop.

The magnitude of the A.C. component in the output is a function of the number of stages of low pass filters used and the values of the components in each stage.

The voltage at the output of the power supply varies to the same extent as the input providing that the transformer has no regulating effect. For many applications this is quite satisfactory; however, other applications need more stringent control.

For moderate current requirements one could resort to a Zener diode for regulation. The Zener diode is a solid state device for which the current-voltage relationship, as shown in Figure 22, has transition at some point called the Zener potential.

Once the Zener potential is attained, additional current flow does not produce an appreciable change in potential. The diode is utilized in a circuit similar to that in Figure 21.

As the current requirements of the load varies, or the input voltage varies, the current passing through the diode compensates for these changes. These diodes are

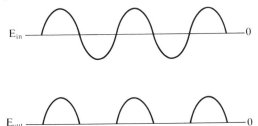

Figure 19 *Wave forms of half-wave rectifier.*

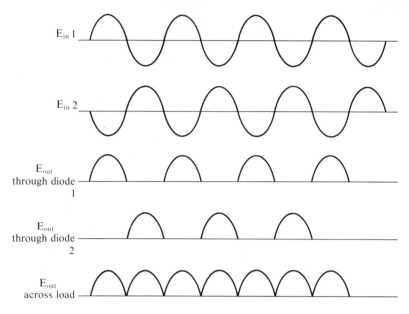

$E_{in}\,1$

$E_{in}\,2$

E_{out}
through diode
1

E_{out}
through diode
2

E_{out}
across load

Figure 20   *Wave forms of full-wave rectifier.*

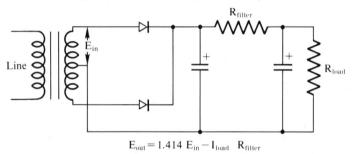

R_{filter}

E_{in}

Line

R_{load}

$E_{out} = 1.414\ E_{in} - I_{load}\ R_{filter}$

Figure 21   *Full wave rectifier with filter.*

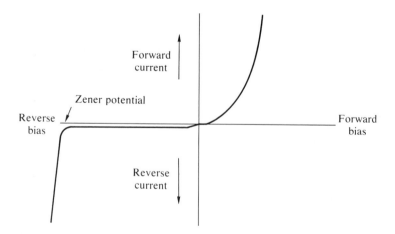

Forward
current

Zener potential

Reverse
bias

Forward
bias

Reverse
current

Figure 22   *Current-voltage characteristic of a pn junction.*

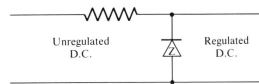

Figure 23 *Simple Zener regulator.*

available in voltages from a few volts to 200 volts and in power ratings to 50 watts. The Zener diode exhibits a distinct temperature coefficient of voltage, which reaches zero and changes sign in the vicinity of 5 or 6 volts. By selecting diodes in this region, one can construct a device that has a very low temperature coefficient approaching 5 ppm degree per Centigrade. This temperature coefficient can be maintained only at one specific current.

If the current requirements of the load exceed the regulating ability of the diode, or if the output voltage must be continuously variable, an electronically regulated power supply is most advantageous. This is shown in Figure 24. The output of the power supply is compared to a reference voltage and the difference amplified. The reference may be derived from a temperature compensated Zener diode. The amplified error signal is connected to a transistor or vacuum tube in such a way that the error is reduced and the voltage becomes stable. Power supplies are available that have a resolution of 0.1 ppm, a regulation of 5 ppm, and a temperature coefficient of 2 ppm.

AMPLIFIERS

Most electronic instruments require a modification of the input voltage derived from a detector or transducer. This modification can take the form of voltage amplification, impedance transformation, or power amplification. All of these are handled using various forms of amplifiers. Only a limited description of internal construction of the amplifier will be given, and therefore the amplifier will be generally considered to be a black box. It is best to discuss a few terms related to amplifiers.

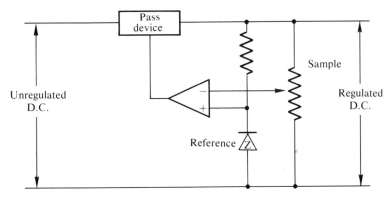

Figure 24 *Regulated power supply.*

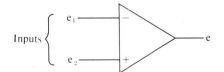

Figure 25 *Representation of an operational amplifier: e is the output voltage; e_1 is the signal voltage at the negative input; e_2 is the signal voltage at the positive input.*

Although this discussion is primarily concerned with operational amplifiers the terms are not necessarily limited to them.

Differential Input. When the difference between rather than the absolute value of two voltages is amplified, the device is called a differential amplifier, which is symbolically shown in Figure 25.

The minus sign corresponding to input e_1 refers to a signal that is amplified with an inversion of phase. In other words, a signal will be amplified with a 180 degree inversion of phase. The plus sign corresponding to input refers to amplification without phase inversion. If one input (usually the positive input) is connected to the common or reference junction, the device is called a single-ended amplifier. Certain modified models have internal connections that render the positive input unavailable for use. Other amplifiers are designed with one input.

Input Impedance. The rigid analysis of the input circuit of an amplifier shows it to be rather complex. A simple equivalent circuit, as shown in Figure 26, is a resistor and capacitor in parallel. All amplifying devices (transistor and vacuum tubes) require a small current to facilitate the control of the device. Besides this, leakage currents exist within the mechanical arrangement of the inputs. If one considers a fixed voltage, these currents may be reported as resistance. Modern technology has provided us with input resistances of 10^{14} ohms for solid state devices and 10^{16} ohms for special electrometer tubes.

The capacitive component has two primary sources. One is the proximity of the input leads to the chassis and other components, whereas the second is the inherent capacitance electrically associated with the input transistor or vacuum tube.

Voltage Gain. The gain of an amplifier is defined as the ratio of the change in the output voltage to the change in the input voltage. Since no restrictions are placed on direction of change, the gain may be positive or negative. Amplifiers with gains of 10^8 are available. When one speaks of gain for an amplifier it is understood to mean voltage gain unless otherwise noted.

Power Gain. The power input to an amplifier is normally in the order of picowatts (10^{-12} watts) to microwatts. The output is capable of providing milliwatts to watts of power. The ratio of output power to input power is called power gain. Amplifiers are available with power gains from unity to 10^{13}.

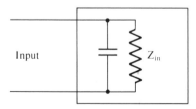

Figure 26 *Simplified schematic diagram of the input of an amplifier.*

Common Mode Rejection Ratio. For a differential amplifier to function properly, it must respond to the difference between two signals rather than to the signals. The gain should be small for the common signal and large for the signal representing the difference. The characteristic that indicates the effectiveness of rejecting the common signal is called the common mode rejection ratio; it is mathematically expressed as

$$\text{CMRR} = \frac{\text{Gain for } \Delta e_{in}}{\text{Gain for } e_{in}} \tag{28}$$

The common mode rejection ratio of most differential amplifiers is at least 1000 and in some instances exceeds 10^6.

Output Impedance. When the load attached to the output of an amplifier varies, a change in the load current occurs. The output impedance is calculated from the change in output voltage for a given change in output current.

$$Z_{out} = \frac{\Delta e}{\Delta I_{out}} \tag{29}$$

Most high quality operational amplifiers have output impedance in the order of hundreds of ohms or less.

Frequency Response. There is a limit to the frequencies to which an amplifier will respond. There is always an upper limit, which is generally taken to be the frequency at which the voltage gain drops by a value of two.

For D.C. amplifiers there is no lower frequency limit, but for many A.C. amplifiers there is a controlled lower limit established by the value of components used.

The present trend is to manufacture amplifiers with as broad a frequency response as possible. The increased bandwidth includes all noise within that frequency range. It is normally possible to sacrifice frequency response for reduced noise. It is sometimes advantageous to reduce the bandwidth to such an extent that only one specific frequency is amplified; this in turn reduces the noise content of the output signal. Such amplifiers are called tuned amplifiers.

Noise. The output of all amplifiers contains some spurious signals that are unrelated to the signal being studied. This is called noise. The noise is normally amplified with the signal being investigated and it is the noise that places a lower limit on signal voltage. The source of noise may be the components or the surroundings. Proper design techniques can greatly limit the magnitude of the noise, but complete elimination is virtually impossible.

The noise that affects an amplifier may be classified into several categories:

WHITE NOISE. Thermal white noise is characterized as a one-dimensional black body radiation produced by resistors. It is usually uniform in magnitude through the frequency spectrum and is a function of resistance, temperature, and bandwidth. Most amplifiers normally encountered in chemistry involve the region of the spectrum in which white noise is applicable.

SHOT NOISE. In vacuum tubes there is random emission and a transient time for the electron flow, which produces a form of white noise called shot noise. The magnitude of this noise is normally lower than that of thermal white noise. The greater the electron flow the larger will be the value of this noise. A similar type of noise exists in transistors as a result of generation and recombination of holes and free electrons.

PICK UP. In almost every laboratory there is a source of A.C. power. This power is distributed throughout the laboratory through wires. Since current flowing through wires produces an electromagnetic field, a source of noise is always present. Most amplifiers are capable of responding to this frequency range. Provisions must be made to limit insertion of A.C. line voltage into the input. There are a number of procedures for reducing pick up. Wires that carry A.C. power within the instrument are usually twisted. Leads going to the input are covered with a braided wire, called a shield, which is connected to the chassis ground. In addition to this, all leads going to sensitive inputs are kept short.

This form of noise is by far the most common encountered in the laboratory. Within limits, it is also the one that can be most easily controlled.

Feedback. The gain of most operational amplifiers is quite high, being somewhere between 2×10^4 and 1×10^7. This value is termed the open loop gain, and is given the symbol A. In most applications, it is desirable to have a gain that is considerably less than the open loop gain. This new term is called the closed loop gain. The open loop gain is altered by coupling some of the output voltage back to the input. This coupling is called feedback.

Feedback can be classified as positive or negative. Positive feedback occurs when the output signal is in phase with the input signal. Since this must cause the closed loop gain of the amplifier to increase, it produces a rather unstable condition. Most oscillators are designed around positive feedback.

The second and more important type is negative feedback. In this form, the output is 180 degrees out of phase with the input. This has the effect of reducing the overall gain of the amplifier. It increases the bandwidth and the general stability of the amplifier at the cost of having a closed loop gain less than the open loop gain.

For an operational amplifier two conditions are maintained to a degree dictated by the design of the amplifier. The first condition is that the input current into the amplifier is considered to be zero. The second condition is that the voltage at the input is maintained at zero volts in a single input amplifier, or the difference in potential for a differential amplifier is zero.

The latter condition is of paramount importance to the theory of the operational amplifier. Figure 28 can be used to elucidate this point. The junction of R_{in}, R_f, and the inverting input is called the summing point. The difference in voltage between the summing point and the noninverting input is symbolized by e_s. When e_s is more

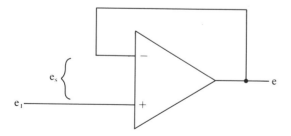

Figure 27   *Simple follower.*

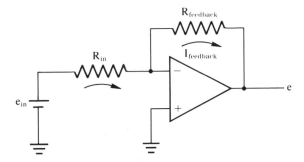

Figure 28 *Typical hookup for voltage amplification.*

than a fraction of a millivolt, the output is very large, (i.e., $e = -e_sA$). By applying Ohm's law, and under the limiting assumption that no appreciable current flows into the amplifiers, it can be shown that

$$e_s = \frac{e_{in}R_f/R_{in}}{1 + R_f/R_{in} + A} \tag{30}$$

Since A is usually the largest term in the expression, e_s is quite small.

Figure 27 is an example of one of the most simple and yet most useful configurations for an operational amplifier. It is called the follower. In this circuit the output is fed directly back to the inverting input. The output is a reproduction of the input as long as the error voltage (difference between the two inputs) is small. Furthermore, the input impedance is high, being somewhere between 10^7 and 10^{12} ohms with a corresponding very low output impedance (normally less than 1 ohm).

In Figure 28 the noninverting input is connected to the chassis ground and therefore the voltage at the inverting input must be zero. The current into the summing point must be

$$I_{in} = \frac{e_{in}}{R_{in}} \tag{31}$$

But, the net current at the input must be zero so that the current flows toward the output, making the output more negative than zero.

Since I_{in} has been fixed by the input voltage and R_{in}, the output voltage must vary. Accordingly, it is given a minus sign because the current is moving away from the input.

$$e = -I_{in}R_{feedback} \tag{32}$$

$$\text{Gain} = \frac{e}{e_{in}} = \frac{-I_{in} \cdot R_{feedback}}{I_{in} \cdot R_{in}} = \frac{-R_{feedback}}{R_{in}} \tag{33}$$

The maximum value of R_f/R_{in}, of course, cannot exceed the limits of the open loop gain of the amplifier. More severe restrictions on the maximum value of R_f/R_{in} are imposed by other considerations, such as drift and frequency response.

The operational amplifier can be used to sum voltages or to take the difference between voltages. In Figure 29, an adder circuit is shown. The output voltage is

$$e = \frac{e_1' \cdot R_f}{R_{in}} - \frac{e_1'' \cdot R_f}{R_{in}} \tag{34}$$

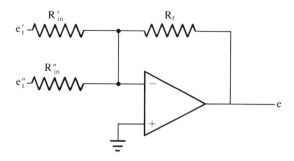

Figure 29 *Adder circuit.*

Under the specific condition when $R_f = R'_{in} = R''_{in}$ the output voltage will be

$$e = -(e'_1 + e''_1) \tag{35}$$

Since the amplifier responds to the difference between the two inputs, a circuit may be designed that is a subtractor. This is shown in Figure 30. The output voltage is

$$e = -e'_1 \frac{R'_f}{R'_{in}} + e''_2 \frac{R''_f}{R''_{in}} \tag{36}$$

Under the specific conditions when $R'_f = R'_{in} = R''_f = R''_{in}$

$$e = -e_1 + e_2 \tag{37}$$

An operational amplifier can also be used as a current amplifier or, to be more exact, a current-to-voltage converter. Such a circuit is shown in Figure 31. This circuit is not very practical as a voltage amplifier because the input impedance is very low.

Since the current entering the amplifier must go through R_f the output signal is

$$e = -I_{in}R_f \tag{38}$$

If R_f is large, a substantial voltage is developed for a small input current and the amplifier is able to present this output at a modest power level.

Several other circuits are of interest. An integrator circuit is shown in Figure 32.

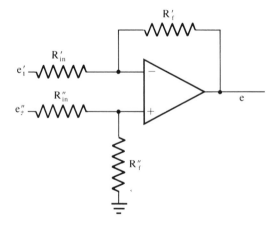

Figure 30 *Subtractor circuit.*

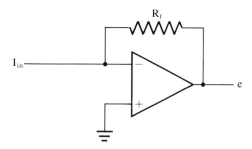

Figure 31 *Amplifier as a current-to-voltage converter.*

Since the input current is e_{in}/R_{in} the feedback current must be $-I_{in}$. The voltage on a capacitor is given by

$$e = \frac{1}{C} \int I \, dt \tag{39}$$

Substituting in for I_{in}

$$e = \frac{-1}{R_{in}C} \int e_{in} \, dt \tag{40}$$

If the resistor and capacitor are interchanged, the circuit becomes a differentiator. The output potential is given by equation 41, whereas the circuit is shown in Figure 33.

$$e = -R_f C \frac{de}{dt} \tag{41}$$

An interesting circuit is that of the comparator shown in Figure 34A. The output voltage can be at either extreme depending upon which input is most positive. When input 1 is more positive than input 2, the output will be fully negative. When input 1 is less positive than input 2, the output will be fully positive. The circuit is quite sensitive but has the limitation that the extremes of voltages applied to the input cause changes in the exact offset between inputs, and consequently the point of transition changes. To eliminate this difficulty, bounds are put on the output so that the circuit does not limit from one extreme to the other. Figure 38B shows such a circuit.

Comparison is made by sensing the direction of flow of current. Since the input is connected to ground through diodes, neither can deviate from the common potential by more than the forward breakdown voltage of the diode (which is about

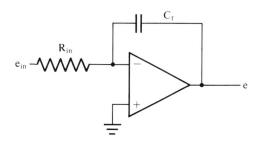

Figure 32 *Integrating circuit.*

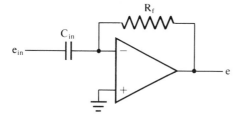

Figure 33 *Differentiator circuit.*

0.6 V for a silicon diode). Positive feedback is used if a snap action is desired for slowly changing signals.

SPECIAL PURPOSE AMPLIFIERS

Besides operational amplifiers, there are other classes of amplifiers that perform specific functions. Among these are servo amplifiers, such as those in recorders, tuned amplifiers, such as those in conductance bridges, and pulse amplifiers, such as those in radiation detection equipment.

INSTRUMENTS

Most physical chemists encounter many instruments in the course of their study. The basic operation of several will be discussed here.

Vacuum Tube Voltmeter. This is one of the most common instruments and its basic principles are incorporated in many other instruments. It is a form of difference amplifier. A simplified drawing is shown in Figure 35. A constant input resistance is attached to the grid of the input tube. As the potential at the grid changes, the potential at the corresponding cathode also changes. The second tube serves as a reference. This tube could be used as the zero position adjustment; however, in most circuits the zero adjustment is in the cathode circuit common to both tubes. The variable resistor in series with the meter is used to calibrate the circuit. By substituting a voltage divider for the grid resistor, it is possible to obtain multiple ranges. The input impedance of the device is limited by the current to the tube and

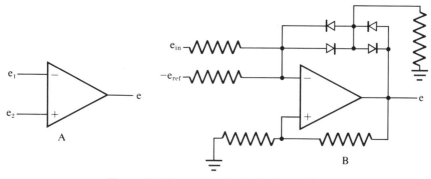

Figure 34 *Comparator: A, simple; B, precision.*

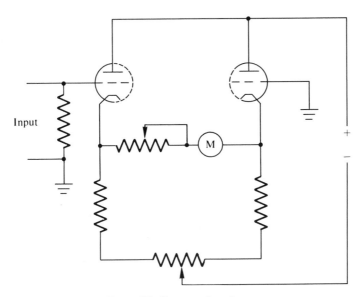

Figure 35   *Vacuum tube voltmeter.*

the size of the grid resistor. In normal practice the grid resistor is 10 megohms, while the grid current is 0.01 microamperes or smaller.

pH Meter.   The ordinary vacuum tube voltmeter, using conventional vacuum tubes, has an input impedance of 10 megohms. Most pH meters use a glass electrode that has a resistance of 10^8 to 10^{10} ohms. The equivalent circuit is shown in Figure 36. The voltage produced by the electrochemical cell should be dissipated across the resistance of the measuring device rather than the internal resistance of the cell.

For an electrode resistance of 10^8 to 10^{10} ohms, it is necessary to have an input resistance of 10^{10} to 10^{12} ohms in order to have a measuring accuracy of 1 per cent. If the required accuracy is greater, the corresponding input resistance must be greater. This cannot be accomplished with conventional vacuum tubes unless the tubes are selected and operated under specific conditions (i.e., low plate and heater voltages.) The problem can also be circumvented by employing 100 per cent negative feedback in the input. However, it is simpler to use specially prepared vacuum tubes, called electrometers, which have input resistances from 10^{13} to 10^{16} ohms. Some solid state devices, metal-oxide-semiconductor field effect transistors, have input resistances of 10^{14} ohms. By using either device a modified difference amplifier can be designed

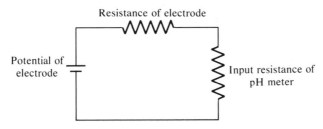

Figure 36   *Equivalent circuit of a pH meter input.*

to meet the requirements of pH measurements. Since response time is normally not of paramount interest, the input is generally well filtered with a low pass filter.

A direct reading pH meter could be built around a high quality operational amplifier. Commercial, low drift, operational amplifiers are available with input impedances in the order of 10^{12} ohms or higher. In order to design such an instrument, certain design requirements should be established: (1) as high an input impedance as possible, (2) a meter output graduated in pH units, (3) a means of calibrating the meter, (4) provision for zero adjustment, and (5) a means of compensating for variations in temperature.

Since it is desired to have maximum input impedance, the electrodes should be connected to an input that is not encumbered by feedback networks. The non-inverting input is suited for this application. The indicator is a meter with mechanical zero at center scale; therefore, the zero control must be able to adjust the output of the amplifier to zero volts at pH 7. For this reason, the zero adjustment is placed in the noninverting input. Temperature compensation and standardization could be incorporated in the metering circuit. A simplified diagram is shown in Figure 37.

The range of the zero control is arbitrarily selected as 0 ± 100 millivolts to compensate for any amplifier offset and also for the fact that the electrode voltage is probably not zero at pH 7. Two mercury electrodes (1.3 volts each) are connected in series and the junction between them returned to chassis ground. This allows control of both plus and minus deviations from zero. The electrodes do not draw appreciable current; therefore the resistance of the zero adjustment can be quite high without introducing voltage errors. This high resistance has the added advantage of maintaining a long life expectancy for the batteries. The adjustment control could be a 1 megohm potentiometer over which 200 millivolts is impressed. By Ohm's law, the current is calculated to be 0.2 microampere. Since this will be a series circuit, the entire 2.6 volts of the battery must be dissipated with a current flow of 0.2 micro-ampere, which in turn requires a total resistance of 13 megohms. This total resistance can be set in such a manner that the zero potentiometer is in a position that yields a control of 0 ± 100 millivolts.

A zero center meter with a full scale sensitivity of ± 100 microamperes and a resistance of 1000 ohms requires 200 millivolts to deflect the meter from one extreme to another. Normally pH 7 is selected as the center of the scale so that the deflection will be ± 100 millivolts. One wishes to present 0 to 14 pH units on a 200 microampere scale so that every 14.3 microamperes correspond to one pH unit. The meter scale could be marked in a manner that yields 14 pH units. At 25° C., using a glass

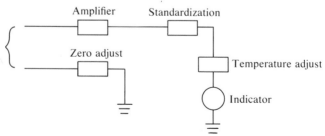

Figure 37 *Basic pH meter.*

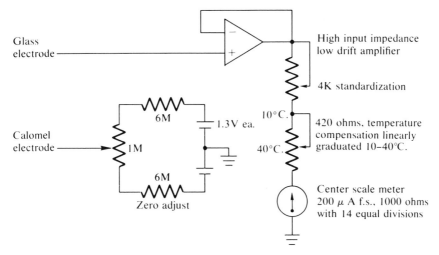

Figure 38 *Detailed diagram of a pH meter.*

electrode–calomel electrode combination, 59 millivolts are produced for each pH unit change. This corresponds to 826 millivolts for the pH range of 0 to 14. Since only 200 millivolts are required to fully deflect the meter, the gain of the amplifier need only be 0.242. Since this is less than one, it would be more advantageous to have a gain of exactly one and reduce voltage sensitivity of the meter. One could then employ the voltage follower configuration for the amplifier.

It is now required to deflect the meter 200 microamperes for an impressed voltage of 826 millivolts. Applying Ohm's law, one obtains a total resistance of 4.13 kilohms. Because the meter resistance is 1000 ohms, a total of 3.13 kilohms must be placed in series with the meter. However, this does not provide any margin of error or allow control; a suitable compromise would be a 4-kilohm potentiometer.

The voltage per pH unit changes at a rate of +0.2 millivolt for every one degree centigrade increase in temperature if the temperature is approximately 300° K.

Most pH meters have a temperature range of 25 ± 15° C. corresponding to 59 ± 3 millivolts per pH unit. The full scale voltage would then be 826 ± 42 millivolts. Hence, a change of 42 millivolts, due to a temperature change of 15° C., should cause no change in the pH reading. By Ohm's law, one finds that an addition of 210 ohms in series with the meter would compensate for a 15° C. increase in temperature. The converse is also true. A more detailed diagram is presented in Figure 38.

The instrument is adjusted to the value of a known buffer (in the region of pH 7), by using the zero adjustment. A second buffer is employed and the standardization control is used to set the pH value on the meter.

Conductance Bridge. In studies of the conductance properties of solutions, the resistance measurements are complicated by the parallel capacitance associated with the cell. Most conductance measurements are performed using a modified version of the Wheatstone bridge. Since conductance measurements cannot be performed at D.C. because of the irreversible effect of electrolysis, it is necessary to conduct the measurements using A.C. voltages. The parallel capacitance of the cell

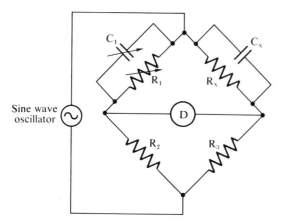

Figure 39 *Conductance bridge.*

causes phase shifts in the arm of the bridge that contains the cell. It is necessary to cause a similar phase shift in the measuring arm in order to balance the bridge. This can be accomplished by employing a capacitor in the measuring arm. The detector must be A.C. voltmeter. A simplified circuit is shown in Figure 39, where R_1 is the measuring resistor, R_2 and R_3 are fixed resistors, R_x is the unknown resistance, C_x is the capacitance of the unknown, C_1 is the compensating capacitor, and D is the detector.

Most conductance measurements are conducted at low frequencies in the range of 1000 hertz. Elimination of pick up from power lines is one of the greatest problems, and proper grounding is essential.

Oscilloscope. The heart of an oscilloscope is the cathode ray tube, which consists of an electron beam passing through two sets of deflecting plates. The plates form the sides of a square. A typical cathode ray tube is shown in Figure 40.

A beam of electrons is generated at a heater, passes through a grid (which controls the intensity), a focusing anode, an accelerating anode, and two pairs of parallel plates, and finally impinges on the fluorescent screen. When the potentials on the deflecting plates are steady, the beam rests in one position. If a varying potential is applied to the vertical plates, the electron beam is deflected up or down depending upon the polarity of the signal. If a potential, increasing in a linear manner, is applied to the horizontal plates, the beam can be made to move from left to right. Since the period of time required to move the beam from left to right depends upon how fast the potential is increased, this can be calibrated in time units, and is called the sweep.

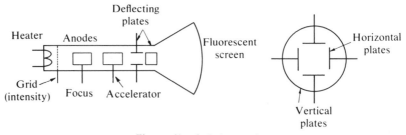

Figure 40 *Cathode ray tube.*

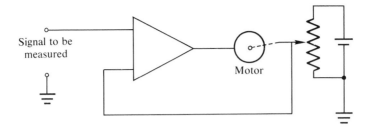

Figure 41 *Servo recorder.*

The beam is then not stationary in the horizontal plane. As the signal in the vertical plane changes, a pattern is traced out, which is related to vertical signal. The spread of that pattern is dependent upon how fast the signal is swept across the face of the tube. In a well designed oscilloscope the sweep is returned to its starting point in such a manner that there is no visible trace of the face of the tube; this is called blanking.

It is also possible to use the horizontal plates for signal display if a sweep is not required. The design of the amplifiers must be such that the pattern presented is an exact representation of the signal to be measured. Oscilloscopes are available having capabilities of presenting signals from D.C. to 100 megahertz or greater.

Recorders. Oscilloscopes are capable of presenting patterns of signals with rapidly changing characteristics. When it is required to measure signals that change slowly, in seconds or minutes, it is generally more advantageous to have a permanent record.

Recorders consist of a special form of difference amplifier, called a servo amplifier, capable of driving a motor. A simplified drawing is shown in Figure 41. The potentiometer-battery combination provides a source of voltage the value of which is dependent upon the position of the movable contact. A signal applied to the input of the amplifier causes an error between inputs. This in turn causes the motor to rotate, changing the position of the movable contact. This continues until the difference in potential between the inputs is reduced to zero. If the motor also moves a pen, writing on graph paper, there will be a presentation of the input signal.

When the paper moves at a constant rate, the recorder is a X vs. time recorder. Modern recorders may have more than one channel presented on the same graph. If the time axis is replaced with another amplifier, the presentation can be the result of two voltages or an X vs. Y recorder.

Digital Voltmeter. In most classic voltmeters, the meter movement is not completely linear. When accuracy is required, one can resort to a high quality potentiometer, but this is time consuming. A second approach is the digital voltmeter, which has a high degree of accuracy, approaching 0.01 per cent in some cases. There are a number of classes of digital voltmeters. The ramp type and the integrating type will be discussed here.

In the ramp type of digital voltmeter, a linearly increasing voltage is generated in a circuit similar to the operational integrator. This ramp is compared to the unknown voltage. When the ramp has a zero voltage, a timer is turned on; when the ramp and the unknown voltage are of equal value, the timer is turned off. If the

TABLE 4.

RANGE	SOURCE	OPTICAL MATERIAL	DETECTOR
Ultraviolet	H_2, D_2, Xe, A lamps	Quartz	Phototube with quartz window
Visible	Tungsten lamp	Glass	Phototube
Infrared	Nernst glower, globar	Sodium chloride, cesium iodide, lithium fluoride	Bolometer, thermocouple

ramp is calibrated in terms of volts/time, all that is needed is the time factor to determine the voltage.

In the integrating type of digital voltmeter, the average rather than the absolute voltage is measured. When a positive voltage is applied to an operational amplifier hooked in an integrating configuration, the output is a negative going ramp. The ramp continues until a level is reached at which a comparator is activated, in turn triggering a pulse generator. The output of the pulse generator is a rectangular wave with well defined characteristics, both in pulse width and amplitude. This pulse discharges the integrating capacitor back to zero volts. The net result is the generation of a series of pulses that can be counted and calibrated in terms of voltage.

Spectrophotometers. The spectrophotometer is one of the most useful instruments available to the chemist. Aspects of the nature of the chemical bond, as well as structure and concentration, are investigated with this instrument. The subject is so broad that only the most cursory discussion can be made here. One can break down the field of optical spectrophotometers into three general spectral ranges— ultraviolet, visible, and infrared spectra. The generation, focusing, and dispersion of light in these three ranges is covered in any text on instrumental analysis. Table 4 summarizes the sources, optical materials, and detectors available for analysis of the three ranges. This section will deal only with the detectors and the electronics.

Detectors. ULTRAVIOLET AND VISIBLE RANGE. The electronics of spectrophotometers generally starts with the detector and continues to the amplifying circuits. For studies in the ultraviolet and visible ranges, the phototube is frequently employed. The vacuum phototube consists of a photosensitive cathode and an anode (to collect electrons) encased in an evacuated tube. The photosensitive surface is coated with a substance that is capable of emitting photoelectrons. Alloys or compounds of antimony, cesium, selenium, bismuth, and silver are commonly employed. The

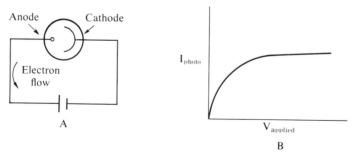

Figure 42   Phototube: A, schematic; B, current-voltage curve.

energy of the electron ejected by impinging photons is given by Einstein's equation. These ejected electrons are called photoelectrons.

$$E = hv - w \tag{42}$$

where E is the energy of the photoelectron, h is Planck's constant, v is the frequency of the photon, and w is the work function of the photocathode.

The work function of most photocathodes exceeds hv when the wavelength reaches or exceeds 1.4 microns. This places an upper wavelength limit on the spectral range of phototubes. The lower limit is established by the optical transmission of the tube envelope and the transparency of the photocathode to light at lower wavelengths. Phototubes with special lithium fluoride windows may be obtained with cut-off wavelength to 0.12 micron.

The vacuum phototube has a current-voltage relationship at a fixed light level, which is constant once the applied potential becomes sufficiently high. This results because the electrons are generated by a photo process and the applied potential only collects these electrons.

At any fixed wavelength the photocurrent is a linear function of light intensity. At a fixed light intensity, the photocurrent exhibits a bell shaped relationship with changing wavelength. For this reason instruments must be calibrated for each change in wavelength.

One of the limitations of a phototube is the lack of sensitivity, being in the region of 10 to 30 microamperes per lumen. In an effort to overcome this lack of sensitivity, the photomultiplier tube was developed. In this tube the photoelectrons are electrostatically directed to a secondary emitting surface, called a dynode. Under normal operating conditions, three to six secondary electrons are emitted per primary electron. The process of electrostatically focusing the emitted electrons on to the next dynode is continued for seven to thirteen stages until the electron stream finally reaches the anode. The net result is the production of tubes with photosensitivities in the region of 100 to 10,000 amperes per lumen.

Electron multiplication is not without its pitfalls because the increased electron multiplication also amplifies any noise. By proper selection of operating conditions, the best ratio of signal to noise can be utilized. Since much of the spurious signal is thermally generated, it is often advantageous to cool the phototubes as low as the temperatures of liquid nitrogen. One other problem is that the efficiency of the electron multiplication process is dependent upon the applied dynode string potential; hence the overall gain is quite sensitive to voltage.

INFRARED RANGE. The basic principle in the detection of infrared radiation is to observe the heat generated. However, this is in the microwatt or smaller range. One can resort to a bolometer, which consists of a coil of wire with a well defined temperature coefficient. If the wire is blackened it will absorb energy and its resistance will change. One need only measure this resistance change.

A second and more important detector is the thermocouple. This device thermally generates a potential, which is then measured. The mass of the thermocouple must be small so that it has appreciable sensitivity. Both of these devices lack speed of response; most infrared spectrophotometers are slower than their ultraviolet counterparts.

A third type of detector that is becoming available is solid state junctions that have sensitivities in the infrared range. These offer promise in that they respond to photons rather than heat generated over a fixed period of time.

Electronics. Once the photosignal is generated, the function of the electronic package is to transform that signal into useful information. Most modern spectrophotometers are of the dual beam variety, with one channel acting as the reference and the other measuring samples. This does not mean that two detectors are necessary because some instruments alternately shine the reference beam and the sample beam onto the same detector. This process has the advantage that one does not need two exactly matched detectors.

In essence, the difference in signal intensity between these two channels is amplified. The amplified signal is then presented on a recorder. In many instances, the log of the signal is recorded.

As mentioned previously, the response wave of the detector is not linear; as the wavelength changes, so does the efficiency of the detector. To compound this problem, the light source is almost never uniform in intensity through the spectral range. The spectrophotometer electronics must in some manner accommodate these difficulties. It is normally done by monitoring the reference channel light and closing a set of slits so that the detector receives only a constant light intensity. This also means that the spectral response curve of the detector must be trimmed so that it is relatively flat.

REFERENCES

1. H. Malmstadt and C. Enke. Digital Electronics for Scientists. W. A. Benjamin, Inc., New York, 1969.
2. H. Malmstadt, C. Enke and E. Toren. Electronics for Scientists. W. A. Benjamin, Inc., New York, 1963.
3. Application Manual for Computing Amplifiers for Modelling, Measuring, Manipulating and Much Else. Philbrick/Nexus Researches, Inc., Dedham, Mass., 1968.
4. J. Sharpe. Photomultipliers. EMI/US Document Ref. No. C.P. 5306. U.S. Government Printing Office, Washington, D.C., 1961.
5. H. Strobel. Chemical Instrumentation. Addison-Wesley Publishing Co., Inc., Reading, Mass., 1962.
6. H. Willard, L. Merritt and J. Dean. Instrumental Methods of Analysis. 4th Edition, D. Van Nostrand Co., Inc., Princeton, N.J., 1965.

XV

APPENDIX

COMPUTER PROGRAMMING

Many problems in physical chemistry can now be solved very quickly with the aid of computers. The saving in time can be illustrated with the example of x-ray diffraction studies. Before the use of digital computers, it took approximately 1 to $1\frac{1}{2}$ years of calculation to determine the atomic positions in a crystal after all the diffraction patterns were obtained. Now diffraction data is fed directly into computers and the total structure can be determined in a couple of days.

Books on computer programming and its application to chemical problems would now fill a small library. It is impossible to sum up in a chapter what others try to accomplish in a book. Therefore, the purpose of this appendix is to introduce the student to programming not by describing in detail the rules of using Fortran language but by illustrating a relatively simple computing problem and how the step-by-step programming is accomplished in this special case. The expectation is that once the student understands how simple such a task is, he will have the initiative either to go to the proper textbooks or to take an introductory course in programming that will enable him to do more complicated problems.

FIRST EXAMPLE

The problem selected here is the light-scattering data of polymer solutions that can yield three molecular parameters: (a) molecular weight, (b) radius of gyration (dimension of the molecule), and (c) an interaction parameter between solvent and solute expressed as the second virial coefficient.

The experimental procedures are described in Experiment 45, and the calculations to be done are summarized as follows:

Light-scattering intensities of a polymer-solvent system are measured as a function of scattering angle and concentration. Light-scattering measurements are also performed on the solvent. In all cases the light-scattering intensities are determined as an instrument response containing two parameters—galvanometric reading (g) and resistance reading (r). The relationship of these readings to the scattering intensity,

I_θ, hence to the Rayleigh ratio,

$$R_0 = (I_0/I_0) \, d^2 \tag{1}$$

is determined by using a calibrating material, the absolute scattering of which has been established. (The R_{90} of benzene is 16.3×10^{-6} with the 5461 Å mercury line as a monochromatic light source.) In equation 1, I_0 is the intensity of the incident beam, and d is the distance of the photomultiplier from the scattering center. Therefore, the galvanometric reading for benzene, gb, and the resistance reading, rb, are also obtained at a 90-degree scattering angle.

The instrument constant, q, is calculated as

$$q = \frac{16.3 \times 10^{-6} rb}{gb} \tag{2}$$

With the aid of this instrument constant, a pair of readings (g/r for a polymer solution of concentration c at a set angle of θ, and gs/rs for the solvent at the same angle) can be converted into the proper Rayleigh ratio.

$$R_0 = (g/r - gs/rs) \cdot q \tag{3}$$

In order to determine the molecular weight and other data, a Zimm plot is graphed. In this we plot c/R'_0 against $\sin^2(\theta/2) + 200c$ where

$$R'_0 = R_\theta \frac{\sin \theta}{1 + \cos^2 \theta} \tag{4}$$

and 200 is an arbitrary number to spread the plot. (One can select any number from 10 to 1000, depending on the sample.)

Thus c/R'_0 is calculated for every θ angle and c concentration, and it should be tabulated together with the corresponding $\sin^2 \theta/2 + 200c$.

In the following discussion, the program written in Fortran language to accomplish this is explained line by line.

PROGRAM ZIMMPL

Heading the program is the title in this case, program for the Zimm plot.

dimension theta (100), s (100), gs (100), rs (100), y (100), $t\sin$ (100)

The first line gives the *maximum* dimension of the program. The maximum number of angles, θ, will be 100 and so will be the maximum number of other parameters that are *functions of* θ: $s = (1 + \cos^2 \theta)$; $gs = $ galvanometric readings for the solvent; $rs = $ resistance readings for the solvent;

$$y = q \frac{\sin \theta}{1 + \cos^2 \theta}$$

where q is the instrument constant as in equation 2, and $t\sin = (\sin^2 \theta/2)$.

32 read 600, gb, rb, n, na

The number 32 is the code of this command. The computer is commanded to read the following: *first* 600, which the format of the command explained in the next line; *second, gb,* the galvanometric reading for benzene at an angle of 90 degrees; *third, rb,* the resistance reading for benzene at 90 degrees; *fourth, n,* the total number of angles at which measurements were taken on each sample; and *fifth, na,* the total number of samples (i.e., solutions with different concentrations).

600 format (2F 10.5, 2I 10)

The number 600 refers to the format of the preceding command. It states that the first *two* parameters, *gb* and *rb,* are *fixed numbers,* (2F). Each may have a total number of ten digits of which five are reserved for digits after the decimal point (2F 10.5). The other *two* parameters (i.e., the number of angles and the number of concentrations) are *integer* numbers, *I,* and each may have ten digits (2I 10).

$q = 16.3\ E - 6 * rb/gb$

This commands the computer to calculate the instrument constant, q. Comparing equation 2 with this formula, one sees that the number, 16.3×10^{-6}, is written in Fortran language as $16.3E - 6$ (E standing for 10 on the power of) and the multiplication sign is $*$.

33 read 605 [theta (i), gs (i), rs (i)), i = 1, n]

The number 33 is the code for this command. It states that the format will be specified below with code 605. It commands the computer to read the theta (θ) angles from 1 to n and the corresponding galvanometric, gs, and resistance, rs, values at each θ for the solvent.

605 format (3F 10.5)

In the previous command there were three parameters to be read for each angle: θ, gs, and rs. All *three* parameters have the same format being a *fixed number* (F) and allocating ten digits for each of which five are reserved for digits after the decimal point (3F 10.5).

do 34 i = 1, n

This commands the computer to do the following calculations specified in the next three lines for each angle θ from 1 to n.

s(i) = 1. + cos f (theta (i)) ** 2

This commands the calculation of the function $1 + \cos^2 \theta$ for all θ angles.

y(i) = q * sin f (theta (i))/s(i)

This command orders the calculation of $y(i)$ for all θ angles.

$$y(i) = q\ \frac{\sin \theta}{1 + \cos^2 \theta}$$

34 tsin (i) = sin f [theta (i)/2)] ** 2

This command defines the $tsin\ (i) = \sin^2 (\theta/2)$ to be calculated for each θ.

do 35, $JJ = 1$, *na*

This command indicates the closing of a loop. Up to now the computer read the data for the blank (solvent) and calculated the functions dependent on angles only. From here on the computer will read in the data obtained for different polymer solutions, hence the concentration dependent part. This command therefore indicates that after the data for one concentration is read by the computer, it should calculate the desired values and return here to start the same process for the data for the next concentration. The *JJ* means that the concentrations should be taken for numbers 1 to *na*.

36 read 602, *c*

This commands the computer to read the concentration, *c*, which has a format 602 to be given.

602 format (*F* 10.5)

The concentration is given in fixed number (*F*) having a maximum of ten digits of which five may be after the decimal point.

do 35 $i = 1$, *n*

This commands the computer to do the following calculations for each angle from 1 to *n* for a given concentration.

read 603, *g*, *r*

This commands the computer to read the galvanometric, *g*, and the resistance, *r*, values for the polymer solutions for all angles and concentrations. The format of these values will be given as 603.

603 format (2*F* 10.5)

Both the *g* and *r* values are fixed numbers for each ten digits and are allocated with five digits after the decimal point.

$b = g/r - gs(i)/rs(i)$

The computer is commanded to calculate the quantity *b* for each angle and concentration. The quantity *b* is simply the instrument response for a solution minus the instrument response for the solvent.

$a = c/(b * y(i))$

The quantity *a*, which is equal to c/R_θ', is to be calculated. This will be the ordinate of the Zimm plot.

$x = t\sin(i) + 200 * c$

The quantity $x = \sin^2(\theta/2) + 200c$ is to be calculated and this will be the abscissa of the Zimm plot.

35 print 601, theta (*i*), *c*, *a*, *x*

This commands the computer to print out the results in the format 601 to be specified in the tabulated form of angle, concentration, and the calculated quantities of *a* and *x*.

601 format (1*x*, *F* 20.5, 5*x*, *F* 20.5, 5*x*, *F* 20.5, 5*x*, *F* 20.5 | | |)

This format specifies that 1*x* means that one space is left open, and then the theta should be printed as a fixed number for which 20 digits are allocated with five digits after the decimal point (*F* 20.5): then five spaces are left open (5*x*), and the concentration should also be printed in (*F* 20.5) format, another five spaces left open before *a* is printed out in the same format, another five spaces left open and *x* is also printed out in (*F* 20.5) format. The | | | sign at the end indicates three carriage returns (spacings) before the same data is printed out for the next concentration.

Now that the program is available, we shall give an example of how the data has to be provided to the computer. The following data is for the Zimm plot program. After each number the *symbol of the parameter* as it appears in the program is printed *in parenthesis* for the student to identify these quantities. *However, these symbols should not be typed or punched into cards when the data is presented to the computer.*

56.	(*gb*)	1000.	(*rb*)	6	(*n*)	6	(*na*)
1.22173	(70° in radians)	11.	(*gs*)	1000.	(*rs*)		
0.87266	(50° in radians)	17.	(*gs*)	1000.	(*rs*)		
0.78540	(45° in radians)	25.	(*gs*)	1000.	(*rs*)		
0.69813	(40° in radians)	42.	(*gs*)	333.3	(*rs*)		
0.61087	(35° in radians)	65.	(*gs*)	100.	(*rs*)		
0.52360	(30° in radians)	41.	(*gs*)	33.33	(*rs*)		

0.00387	(concentration c_1)		
28.	(*g*)	10.	(*r*)
46.	(*g*)	10.	(*r*)
53.	(*g*)	10.	(*r*)
64.	(*g*)	10.	(*r*)
78.	(*g*)	10.	(*r*)
34.	(*g*)	3.33	(*r*)

0.00290	(concentration c_2)		
22.	(*g*)	10.	(*r*)
35.	(*g*)	10.	(*r*)
41.	(*g*)	10.	(*r*)
50.	(*g*)	10.	(*r*)
64.	(*g*)	10.	(*r*)
84.	(*g*)	10.	(*r*)

and so forth.

SECOND EXAMPLE

As a second example, a computer program is presented here that was written by one of my former students. Kenneth Krul took the regular undergraduate physical chemistry course at Adelphi University in 1967 and has written the following program for indexing the reflections obtained from x-ray powder diagrams.

The experiment is Experiment 22 in this book and it deals with the *x*-ray diffraction patterns of dimethyl sulfoxide (DMSO) (or dimethylformamide [DMF], and so forth) and water. In that experiment the student has to determine the dimensions of the concentric rings (diffraction lines) obtained on a film that was exposed to diffracted *x*-rays in a powder camera. The measured lines must be corrected to account for film shrinkage in development. From the corrected data the *d* spacings are calculated using Bragg's equation. These *d* spacings are indexed using the given symmetry and the dimensions of the appropriate unit cell. The program is written to accomplish this. In the first part the measured dimensions of the reflection lines are corrected for shrinkage and the *d* values are calculated. In the second part subroutines are written for different unit cells. The proper unit cell will be selected and the computer will calculate *d* values for different *hkl* indexing.

CALCULATION OF *D*-SPACINGS AND INDEXING POWDER DIAGRAMS (PROGRAM INDEX)

For run:

CARDS NECESSARY

I. Control Cards (Master Instructions).

II. Data Cards. (See Variable Definitions.)
   1. Print in first two spaces of data card. For example: 01, 02, ... , 10, 11, 12, and so forth.
   2. M, IC, IT, IH, IO, IM: M in spaces 1 and 2, i.e., 01, 02, ... , 11, 12, ... IC, IT, IH, IO, and IM in spaces 3, 4, 5, 6, and 7.

   3. A, B, C, BETA:
      A in spaces   1–10 ⎫
      B in spaces 11–20 ⎪
      C in spaces 21–30 ⎬ include all decimal points on data cards.
      BETA in spaces 31–40 ⎭

   4. SA1, SA2, SB1, SB2:
      SA1 in spaces   1–10 ⎫
      SA2 in spaces 11–20 ⎪
      SB1 in spaces 21–30 ⎬ include all decimal points on data cards.
      SB2 in spaces 31–40 ⎭

   5. ALPHA (Jo), BRAVO (Jo):
      ALPHA (Jo) in spaces   1–20 ⎫ include all decimals points on data cards.
      BRAVO (Jo) in spaces 21–40 ⎭
      e.g.   Readings are
                 2.843   4.692
                 2.961   4.701
                 3.001   4.900

Then three cards are necessary for ALPHA-BRAVO data

ALPHA(1) = 2.843   BRAVO(1) = 4.692

ALPHA(2) = 2.961   BRAVO(2) = 4.701 and so forth

/ 2.843		4.692	
	20, 21		40

III. Control Cards.

1. End of file cards. If IRUN is greater than 1, Data cards 2, 3, 4, and 5 are repeated in that order for each set.

VARIABLE DEFINITIONS (INPUT)

I. J, K, L: the whole number coordinates corresponding to the h, k, l indices in Experiment 22.

II. IRUN: the number of data sets. For example, if data from water, DMSO, and DMF are to be run, IRUN = 3.

III. M: the number of data in a set. If, for example, 27 lines are recorded from an x-ray plate, M = 27.

IV. IC, IT, IH, IO, IM: control variables governing which theoretical subroutines are to be used. They are always either zero or one.

Zero: The subroutine is not used.

One: The subroutine is used.

They correspond thus:

Variable	Cell	Subroutine
IC	Cubic	CUBIC
IT	Tetragonal	TETRA
IH	Hexagonal	HEXAG
IO	Orthorhombic	ORTHO
IM	Monoclinic	MONO

V. A, B, C, BETA: correspond to a, b, c, and β in Experiment 22.

VI. SA1, SA2, SB1, SB2: correspond to the S values from the forward reflection (SA1, SA2) and the backward reflection (SB1, SB2) used to determine the shrinkage correction factor.

VII. ALPHA(Jo), BRAVO(Jo): the M sets of S values, BRAVO being the larger.

VARIABLE DEFINITIONS (OUTPUT)

I. CORFAC: shrinkage correction factor.

II. DSPACE: the experimental d-spacings.

III. DC, DT, DH, DOT, DM: theoretical d-spacings calculated for the following cells:

> DC-cubic
> DT-tetragonal
> DH-hexagonal
> DOT-orthorhombic
> DM-monoclinic

IV. J, K, L: correspond to h, k, l.

```
        PROGRAM X-RAY DIFFRACTION
        DIMENSION A(100), B(100), D(100), DP(100), DA(100)
c       M is the number of pairs of readings
        M = 27
        DO 5  J = 1, M
        READ 6, A(J), B(J)
  6     FORMAT (2F20.8)
c       A(J) is the low reading and B(J) is the high
  5     CONTINUE
        ALAM = 1.5418
        ALAM is the wavelength
        DO  7  J = 1, M
        D(J) = (ALAM)/2.0 * SIN((B(J) − A(J))/(2.0 * 57.296)))
        DA(J) = D(J)
  7     CONTINUE
        N = 1
        DO 49  J = 1, M
        K = 1
 75     If (D(N) − D(K)) 1, 2, 2
  2     K = K + L
        If (K − M) 75, 75, 38
  1     D(N) = D(K)
        If (K − M) 74, 38, 38
 74     K = K + 1
        GO TO 75
 38     DP(N) = D(N)
        PRINT 100, DP(N)
100     FORMAT (F20.8)
        DO 48  L = 1, M
        D(L) = DA(L)
 48     CONTINUE
        I = 1
        K = 1
 22     If (D(I) − DP(KN)) 66, 67, 66
 66     I = I + 1
        GO TO 22
 67     D(I) = 0.0
        If (K − N) 83, 51, 51
 83     K = K + 1
        I = 1
        GO TO 22
 51     N = N + 1
 49     CONTINUE
        STOP
        END.
```

```
FORTRAN (3.1)/MASTER
        PROGRAM INDEX
        DIMENSION ALPHA(100),BRAVO(100),DELTA(100),DP(100),DA(100),
        J(214),K(214),L(214)
            COMMON J,K,L,A,B,C,BETA,ALPHA,BRAVO,DELTA,DP,DA,I
        DO 13 I=1,214
        READ 14,J(I),K(I),L(I)
   14   FORMAT (3I1)
   13   CONTINUE
        JER=1
        READ 9,IRUN
    9   FORMAT (I2)
   12   READ 15,M,IC,IT,IH,IO,IM
   15   FORMAT (I2,5I1)
        READ 16, A,B,C,BETA
   16   FORMAT (4F10.8)
        READ 100,SA1,SA2,SB1,SB2
  100   FORMAT (4F10.8)
        DO 18 JO=1,M
        READ 19, ALPHA(JO),BRAVO(JO)
   19   FORMAT (2F20.8)
   18   CONTINUE
        ZED=((SB1+SB2)/2.)-((SA1+SA2)/2.)
        CORFAC=180./ZED
        PRINT 101
  101   FORMAT (10x,6HCORFAC,/)
        PRINT 102, CORFAC
  102   FORMAT (F20.8)
        ALAM=1.5418
        PRINT 17
   17   FORMAT (10X,6HDSPACE,/)
        DO 20 JO=1,M
        GAM=(BRAVO(JO)-ALPHA(JO))*CCREAC
        DELTA(JO)=(ALAM)/2.0*SIN(GAM/(2.0*57.296))
        DA(JO)=DELTA(JO)
   20   CONTINUE
        N=1
        DO 21 JO=1,M
        KO=1
   22   IF (DELTA(N)-DELTA(KO))1,2,2
    2   KO=KO+1
        IF (KO-M)22,22,23
    1   DELTA(N)=DELTA(KO)
        IF (53-M)24,23,23
   24   KO=KO+1
        GO TO 22
   23   DP(N)=DELTA(N)
        PRINT 736,DP(N)
  736   FORMAT (F20.8)
        DO 25 LO=1,M
        DELTA(LO)=DA(LO)
   25   CONTINUE
        IZ=1
        KZ=1
   26   IF (DELTA(IZ)-DP(KZ))27,28,27
   27   IZ=IZ+1
        GO TO 26
```

```
28   DELTA(IZ)=0.0
     IF (KZ-N)29,30,30
29   KZ=KZ+1
     IZ=1
     GO TO 26
30   N=N+1
21   CONTINUE
     IF (IC-1)32,31,31
31   CALL CUBIC(J,K,L,CGAM,DC,A,B,C,BETA)
32   IF (IT-1)34,33,33
33   CALL TETRA(J,K,L,TGAM,DT,A,B,C,BETA)
34   IF (IH-1)36,35,35
35   CALL HEXAG(J,K,L,HGAM,HEX,DH,A,B,C,BETA)
36   IF (IO-1)38,37,37
37   CALL ORTHO(J,K,L,OR1,OR2,OR3,ORTH,DOT,A,B,C,BETA)
38   IF (IM-1)40,39,39
39   CALL MONO(J,K,L,MON1,MON2,MON3,MON,A,B,C,BETA,DM)
40   CONTINUE
     IF (JFR-IRUN)42,41,41
42   JER=JER+1
     GO TO 12
41   CONTINUE
     STOP
     END


     SUBROUTINE CUBIC (J,K,L,I,CGAM,DC,A,B,C,BETA)
     DIMENSION ALPHA(100),BRAVO(100),DELTA(100),DP(100),DA(100),
     J(214),K(214),L(214)
        COMMON                    ALPHA,BRAVO,DELTA,DP,DA,
     PRINT 199
199  FORMAT (16X,2HDC,/)
     DO 200 I=1,214
     CGAM=(J(I)**2)+(K(I)**2)+(L(I)**2)
     DC=A/SORT(CGAM)
     PRINT 201,DC,J(I),K(I),L(I)
201  FORMAT (10X,F20.8,5X,I1,5X,I1,5X,I1)
200  CONTINUE
     RETURN
     END


     SUBROUTINE TETRA(J,K,L,I,TGAM,DT,A,B,C,BETA)
     DIMENSION ALPHA(100),BRAVO(100),DELTA(100),DP(100),DA(100),
     J(214),K(214),L(214)
        COMMON                    ALPHA,BRAVO,DELTA,DP,DA,
     PRINT 203
203  FORMAT (16X,2HDT,/)
     DO 204 I=1,214
     TGAM=(C**2)*(J(I)**2)+(K(I)**2)*(C**2)+(A**2)*(L(I)**2)
     DT=(A*C)/SORT(TGAM)
     PRINT 205,DT,J(I),K(I),L(I)
205  FORMAT (10X,F20.8,5X,I1,5X,I1,5X,I1)
204  CONTINUE
     RETURN
     END
```

```
        SUBROUTINE HEXAG(J,K,L,I,HFX,HGAM,DH,A,B,C,BETA)
        DIMENSION ALPHA(100),BRAVO(100),DELTA(100),DP(100),DA(100),
        J(214),K(214),L(214)
            COMMON                    ALPHA,BRAVO,DELTA,DP,DA,
        PRINT 206
206     FORMAT (16X,2HDH,/)
        DO 207 I=1,214
        HEX=(J(I)**2)+(J(I)*K(I))+(K(I)**2)
        HGAM=(4.*(C**2)*HEX)+3.*(A**2)*(L(I)**2)
        DH=(1.732*A*C)/SORT(HGAM)
        PRINT 208,DH,J(I),K(I),L(I)
208     FORMAT (10X,F20.8,5X,I1,5X,I1,5X,I1)
207     CONTINUE
        RETURN
        END


        SUBROUTINE ORTHO (J,K,L,I,A,B,C,BETA,OR1,OR2,OR3,ORTH,DOT)
        DIMENSION ALPHA(100),BRAVO(100),DELTA(100),DP(100),DA(100),
        J(214),K(214),L(214)
            COMMON                    ALPHA,BRAVO,DELTA,DP,DA,
        PRINT 209
209     FORMAT (16X,3HDOT,/)
        DO 210 I=1,214
        OR1=(J(I)**2)*(B**2)*(C**2)
        OR2=(K(I)**2)*(A**2)*(C**2)
        OR3=(L(I)**2)*(A**2)*(B**2)
        ORTH=OR1+OR2+OR3
        DOT=(A*B*C)/SORT(ORHT)
        PRINT 211,DOT,J(I),K(I),L(I)
211     FORMAT (10X,F20.8,5X,I1,5X,I1,5X,I1)
210     CONTINUE
        RETURN
        END


        SUBROUTINE MONO(J,K,L,A,B,C,I,BETA,MON1,MON2,MON3,MOND,
        DM)
        DIMENSION ALPHA(100),BRAVO(100),DELTA(100),DP(100),DA(100),
        J(214),K(214),L(214)
            COMMON                    ALPHA,BRAVO,DELTA,DP,DA,
        PRINT 212
212     FORMAT (16X,2HDM,/)
        DO 213 I=1,214
        MON1=(J(I)**2)/(A**2)
        MON2=(L(I)**2)/(C**2)
        MON3=((2*J(I)*L(I))/(A*C))*COS(BETA/57.296)
        MOND=(B**2)*(MON1+MON2+MON3)+(K(I)**2)*((SIN(BETA/
        57.296))**2)
        DM=B*SIN(BETA/57.296)/SORT(MOND)
        PRINT 214,DM,J(I),K(I),L(I)
214     FORMAT (10X,F20.8,5X,I1,5X,I1,5X,I1)
213     CONTINUE
        RETURN
        END
```

THIRD EXAMPLE

Finally, a program is presented here for curve fitting. This is a common problem in experimental physical chemistry. The problem selected here deals with only one independent variable (x) and the dependent variable (y) is calculated either as an exponential or as a polynomial function of Kth degree of the independent variable.

POLYNOMIAL AND EXPONENTIAL CURVE FIT IN ONE INDEPENDENT VARIABLE

I. Purpose: given a set of N points (X_i, Y_i), $N \leq 1000$, to find the least square curve fits of the form

$$Y_{\text{pred}} = Ae^{BX} \tag{1}$$

or

$$Y_{\text{pred}} = \sum_{i=0}^{K} A_i X^i, \qquad K \leq 20 \tag{2}$$

II. Method: classic least square techniques are used. The *standard deviation* of the prediction is given by

$$s^2 = \frac{1}{N-1} \sum_{i=1}^{N} (Y_{\text{pred}} - Y_i)^2 \tag{3}$$

The user has the following options:

Option 1—to fit the points to an exponential curve of form 1.

Option 2—to fit the points to a polynomial of form 2 for a single $K \leq 20$.

Option 3—to fit the points to polynomials of form 2 for a sequence $K_1, \ldots, K_M$.

Option 4—to fit the points to polynomials of form 2 for a sequence $K_1, \ldots, K_M$, but stopping when $s < \text{ACC}$, a preassigned quantity.

Option 5—to fit the points to polynomials of form 2 for all K, $\text{KLOWER} \leq K \leq \text{KUPPER}$.

Option 6—to fit the points to polynomials of form 2 for all K, $\text{KLOWER} \leq K \leq \text{KUPPER}$, but stopping when $s < \text{ACC}$, a preassigned quantity.

III. Input (also described in the comments at the beginning of the FORTRAN source program).
  1. Read ITEST, N in FORMAT(2I5).
      ITEST = the number of the option described in METHOD.
      N = number of data points, limited to 1000.
  2. If ITEST = 2, read K in FORMAT(I5).
      If ITEST = 3, read M, $(K(I), I = 1, M)$ in FORMAT(16I5).
      If ITEST = 4, read ACC, M, $(K(I), I = 1, M)$ in FORMAT(E10.0, 14I5/(16I5)).
      If ITEST = 5, read KLOWER, KUPPER in FORMAT(2I5).
      If ITEST = 6, read ACC, KLOWER, KUPPER in FORMAT(E10.0, 2I5).

3. Read data points $(X(I), Y(I), I = 1, N)$ in FORMAT(8E10.0).
4. Read IPR in FORMAT(I5)
IPR $= 1$ if deviations at each point are to be printed.
IPR $= 0$ if deviations at each point are not to be printed.
Each set of data immediately follows the preceding set. Place two end-of-file cards after the last set of data.

IV. Output. The coefficients of the curve fits are printed. The standard deviation of the prediction, and, if requested, the deviations at each point are printed.

V. Limitations. The degree of any polynomial curve fit is restricted to 20. The number of data points cannot exceed 1000.

VI. Examples.
Example 1:

Point	X	Y
1	0	2
2	1	1
3	2	2
4	3	1
5	−4	2
6	−3	3
7	−2	1
8	−1	3

Find the exponential curve fit of form 1 with deviations at each point to be printed. Here,
$$\text{ITEST} = 1, N = 8, \text{IPR} = 1$$

Example 2:

Point	X	Y
1	0	0
2	1	1
3	−1	0
4	2	1
5	−2	−1
6	3	0

Find the polynomial curve fit of form 2 for $K = 2$, with no deviations printed. Here,
$$\text{ITEST} = 2, K = 2, N = 6, \text{IPR} = 0$$

Example 3:
For the data of example 2, find the polynomial curve fits of form 2 of degrees 2 and 4, with no deviations printed. Here,
$$\text{ITEST} = 3, M = 2, K_1 = 2, K_2 = 4, N = 6, \text{IPR} = 0$$

INPUT SHEETS

FORTRAN CODING FORM

IBM

Program				Punching Instructions		Page	of
Programmer		Date		Graphic	Card Form #		Identification
				Punch			73 ... 80

C FOR COMMENT

STATEMENT NUMBER | FORTRAN STATEMENT

Column markers: 1 ... 5 6 7 ... 10 ... 15 ... 20 ... 25 ... 30 ... 35 ... 40 ... 45 ... 50 ... 55 ... 60 ... 65 ... 70 72

* A standard card form. IBM electro 888157. is available for punching source statements from this form.

Example 4:

For the data of example 3, find the polynomial curve fits of form 2 for all K, $5 \leq K \leq 7$, stopping when $s < 0.1$, with no deviations printed. Here,

ITEST $= 6$, KLOWER $= 5$, KUPPER $= 7$, ACC $= 0.1$, $N = 6$, IPR $= 0$

Printout. (Page ejections are omitted, and four decimal places rather than 9 are shown here.)

EXPONENTIAL CURVE FIT

$$Y = 1.6343\text{E}00 \text{ EXP} (-8.6724\text{E-}02 \text{ X})$$

PT.	X	Y	CALC. Y	DEV.
1	0	2.0000E 00	1.6343E 00	3.6569E-01
2	1.0000E 00	1.0000E 00	1.4985E 00	−4.9855E-01
3	2.0000E 00	2.0000E 00	1.3741E 00	6.2594E-01
4	3.0000E 00	1.0000E 00	1.2599E 00	−2.5992E-01
5	−4.0000E 00	2.0000E 00	2.3120E 00	−3.1201E-01
6	−3.0000E 00	3.0000E 00	2.1199E 00	8.8005E-01
7	−2.0000E 00	1.0000E 00	1.9438E 00	−9.4385E-01
8	−1.0000E 00	3.0000E 00	1.7824E 00	1.2176E 00

STANDARD DEVIATION $= 7.6410\text{E-}01$

POLYNOMIAL CURVE FIT OF DEGREE 2

COEFFICIENTS—A(I) IS COEFFICIENT OF X TO POWER I

$A(0) = 5.1429\text{E-}01$

$A(1) = 4.3571\text{E-}01$

$A(2) = -1.7857\text{E-}01$

STANDARD DEVIATION $= 3.1168\text{E-}01$

POLYNOMIAL CURVE FIRST OF DEGREE

2   4

DEGREE $= 2$

COEFFICIENTS—A(I) IS COEFFICIENT OF X TO POWER I

$A(0) = 5.1429\text{E-}01$

$A(1) = 4.3571\text{E-}01$

$A(2) = -1.7857\text{E-}01$

STANDARD DEVIATION $= 3.1168\text{E-}01$

DEGREE = 4

COEFFICIENTS—A(I) IS COEFFICIENT OF X TO POWER I
A(0) =   2.3810E-01
A(1) =   4.1270E-01
A(2) =   1.2500E-01
A(3) =   2.7778E-02
A(4) = −4.1677E-02

STANDARD DEVIATION = 1.6903E-01

POLYNOMIAL CURVE FIRST OF DEGREE
   5   6   7

STOP WHEN ST. DEV. IS LESS THAN 1.0000E-01

DEGREE = 5

COEFFICIENTS—A(I) IS COEFFICIENT OF X TO POWER I
A(0) = −3.6453E-09
A(1) =   7.0000E-01
A(2) =   6.6667E-01
A(3) = −2.5000E-01
A(4) = −1.6667E-01
A(5) =   5.0000E-02

STANDARD DEVIATION = 2.1660E-09

ACCURACY ATTAINED

REFERENCES

1. D. D. McCracken. Digital Computer Programming. Wiley & Sons, Inc., New York, 1957.
2. P. A. D. de Maine and R. D. Seawright. Digital Computer Programs for Physical Chemistry. Vol. I and II, The Macmillan Co., New York, 1963 and 1965.
3. T. R. Dickson. The Computer and Chemistry. W. H. Freeman & Co., San Francisco, 1968.

INDEX

ELEMENTS AND ATOMIC WEIGHTS, BASED ON CARBON-12 12.000

	Symbol	Atomic No.	Atomic Weight		Symbol	Atomic No.	Atomic Weight
Actinium	Ac	89		Mercury	Hg	80	200.59
Aluminum	Al	13	26.9815	Molybdenum	Mo	42	95.94
Americium	Am	95		Neodymium	Nd	60	144.24
Antimony	Sb	51	121.75	Neon	Ne	10	20.183
Argon	Ar	18	39.948	Neptunium	Np	93	
Arsenic	As	33	74.9216	Nickel	Ni	28	58.71
Astatine	At	85		Niobium	Nb	41	92.906
Barium	Ba	56	137.34	Nitrogen	N	7	14.0067
Berkelium	Bk	97		Nobelium	No	102	
Beryllium	Be	4	9.0122	Osmium	Os	76	190.2
Bismuth	Bi	83	208.980	Oxygen	O	8	15.9994
Boron	B	5	10.811[a]	Palladium	Pd	46	106.4
Bromine	Br	35	79.909[b]	Phosphorus	P	15	30.9738
Cadmium	Cd	48	112.40	Platinum	Pt	78	195.09
Calcium	Ca	20	40.08	Plutonium	Pu	94	
Californium	Cf	98		Polonium	Po	84	
Carbon	C	6	12.01115[a]	Potassium	K	19	39.102
Cerium	Ce	58	140.12	Praseodymium	Pr	59	140.907
Cesium	Cs	55	132.905	Promethium	Pm	61	
Chlorine	Cl	17	35.453[b]	Protactinium	Pa	91	
Chromium	Cr	24	51.996[b]	Radium	Ra	88	
Cobalt	Co	27	58.9332	Radon	Rn	86	
Copper	Cu	29	63.54	Rhenium	Re	75	186.2
Curium	Cm	96		Rhodium	Rh	45	102.905
Dysprosium	Dy	66	162.50	Rubidium	Rb	37	85.47
Einsteinium	Es	99		Ruthenium	Ru	44	101.07
Erbium	Er	68	167.26	Samarium	Sm	62	150.35
Europium	Eu	63	151.96	Scandium	Sc	21	44.956
Fermium	Fm	100		Selenium	Se	34	78.96
Fluorine	F	9	18.9984	Silicon	Si	14	28.086[a]
Francium	Fr	87		Silver	Ag	47	107.870[b]
Gadolinium	Gd	64	157.25	Sodium	Na	11	22.9898
Gallium	Ga	31	69.72	Strontium	Sr	38	87.62
Germanium	Ge	32	72.59	Sulfur	S	16	32.064[a]
Gold	Au	79	196.967	Tantalum	Ta	73	180.948
Hafnium	Hf	72	178.49	Technetium	Tc	43	
Helium	He	2	4.0026	Tellurium	Te	52	127.60
Holmium	Ho	67	164.930	Terbium	Tb	65	158.924
Hydrogen	H	1	1.00797[a]	Thallium	Tl	81	204.37
Indium	In	49	114.82	Thorium	Th	90	232.038
Iodine	I	53	126.9044	Thulium	Tm	69	168.934
Iridium	Ir	77	192.2	Tin	Sn	50	118.69
Iron	Fe	26	55.847[b]	Tinanium	Ti	22	47.90
Krypton	Kr	36	83.80	Tungsten	W	74	183.85
Lanthanum	La	57	138.91	Uranium	U	92	238.03
Lawrencium	Lw	103		Vanadium	V	23	50.942
Lead	Pb	82	207.19	Xenon	Xe	54	131.30
Lithium	Li	3	6.939	Ytterbium	Yb	70	173.04
Lutetium	Lu	71	174.97	Yttrium	Y	39	88.905
Magnesium	Mg	12	24.312	Zinc	Zn	30	65.37
Manganese	Mn	25	54.9380	Zirconium	Zr	40	91.22
Mendelevium	Md	101					

[a]The atomic weight varies because of natural variations in the isotopic composition of the element. The observed ranges are boron, ±0.003; carbon, ±0.00005; hydrogen, ±0.00001; oxygen, ±0.0001; silicon, ±0.001; Sulfur, ±0.003.

[b]The atomic weight is believed to have an experimental uncertainty of the following magnitude: bromine, ±0.002; chlorine, ±0.001; chromium, ±0.001; iron, ±0.003; silver, ±0.003. For other elements the last digit given is believed to be reliable to ±0.5.